GREAT SOLID STATE PHYSICISTS OF THE 20th CENTURY

GREAT SOLID STATE PHYSICISTS OF THE 20th CENTURY

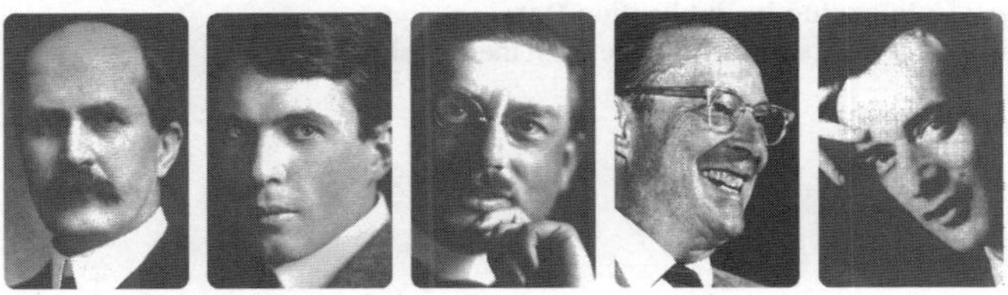

editors

Julio A. Gonzalo & Carmen Aragó López

Universidad Autónoma, Madrid, Spain

World Scientific

New Jersey • London • Singapore • Hong Kong

Published by

World Scientific Publishing Co. Pte. Ltd.

5 Toh Tuck Link, Singapore 596224

USA office: Suite 202, 1060 Main Street, River Edge, NJ 07661

UK office: 57 Shelton Street, Covent Garden, London WC2H 9HE

British Library Cataloguing-in-Publication Data
A catalogue record for this book is available from the British Library.

GREAT SOLID STATE PHYSICISTS OF THE 20TH CENTURY

ISBN 981-238-336-0

Printed in Singapore by World Scientific Printers (S) Pte Ltd

Foreword

In the year 2000 the world resembles very little the world in 1900. It is true that one century ago the first industrial revolution had already taken place in Europe and America. Steam machines, chemical industries, railroad networks and even hydroelectric power generation were becoming widespread realities. But only a few of the most advanced large cities of the world were beginning to utilize electrical illumination. Telephones and automobile transportation were in their infancy. The changes, on the other hand, which have taken place during the 20th century have been much more spectacular and more global.

By the second half of the 19th century the main branches of classical physics (mechanics, optics, electricity, thermodynamics) had achieved maturity. The technologies based on them, of course, took more time to develop. Just at the beginning of the 20th century the primitive quantum theory, special relativity and the atomistic character of matter as made up of nuclei, containing protons, neutrons and electrons, were being formulated. Quantum mechanics was set forth twenty-five years later. Within another twenty years the basic conceptual framework for modern solid state physics, or condensed matter physics or materials physics had been established, taking advantage of experimental findings of fundamental importance, like X-ray diffraction by crystalline lattices discovered by von Laue, son-in-law of Max Planck and exploited systematically by the Bragg's to determine crystal structures. Debye's and Einstein's contributions to explain the low temperature behavior of the specific heat of solids required laying, to begin with, the foundations of lattice dynamics.

By the middle of the 20th century everything was done to make the decisive step from vacuum tube electronics to solid state electronics, based on doped semiconductors. It was performed by Bardeen, Shockley and Brattain. This decisive step, compounded with the trend for miniaturization required by the space race and with the need to power and control artificial satellites, set in motion developments which, by the end of the century, had caused the computer revolution, large scale industrial automatization and the late revolution in communications carried out in good measure using solid state lasers, among other devices.

It has been said that the 20th century has been the century of physics. Certainly, the advances made in the realms of Elementary Particle Physics and Cosmology have been really spectacular. But these advances would not have been possible without the previous extraordinary advances in Materials and Solid State Physics. The large particle accelerators used in High Energy Physics require special superconducting magnets to produce the very large magnetic fields needed. The sophisticated particle detectors being used nowadays require also large arrangements of sensitive semiconductor components. And the fine instruments used in observational Cosmology, capable of detecting, for instance, minor anisotropies in the cosmic microwave background radiation, require also very special materials and very special solid state devices. So it may be said, perhaps more appropriately, that the 20th century has been the century of Solid State Physics.

Evidently, the experimental and the theoretical components are both essential in Physics. But in Elementary Particles one is at the limit of very high energies and very small sizes, while in Cosmology one is at the opposite limit of very distant and very large cosmic objects. The obliged reference for us to combine properly experiment and theory in the right proportion is however the distinctive characteristic of Materials Physics or Solid State Physics. The conceptual framework appropriate to formulate mechanics, electricity, optics and thermodynamics was actually that of meso-physics, rather than micro or macro-physics, which could be approached later, after physics had achieved maturity describing real facts and events more accessible to experimentation.

The phenomenal success of semiconductor materials in the second half of the 20th century has been accompanied by the very significant success of magnetic, piezoelectric, electrooptic materials used in a great variety of transducers and devices (electromechanical, thermoelectric, optoelectronic, etc.). The investigation of cooperative effects and phase transitions in ferromagnetic, ferroelectric, ferroelastic, superconducting materials has been one of the main efforts and it is now still very active. By the beginning of the 21th century Solid State Physics faces the challenge of going from microelectronic materials and composite materials to the artificial, specially designed materials required for the new and promising nanotechnologies.

In this book, based upon a plenary Session at the 10th International Meeting on Ferroelectricity, held in Madrid on September 5, 2001, a few outstanding Solid State physicists of the 20th century are selected. William H. Bragg and William L. Bragg, father and son, Peter Debye, John Bardeen, and Lev Landau are household names. They were great scientists, true gentlemen, and models of integrity for their students and their colleagues. The men and their scientific achievements are presented by distinguished men of science, such as Anthony Mike Glazer (Oxford), Eric Courtens (Montpellier), Fernando Sols (UAM, Madrid) and Vitaly Ginzburg (Moscow). Finally, Prof. Stanley L. Jaki (New Jersey) gave an overview of the relevance of Materials Physics.

Dr. Gen Shirane (BNL, New York) and Prof. Alex K. Müller (Physik-Institut, U. Zürich) were the chairman of that plenary session.

Julio A. Gonzalo and Carmen Aragó

Madrid, July 9, 2002

Acknowledgements

We gratefully acknowledge permission from Royal Society (articles by E. N. da C. Andrade, Sir David Phillips and Mansel Davies) and the Nobel Foundation for making available very useful written and graphic material for the outstanding Solid State Physicists taken up in this book.

We are very grateful to our young colleague Jorge García for checking carefully the manuscript and making many useful suggestions.

Finally, we are specially grateful to the Hon. Minister of Education, Culture and Sports of Spain, who presided the Opening Session of the 10th International Meeting on Ferroelectricity and supported this publication through an special grant to prepare properly the full manuscript of the book.

Contents

PART I THE BRAGGS

William Henry Bragg

William Lawrence Bragg

The Braggs

A. M. Glazer

Clarendon Laboratory, Parks Rd., Oxford, OX1 3PU, UK.

The name of "Bragg" is universally known for its connection with the birth of the field of X-ray crystallography. However, many outside this particular field of study do not realise that there was not one Bragg but two, father and son, and this abstract of my talk given at the IMF in Madrid 2001 briefly describes their careers.

The Bragg family has its origins in Cumberland in England, and in the early part of the 19th century, a certain John Bragg married Lucy Brown. They had four children, one of whom was Robert John Bragg. Unfortunately around 1840 the father was lost at sea and the remaining family moved to Birkenhead. Robert John himself went to sea and had some near escapes. Returning to England he married Mary Wood in 1861. In 1862 she gave birth to William Henry Bragg (WHB), plus two more sons later. In 1884 WHB went up to Trinity College Cambridge to read Mathematics, and in 1885, after encouragement by J.J. Thomson, then Cavendish Professor of Physics, he took up an appointment as Professor of Mathematics at the University of South Australia, in Adelaide. In 1889, WHB married Gwendoline Todd, the daughter of Charles Todd, the Government Astronomer of South Australia. From this marriage there were three children, William Lawrence (WLB) in 1890, Robert Charles in 1891 and Gwendoline Mary in 1907.

So much for the early life of the Bragg family, and we must switch our attention to the realm of science. Perhaps the first and most important development that set the two Braggs WHB and WLB on to their future paths was the Roentgen's momentous discovery of X-rays in 1895. Suddenly, scientists throughout the world began experimenting with this mysterious, invisible form of radiation that had such unusual properties, well known to us today. WHB constructed his own X-ray generator and tube, and perhaps some idea of how hair-raising this was can be gleaned from the following account made by his son WLB at the time:

"I was scared stiff by the fizzing sparks and smell of ozone, and could only be persuaded to submit to the exposure after my much calmer brother Bob had his radiograph taken to set me an example"

Experiments of this sort often required the person to submit to up to an hour of exposure, and one wonders at the dosage they received. It is clear that by asking his son to submit to this WHB had no idea about any possible dangers of X-rays. Of course this was in the days before the concept of 'safety regulations' was known, and I wonder if the important discoveries made in the 19th and early 20th centuries would have been possible under modern regulations!

During this period WHB carried out experimental and theoretical work on α particles, publishing many papers on the subject. So it was that in 1907, he was elected to the prestigious Royal Society, an honour normally granted only to the highest rated scientists in Great Britain. In 1908 WHB returned to England to take of the post of Professor of Physics at the University of Leeds, where he remained until 1915. WLB, like his father before him, went to Cambridge in 1909 to read Mathematics, finishing his time there in 1914.

It was during this period, that the next major discovery occurred. In 1912 in Germany, following discussions with Ewald, who at that time was a student working on radiation from dipole arrays, Max von Laue realised that the typical interplanar distances in a crystal might be of such a dimension that they might diffract X-rays. It should be realised that at the time, much was unknown about the atomic nature of crystals. Even the very idea of "atoms" was in dispute. Furthermore, there was the controversy as to whether X-rays consisted of particles or of waves. In order to test the diffraction idea, Laue, together with Friedrich and Knipping shone an x-ray beam at a crystal of copper sulphate (by the way in the erroneous and surprising belief that the presence of copper atoms would itself give rise to x-rays). The result was a crude photograph in which they could clearly see spots on the film displaced from the central beam. They very soon improved the technique and within a few days were producing X-ray diffraction photographs of various crystals including alkali halides and zinc sulphide of a quality that even today would be considered good. The results of this were announced on the 8th June of that year at the Bavarian Academy.

There followed several attempts by the German scientists to interpret the X-ray patterns and work back to some idea about the structures of the crystals. Unfortunately von Laue had decided that the x-rays were all of a particular wavelength, but had he discussed this more fully with Ewald it is likely that he would have realised that they were indeed polychromatic and this would have led him to a proper understanding of the photographs.

In England, WHB and WLB were much excited by the Laue discoveries and decided to understand the patterns obtained. WHB was very much of the "particle" school, whereas WLB had an open mind about this. WLB began a series of experiments of his own to test his father's ideas about the nature of x-rays, typically by designing experiments to show that X-ray particles were channelled down tunnels in the crystal. However these experiments were unsuccessful, and it was WLB who realised that his results could be better interpreted by assuming that the X-rays were wave-like and that they were being scattered by planes of atoms. This led him to formulate the famous Bragg Law relating the interplanar spacings of a crystal to the wavelength of the X-rays and to the angle of scattering. WLB's approach to this was very simple, treating the problem as if it were a form of specular reflection involving the interference of reflected waves. With this formula he could explain the positions of all the spots on Laue's films provided he assumed that the incident X-rays were polychromatic. WLB then understood that the intensities of the spots provided information about the locations of atoms on the planes and with this he was able for the first time to derive the crystal structures of simple ionic solids such as KCl and ZnS, thus succeeding where Laue had failed. This work was published in 1913 by WHB and WLB. Thus began one of the most important and enduring scientific endeavours of the 20th century, namely the study of crystal structures. Also in 1913 WHB built the so-called ionisation spectrometer to study X-ray diffraction using counter techniques, and this was the fore-runner of the modern diffractometer used by crystallographers everywhere today.

At this point in 1914 the First World War erupted and WLB joined the army. He was stationed in France in the Map Section. Part of his involvement was in the design of sound-ranging equipment. This consisted of placing microphones at different places near the battlefield and attempting to use the times at which the sound of enemy gunfire reached the microphones to work out where and how far the guns were placed. It was during 1914 that von Laue was awarded the Nobel Prize in Physics. In 1915, while he was still at the war front at Ypres, WLB learnt that he and his father had been awarded the next Nobel Prize in Physics. This made WLB the youngest ever recipient of the Nobel Prize, an honour unbroken to the present day.

Unfortunately because of the war, neither von Laue nor the Braggs were able to attend an actual prize-giving ceremony. However in the early 20's the Nobel Committee decided to hold a special award ceremony for them. I have been told that WHB refused to go because "there will be Germans there"! So you see that feelings about the war were very deep some years after the event.

It is worth appreciating that despite the excellent work done by the Braggs, it was not without controversy. In particular, all chemists knew that sodium chloride, or common salt, consists of Na atoms and Cl atoms in a strict 1:1 ratio. They therefore were attached to the idea that there must be molecules of NaCl in the crystal. However, WLB's model was not molecular but consisted of an infinite framework of alternating Na and Cl ions in all three dimensions. Many chemists objected to this idea, which today we take for granted, in no uncertain terms. Perhaps the most famous attack on the WLB was made in a letter to Nature by Professor H.E. Armstrong in 1927:

" 'Some books are lies frae end to end' says Burns. Scientific (save the mark) speculation would seem to be on the way to this state!..... Professor W.L. Bragg asserts that 'In Sodium Chloride there appear to be no molecules represented by NaCl. The equality in number of sodium and chlorine atoms is arrived at by a chess-board pattern of these atoms; it is a result of geometry and not of a pairing of the atoms'.

This statement is more than 'repugnant to common sense'. It is absurd to the nth degree, not chemical cricket. Such an unjustified aspersion of the molecular character of our most necessary condiment must not be allowed any longer to pass unchallenged. It were time that chemists took charge once more and protected neophytes against the worship of false gods; at least taught them to ask for something more than chess-board evidence".

It must be wonderful to have made such an important discovery as to stimulate that kind of stinging attack!

After the war, WLB went to Manchester in 1919, staying there until 1937 as Professor of Crystallography. Father and son then decided to divide up the field between them, with WHB taking organic crystallography and WLB the inorganic crystals. Both scientists set up research groups that generated many excellent scientists, some of whom went on to gain Nobel Prizes themselves. They also encouraged women to take up scientific careers, including well known crystallographers such as Kathleen Lonsdale and Dorothgy Hodgkin. This enlightened attitude of the Braggs is probably the reason that so many crystallographers today are women.

It is worth saying something here about a problem that developed around this time between the father and son. WHB was at the time the established international scientist, and it was he who tended to be invited to talk about X-ray diffraction and it was he who tended to be asked to serve on important committees, and so on. However, it is also clear that WLB was the one who had made the initial breakthrough in understanding the X-ray diffraction patterns. WHB was a quiet, kind and thoughtful man, while WLB tended to be more excitable and sensitive in nature. To be fair, WHB always tried to explain the important advances made by "my boy", but unfortunately it is difficult to correct the view of the scientific establishment once an opinion has been formed and taken root. The following letter from WLB to WHB (those were the days when fathers and sons communicated by letter) illustrates something about the feeling between them.

"I have been wondering what you were intending to go on with. I do hope you will never keep from doing any bit of work, Dad, because you think that may be the line I am going on..... If we did happen to do the same thing its all to the family credit, isn't? And I am sure I would never be the loser if people weren't quite sure which of us did a piece of work."

In 1923 WHB became Director of the Royal Institution in London. The RI is world famous as an institute set up in the 19th century in order both the carry out scientific research and to educate the public in scientific matters. WHB followed a distinguished line of Directors who included people like

Michael Faraday. He became President of the Royal Society in 1932, and died at the Royal Institution in 1942.

WLB left Manchester in 1937 to take up the post as Director of the National Physical Laboratory, and then in 1938 he was made Cavendish Professor of Physics in Cambridge. In the Austin Wing of the Cavendish Laboratory he began to assemble a large team of scientists, although the Second World War temporarily halted much of the scientific work. However in 1945 WLB created the Crystallography Laboratory as an entity in its own right within the Cavendish Laboratory. Visitors, students and researchers included W.H. Taylor, Henry Lipson, W. Cochran (well known in Ferroelectrics circles), Max Perutz (later to gain the Nobel Prize for his work on haemoglobin) and later on Crick, Watson and others who worked out the structure of DNA. In this period, WLB became an enabler of science in the sense that he provided the environment within which gifted scientists could work at their best. This was a most fruitful period. It is incredible to think that starting from the Braggs' first solutions of crystal structures containing a mere handful of atoms in the unit cell, today it has become routine for biological crystallographers to solve structures with thousands of atoms per unit cell.

In 1954 WLB was asked to become Director of the Royal Institution like his father before. He agreed to this but not without some trepidation, as the previous Director had left 'under a cloud' having been forced out of office by the membership. WLB was afraid that if he took this post he might receive some of the blame. Nevertheless he was prevailed upon to take the post. To give some idea of the bad feeling that this caused in some circles, I was told the following story by his daughter Margaret Heath. WLB had been invited to the Royal Society to one of their soirees at which the Queen Mother was present. She struck up a conversation with WLB and eventually said: "That is most interesting, Sir Lawrence. Perhaps you can tell me more at dinner". To which WLB responded: "I am sorry Maam, but I have not been invited to dinner".

From 1954 until his retirement in 1966 WLB took particular delight in giving superb lectures to school children at the Royal Institution, and indeed it was as a school child myself at one of those lectures that I first saw him. He had a gift for presenting complex ideas in straightforward and simple terms that anyone could understand. He was a brilliant lecturer, always interesting, clever and above all fun. WLB was a family man, artistic in nature and loved children, and this came out clearly in his lectures.

Perhaps I can end by telling a personal and shameful anecdote. I had met WLB several times while I was a student in the laboratory of Kathleen Lonsdale in London, but to say that he knew me would be an exaggeration. In 1970, while I was in Cambridge, I helped W.H. Taylor to arrange his 80th birthday celebrations. One part of this was a reception at the Royal Society. As I drove my car in haste through London to get to the venue on time, a large black car suddenly cut in front of me forcing me to brake suddenly. With the certitude and rashness of youth, I am sorry to admit that I leant out of the window and hurled abuse at the driver who had dared to cross me. I then saw to my horror that it was the great man himself! Some ten minutes later I found myself in a line of scientists waiting to shake his hand and wish him a Happy Birthday. When it came to my turn, and I had mumbled the customary words, he suddenly looked closely at me, and with a twinkle in his eye and a broad grin, began "Young man, ...". However, before he could finish his sentence he was distracted by someone else who had interrupted him, and I never did discover what he had intended to say to me! WLB died the following year on July 1st 1971 at Waldringfield.

I am indebted to Lady Margaret Heath, daughter of WLB, for many personal insights into WLB and for permission to look through the family album of photographs. For those who wish to read a more in-depth account of WLB in particular I would recommend the Biographical Memoirs of Fellows of the Royal Society, by Sir David Phillips, Volume 25, November 1979.

William Henry Bragg (1862-1942)

E. N. da C. Andrade

Memoirs of Fellows of the Royal Society

In the late fifties Robert John Bragg, a .young man of twenty-five, retired from the sea, where he had been serving as an officer in the merchant navy, and purchased with some monies that had been left to him the farm called Stoneraise Place at Westward, near Wigton, in Cumberland; Here he settled down to a farmer's life. In 1861 he married Mary Wood, the daughter of the Vicar of the parish of Westward, and the next year, on 2 July 1862, William Henry Bragg, later to be President of the Royal Society, was born.

Fortunately there are available certain notes dealing with his early life which Bragg himself prepared some time before his death.[1] These contain many vivid little pictures which, besides giving us intimate glimpses of those early days, will, by their manner of telling, recall the man to those who knew him, and give to those who did not something of his direct and unaffected charm of style. Thus, of his mother he says, 'I do not remember my mother very well, as she died when I was barely seven. Just a few scenes remain. I think she must have been a sweet and kind woman. I remember how one day I was sitting on the kitchen table, and she was rolling pastry, and how suddenly found I could whistle: and how we stared at one another for a quiet moment amazed and proud of the new accomplishment...

Of the school at Market Harborough, in Leicestershire, to which he went at the age of seven, he writes: 'Uncle William had in 1869 succeeded in re-

---

[1] The family has very kindly placed them at my disposal. They were mainly written about 1927, small additions having been made in 1937.

establishing the old grammar school in Market Harborough. It is a quaint structure raised on wooden pillars. The butter market used to be held underneath it. The newly appointed master, Wood by name, was an able man, I believe: and the school grew. I was one of the six boys with which it opened after a long interval. Perhaps it was because of my uncle's connexion with the school that at the end of the first year I was given a scholarship of £8 a year exempting me from fees. At the prize-giving—there were many more than six boys at that, time, so that there was quite an assemblage—my name was called out, and I went up to the desk to get the scholarship, not knowing what it was: I was puzzled and disappointed to go back empty-handed. The school was quite good and I got on quickly enough: in 1873 I went up for the Oxford Junior Locals and was the youngest boy in England to get through: I got a third class; and was told that I -would have done better but that the regulations forbade a higher class to any one who did not pass in Church History; in that I failed, as also in Greek.'

In 1875 he left Market Harborough for King William's College in the Isle of Man. 'The place was a very healthy one and after the first year or two, when the bullying was rather unpleasant, I was happy enough. I stood high in the school and liked my work, especially the mathematics: and fortunately I was fond of all the games and played them rather well. So, though I was a very quiet, almost unsocial boy, who did not mix well with the ordinary schoolboy, being indeed very young for the forms I was in, I got on well enough.' He rose to be head of the school. Early in 1880, when he was seventeen, he went up to try for: a scholarship at Trinity College, Cambridge, and was awarded an Exhibition: he was considered rather young and advised to go back to school for a year. The following year he tried again and did not do so well, but was elected to a minor scholarship on the strength of his previous performance. In his notes Bragg puts his 'stagnation' down to a storm of religious .emotionalism that swept, through the school—the boys were scared of eternal damnation and of hell fire, and very much exercised as to what they should do to be saved. 'It really was a terrible year', says Bragg, who, 'though essentially a religious man, adds, 'But for many years the Bible was a repelling book, which, I shrank from reading'. This from his private notes, but the period evidently left a strong mark on his mind, for in the Riddell Memorial Lecture on 'Science and Faith', given in the year before he died, he says, 'I am sure that I am not the only one to whom when young the literal interpretation of Biblical texts caused years of acute, misery and fear'. What he says in this lecture should be, in the hands of all those, who have to deal with susceptible adolescence.

At Trinity, which he, entered in 1881, he enjoyed himself, although he was a somewhat shy arid lonely lad. He liked the work and the tennis ('my tennis was fairly good') and, the boating: he was keenly alive to the beauty of the place, and to his good fortune in belonging to Trinity. In 1882 he obtained a major, scholarship, which gave him an added status which he appreciated. In 1884 he was I placed third wrangler in the Mathematical Tripos, Part I. 'I had never expected anything so high, not even when I was in my most optimistic mood. I was fairly lifted into a new world. I had a new confidence: I was extraordinarily happy. I can still feel the joy of it! Friends congratulated me: Whitehead (of Harvard now) carne and shook me by the hand, saying, 'May a fourth wrangler, congratulate a third?' He had been fourth the year before. As for the Uncles!'

He continued, at Cambridge, taking part three of the Tripos as it then was and attending lectures, including those being given by J. J. Thomson, who had been appointed Cavendish Professor ai the end of 18&4. It was at the end of 1885 that J. J. Thomson asked him if Sheppard, who had been senior wrangler in Bragg's year, was applying for the professorship in mathematics and physics at Adelaide, which had just become vacant by the resignation of Horace Lamb. Lamb had held the post since the foundation of the university in 1875. Sheppard, was not a candidate, but the query put it into Bragg's head that a man of his own age and qualifications might have a chance for what he had till then regarded, as a post for a senior man. Accordingly he sent in his application. With two others he was interviewed by the electors, who were the Agent General (Sir Arthur Blyth), J. J. Thomson and Horace Lamb, and in due course learn that he had been chosen for the post. The part played by J. J. Thomson in the events that sent Bragg to Australia was described in a letter which he wrote to J. J. on 17 December 1936, to convey congratulations on the Master's eightieth birthday. ' ...I must be allowed to add my own personal congratulations. Just fifty-one years ago, I was walking with you along the K.P. on our way to the Cavendish where you were going to lecture and I was going to be one of the audience. You asked a chance question, which sent me off to the telegraph office after the lecture was over and I applied for the Adelaide post which Lamb was vacating. It was the last day of entry; and of course your remark sent me to Australia. Perhaps you were the one who asked a certain Adelaide man—then visiting London—whether the Council of the University of Adelaide was likely to prefer a senior wrangler who occasionally disappeared under the table after dinner to a young man who had so far shown no signs of indulging in the same way. The Adelaide man was Sir Charles Todd, whose; daughter I married a few years afterwards.'

Bragg was much elated. The salary of £800 a year; a very respectable income in those days, was much beyond what he had expected at his age, and the thought of going to a new country and being his own master excited him. His relations were delighted, in particular his Uncle William, who had always helped him and was very proud of his academic career , although the parting was a great blow to the old man. Bragg enjoyed the voyage out, and was fond of telling in after life how he spent part of his time on the boat in studying physics. The post was that of professor of mathematics *and* physics, and Bragg was wont to declare that, at the time, he had not studied the latter subject and knew nothing of it. The electors apparently attached little importance to his inexperience in this respect, and supposed that he would pick the subject up as he went along. At any rate, he read Deschanel's *Electricity and Magnetism* while outward bound. Apparently the demands on a professor of physics were not very high in those days.

The life in Australia delighted him from the first. 'Going to Australia was like sunshine and fresh invigorating air', he wrote. He at once made friends all round, particularly with the Todds. Charles Todd, the head of the family, was Postmaster General and Government Astronomer of South Australia: he was elected an F.R.S. in 1889 and later became a K.C.M.G. In 1889, three years after he landed, Bragg married his daughter Gwendoline. The three children of this very happy marriage were born in Australia—William Lawrence, the present Sir Lawrence Bragg; Robert, who was killed in the Dardanelles during the First World War; and Gwendolen, now Mrs Alban Caroe.

Bragg's simple, modest, unaffected nature, which had rendered him shy in the conventional Cambridge surroundings of those days, expanded in the free, open, good-natured atmosphere of Adelaide society, which; he said, was a revelation to him after the more formal life of Victorian England. Being a personage agreed with him. It is recorded that in his early days in Australia he was one of the least impressive lecturers, but by careful application he developed, towards that perfection in the art of exposition which he afterwards attained

His social gifts won him a wide popularity. He helped to layout the first golf course in South Australia, at Glenelg, played golf, a game at which he was good, painted in water colours and enjoyed himself. Bicycling came into vogue during his Australian days, and he loved bicycle tours and picnics during the long lazy summer vacations by the sea. It was a pleasant life for a man with a zest for the good simple pleasures of the open air.

More important for science, he became interested in experimenting, of which he had had little or no experience in England. Following Rontgen's great discovery in 1895 he set up the first X-ray tube to operate in Adelaide, possibly the first in Australia. He still, however, made no attempt to carry out any original investigation, and no doubt to those around him it might well have appeared that he was destined to lead a pleasant and useful life as a popular teacher and good friend in the Adelaide community, content with his local fame. At the age, of forty-one he had amassed a long experience of teaching the fundamentals of physics and had developed a keen critical sense: he was a clear and mature, thinker but had produced nothing that could be called research. This maturity and long training in exposition were of immense advantage when he turned to discovery.

In January 1904 the Australian Association for the Advancement of Science met in Dunedin, New Zealand, and Bragg was confronted with the task of giving the presidential address in the section dealing with astronomy, mathematics and physics. At the time the early discoveries in radioactivity were holding the attention of the world of physics and work on the electron was being actively pursued in the Cavendish Laboratory and elsewhere. Exciting discoveries in the new fields were pouring out. In particular, Lenard's second famous paper on the absorption of cathode rays by matter had just appeared. In his address Bragg reviewed the work on radioactivity and on the properties of the electron and, in consequence, he became greatly interested in the question of the penetration of matter by elementary particles. At the time the atom was held to consist either, of electrons widely dispersed in a sphere of positive electricity (J .J. Thomson) or of widely spaced 'dynamids' (Lenard), a dynamid consisting of a positive and a negative charge, closely associated. This conception of a neutral pair was one that Bragg used later in another connexion. In either picture there would be comparatively strong, highly localized electric forces separated by large spaces, in which the forces were feeble. The scattering of the beam of electrons in Lenard's experiments was put down as the effect of a series of random deviations due to passages close to the force centres.

It occurred to Bragg that, if this were so, an alpha particle should, on account of its mass, pass through a thin foil practically undeviated and that, in consequence, the exponential laws of absorption, which held for a beam of electrons, should not apply to a beam of alpha particles. In his own words, taken from his 1904 address[2], 'it cannot be correct to say that the amount of the, radiation which penetrates a distance x is proportional to the expression e^{-ax}: it must rather be proper to say that—

(1) The number of α particles penetrating a given distance does not alter much with the distance until a certain critical value is passed, after which there is a rapid fall.

(2) The enery of the α particles penetrating a given distance gradually decreases as the distance is increased, and dies out at the same critical value.' The passage is quoted, since it shows the typical Bragg clarity and precision, the power to put in the simplest language the essence of a novel problem. Until he had reduced his thoughts to the state where simple and direct expression was possible he was not at ease.

A few months after the address some radium bromide was placed at his disposal, and, assisted by Kleeman, he began his classical researches on the range of the alpha particle and the closely allied questions of the ionization produced by the particle and of the stopping power of substances. Using as detector an electrometer which measured the ionization produced in a thin slab of air between a gauze and a plate, he established the sharply defined range of the alpha particle and the variation of the ionization along the path, obtaining with radium which had come into equilibrium with its products the four ranges, corresponding to the particles from radium, radon, radium A and radium C, according to Rutherford's theory, which was thus confirmed; With Kleeman he found that the 'stopping power' ,of matter was approximately proportional to the square root of the atomic weight. 'Considering the complexity of the phenomena involved in the absorption of an a particle by matter,' it is a matter of interest and also of practical convenience that such a simple rule should hold roughly over the whole range of the elements.'[3] He corresponded freely with Rutherford about the work on the α particle, whose properties they were pursuing independently along different lines.

---

[2] See W.H. Bragg. *Studies in Radioactivity*, London, 1912.

[3] Rutherford, Chadwick and Ellis, *Radiations from Radioactive Substances*, Cambridge, 1930.

He also carried out fundamental work on the ionization produced by α-rays and on the properties and behaviour of the secondary electrons expelled from matter by incident β and γ radiations, electrons to which Bragg applied the name β-rays, or secondary β-rays. In some of the experiments on these β-rays he had the collaboration of Madsen, now Sir John Madsen.

The experiments on the secondary electrons expelled by γ-rays were important for the development of Bragg's theoretical views. They established that the velocity of the β-rays depended not upon the intensity but upon the 'quality' of the 'Y-ray, increasing with the penetrating power of the latter, and that the velocity was independent of the atom from which the β-ray was expelled. These results were shown by Sadler and others to apply to the β-rays produced from matter by incident X-rays. It was largely on experiments of this kind that Bragg based his views that X-rays were corpuscular in nature, views which his own fundamental experiments later showed to require modification. The paradoxical coexistence of particle and wave properties was a favourite theme of Bragg's, to which we shall return.

The fundamental researches carried out, in quick succession, in Adelaide between 1904 and 1908 speedily established his name throughout the world of physics as an original investigator of the first rank. He was elected a Fellow of the Royal Society in 1907, less than three years after the reading of his first original paper, Rutherford being his proposer. It was inevitable that he should be called to a chair in the old world. In 1908 he received the offer of the Cavendish professorship of physics at Leeds, which brought him back to England.

During his twenty-two years in Australia Bragg had identified himself with the life of the community in which he lived and had established himself as a good man, a great teacher and a firm friend. He had many ties with Adelaide, and had thoroughly enjoyed life there. In after days he always spoke with the greatest affection of South Australia.

At Leeds Bragg was, at the beginning, fully occupied with organizing the teaching at the laboratory, and naturally, for a time, he did comparatively little experimental work. He developed and defined his view that X-rays and 'Y-rays were of a corpuscular nature, the lack of deflection in a magnetic or electric field being explained by the hypothesis that an elementary ray was in the-nature of a neutral doublet—'an electron which has assumed a cloak of darkness in the form of sufficient positive electricity to neutralize its charge', to use Bragg's own typically arresting phrase. Although this picture can no longer be deemed satisfactory it did account for many properties of the radiation which are embodied in the quantum theory. In particular Bragg insisted that many of the experimental facts seemed to show that an elementary X- or γ-ray was a definite and concentrated unit rather than a spreading pu1se or wave, as indeed they do. The doublet theory was also successful in explaining in general way certain aspects of the conversion of cathode rays into X-rays and of the release of electrons by X- and γ-rays. It had its usefulness in concentrating attention on the particle aspect of the radiations, which plays so prominent a part in modem theories.

In the course of his work ()n secondary radiation Bragg was led to the view that the ionizing effect of the X-rays is an indirect one, due to the secondary electrons released by the primary X-ray. He was the first to insist on this important fact, and he supported his contention by experiments carried out in conjunction with H. J. Porter, which were published in 1911.

Laue's discovery, with Friedrich and Knipping, announced in June 1912, that X-rays could be diffracted by passing through crystals, caused a sensation, in the physical world. Bragg's interest, was at once captured, and it may be of interest to quote what he wrote in *Nature* in November of that year—'Dr Tutton suggests that the new experiment may possibly distinguish between the wave and the corpuscular theories of the X-rays. This is no doubt true in one sense. If the experiment helps to prove X-rays and light to be of the same nature, then such a theory as that of the 'neutral pair' is quite inadequate to bear the burden of explaining the facts of all radiation. On the other hand, the properties of X-rays point clearly to a quasi-corpuscular theory, and certain properties of light can be similarly interpreted. The problem then becomes, it seems to me, not to decide between two theories of X-rays, but to find, as I have said elsewhere, one theory which possesses the capacity of both.' The theory of the diffraction spots which Laue obtained by consideration of a three-dimensional grating is somewhat complicated, involving as it does the consideration of interfering wavelets in three-dimensional space. The same year W. L. Bragg gave a much, simpler interpretation of the phenomena, by considering the reflection of waves from parallel layers of atoms or diffracting points, each,typical set of parallel crystal planes acting as a reflecting surface for radiation whose wavelength fulfilled the Bragg law $n\lambda = 2d \cdot \sin a$, where d is the distance between parallel crystal planes and a is the glancing angle, i.e. the complement of the angle of incidence. The more densely populated crystal planes gave, speaking generally, stronger reflections, so that planes with higher indices were not strongly represented. W. H. Bragg at once took up, with his son, the experiments on the reflection of X-rays which this interpretation suggested, and early in 1913 there appeared in the Proceedings of the Society the first joint paper, which founded the science of crystal analysis by means of X-rays.

Up to the beginning .of the first Great War in 1914 five further classical papers were produced by Bragg, in one of which, that on the structure of the diamond, he had the collaboration of his son. Among other subjects dealt with were the general technique of the X-ray spectrometer; the characteristic absorption of the different radiations and its effects; the structure of sulphur and quartz; and the general question of intensities. An investigation with S. E. Peirce led to the Bragg-Peirce law, according to which if we keep to frequencies below the band at which the absorption discontinuity takes place, the absorption coefficient per atom is proportional to the fourth power of the atomic number and to the 5/2 power of the wavelength.[4] In these early experiments Bragg made use of the ionization chamber to detect and measure the rays. His earlier work had taught him how to overcome the difficulties connected with this type of measurement and he and his son were strikingly successful with the ionization spectrometer. The photographic method had" already been used by H. G. J. Moseley in his classical researches, but it was only later that Bragg adopted it.

The work of Bragg and his son Lawrence in the two years 1913, 1914 founded a new branch of science of the greatest importance and significance; the analysis of crystal structure by means of X-rays. If the fundamental discovery of the wave aspect of X-rays, as evidenced by their diffraction in crystals, was due to Laue and his collaborators, it is equally true that the use of X-rays as an instrument for the systematic revelation of the way in which crystals are built was entirely due to the Braggs. This was recognized by the award of the Nobel prize for Physics in 1915 to them jointly 'pour leurs recherches sur les structures des cristaux au moyen des rayons de Roentgen', and a further formal acknowledgment was the appearance in Leipzig, in 1928, of a collected reprint, in German translation, of the early papers, under the title *Die Reflexion von Rontgenstrahlen an Kristallen: grundlegende Untersuchungen in den Jahren* 1913, *und* 1914 *von W. H. Bragg und W. L. Bragg*.

The outbreak of war in 1914 found Bragg hard at work at Leeds, where all his early experiments on crystal structure were carried out. He was a leading figure in the university there, and occupied the office of Pro-Vice-

---

[4] The value of the exponent is usually taken as 3, instead of 5/2, to-day.

Chancellor. He continued his X-ray work into 1915, publishing, for instance, a paper on the spinel group of crystals. In this year he was appointed Quain Professor of Physics at University College, London, but by then he had become involved in war work.

The Board of Inventions and Research was instituted in July 1915, for the purpose of giving the Admiralty expert assistance in organizing and encouraging scientific effort in connexion with the requirements of the Naval Service, and Bragg was an original member. The submarine menace was becoming acute and the use of acoustic methods for locating underwater craft came up for discussion before the Board, with the result that Bragg was put in charge of research on the detection and measurement of underwater sounds, within the Anti-Submarine Division of the Admiralty. In the first instance he was, in April 1916, made Resident Director of Research at the Admiralty experimental station at Hawkcraig. After many troubles, largely within the Admiralty (see J. J. Thomson's *Recollections and Reflections*), a laboratory was built for him at Parkeston Quay, Harwich, where Bragg started work in 1917, having under him, among other physicists, A. O. Rankine. The hydrophone or underwater receiver, developed by Bragg and his team, rendered great service in the war against the submarine. The instrument was afterwards simply described by Bragg in his *World of Sound*, a book founded on his first course of Christmas Lectures at the Royal Institution, and in two lectures of which an account is given in *Engineering* for 13 June 1919. During the course of the experiments and research on anti-submarine work principles were established and methods, as well as apparatus, devised which were of great service. It was probably in acknowledgment of his war work, as well as of his scientific eminence, that Bragg was made a C.B.E. in 1917 and was knighted as a K.B.E. in 1920. In, the same year, 1920, he was made an Honorary Fellow of Trinity College, Cambridge, a distinction that gave him great pleasure.

The war over, Bragg took up his work as Quain Professor of Physics at University College, London, and promptly started research there. He gathered about him several young research workers, among whom Backhurst may be mentioned, and founded that school of searchers after the secrets of crystal structure which later flourished at the Royal Institution.

In 1921 Shearer and Astbury joined him, and shortly after Müller and Miss Yardley (Mrs Lonsdale) appeared. In those days Bragg was himself actively experimenting with his own hands, as well as directing research: he was seated at his spectrometer whenever he got the chance. At the beginning of his work at University College he still used his first well-tried weapon, the ionization chamber, as detector, but gradually gave it up, for most purposes, in favour of the photographic plate. The equipment at University College was somewhat scanty at first and Bragg and his students, Müller and Shearer in particular, set to work to develop it. Continuously evacuated X-ray tubes, both hot wire and gas filled, were introduced and a self-rectifying gas tube was evolved, which gave useful service for many years. In those days vacuum pumps had not reached their present state of efficiency and hot wire tubes were rather trouble some. The work was supported by generous grants from the Department of Scientific and Industrial Research, which were more than justified by the results.

The University College period was notable for the first attack on the structure of organic crystals. For much of this work Bragg employed the powder method, whereas hitherto he had worked with single crystals. He embodied his results on naphthalene and naphthalene derivatives in his presidential address to the Physical Society in 1921, having been elected to the presidential chair in the previous year. He worked on the assumption that the benzene or naphthalene ring is an actual structure, which preserves its general form and size from compound to compound and, to the satisfaction of the organic chemist, his results justified this hypothesis. This work was the starting point of the series of investigations on different classes of organic compounds which he afterwards directed at the Royal Institution. He also worked out the probable structure of ice, and at an annual dinner of the Alpine Club exhibited a model, which, being made of soft dental-wax, proved itself by wilting as the evening grew warmer.

In 1919 occurred an event, apparently insignificant, that was to have a profound effect on Bragg' s life—he gave the Christmas Lectures ('A Course of Lectures adapted to a Juvenile Auditory' as the old phrase runs) at the Royal Institution. The title of the course was 'The World of Sound' and it not only established Bragg's name as a popular lecturer but also showed the originality, the personal qualities, which he could bring to an apparently hackneyed subject. He classified his subject on new lines, the six lectures being entitled 'What is Sound?' 'Sound and Music', 'Sounds of the Town', 'Sounds of the Country', 'Sounds of the Sea', 'Sounds in War'. 'Sounds of the Country' in particular, showed his love for, and careful observation of, nature out of doors. His powers of simple exposition, his personal and persuasive style, his affection for young people-all became known to a wide circle.

It may be worth while to quote the opening sentences from the book that embodied these lectures, as they give some idea of his style. All around us are material objects of many kinds, and it is quite difficult to move without shaking some of them more or less. If we walkabout on the floor it quivers a little under the fall of our feet; if we put down a cup on the table, we cannot avoid giving a small vibration to the table and the cup. If an animal walks in thy forest, it must, often shake the leaves or the twigs or the grass, and unless it walks softly with padded feet it shakes the ground. The motions may be very minute, far too small to see, but they are there nevertheless.' The words are simple and colloquial, with no straining after effect, but they are striking and tell us at once that we are in the presence of one who was a lover of nature and a master of simple, direct and captivating exposition.

It was, perhaps, partly as a result of this course of lectures that, when on the death of Sir James Dewar in 1923 the headship of the Royal Institution fell vacant, Bragg was elected to succeed him under the complex of titles Fullerian Professor of Chemistry in the Royal Institution; Director of the Laboratory of the Royal Institution; Superintendent of the House; and .Director of the Davy-Faraday Research Laboratory. The Fullerian Professorship of Chemistry is a traditional title which carries somewhat indefinite duties: the Laboratory of the Royal Institution and the Davy-Faraday Research Laboratory are run more or less as a whole, the Managers of the Institution administering the Davy-Faraday funds and the Resident Professor in the Royal Institution being also the Director of the Davy-Faraday Laboratory: the superintendence of the House entails the general responsibility for the maintenance and smooth running of the Institution which the head would naturally be supposed to exercise: the Davy-Faraday Laboratory was founded by Dr Ludwig Mond under a deed of trust for the carrying out of original research. The whole complicated position is one that could only exist in England, but it works, though whether it works better than a simpler and more unified administration would do has not been proved.

Dewar was over eighty years old at the time of his death and there was much to be done in the way of reorganizing, for .instance, the work in the Davy-Faraday Laboratory, which Bragg, promptly directed to problems of crystal structure. He brought with him from University College Müller and Shearer, who were mainly responsible for the actual installation of the new type of apparatus needed for the X-ray methods. They also continued the work on long chain compounds which they had begun at University

College, while Shearer further worked on the theory of space groups. The work which Müller carried out on the fatty acids was typical of the tendency towards the investigation of organic structures which the laboratory took under. Bragg's direction, while his son at Manchester concentrated on inorganic crystal Structure. Among the early workers in the laboratory may be mentioned J. D. Bemali R. E. Gibbs, who did not pursue organic compounds but worked on the structure of quartz; Miss Yardley, afterwards Mrs Lonsdale, who became one of the pillars of the laboratory; Miss C. F; Elam (Mrs Tipper), known for her work on metals; W. T. Astbury, who came from University College and afterwards investigated the crystal structure of the products of the living body, such as hair and horn; and J. M. Robertson, somewhat later, who worked on the crystal structure of anthracene and naphthalene, and applied the methods of Fourier analysis to deduce crystal structure from intensity measurements. There are other distinguished names among those who spent a short time in the laboratory. It speedily became a world-famous centre of research. Bragg was no longer able to spend as much time as formerly at his own researches, but the work of the laboratory was an informed whole which in all its features gave evidence of his wise guidance and prescience.

One of Bragg's first tasks at the Institution was to give the Chnstmas Lectures. He chose as his title that of the famous poem of Lucretius, 'Conceming the Nature of Things', and talked about atoms, gases, liquids and crystal structure with masterly simplicity and inimitable charm. He always gave the impression that he thoroughly enjoyed addressing the 'Juvenile Auditory' for whom these lectures were traditionally intended. On two other occasions, namely at Christmas 1925 and Christmas 1931, he gave the lectures on 'Old Trades and New Knowledge' and on 'The Universe of Light'. The 1925 discourses gave evidence of his appreciative interest in the industrial aspects of science which he later emphasized by taking 'Craftsmanship and Science' as the subject of his Presidential Address at the British Association at Glasgow in 1928.

Bragg brought to the social life of the Institution a charm and suavity which rapidly won all hearts. Lady Bragg, with a kindheartedness whose sincerity was apparent, and with a graciousness free from all affectation, was an ideal hostess at the mixed gatherings that thronged the Braggs' private apartments after the Friday night discourses. Their daughter" Gwendolen, an accomplished artist, some of whose drawings appeared in Bragg's Universe of Light, lived with them, and Lawrence Bragg was a frequent visitor. The unity of the family was felt outside the walls of their official residence. Lady Bragg's death in 1929 after much suffering was indeed a heavy blow to Bragg and a source of sincere grief to all those who had come to know her at the Royal Institution.

After her mother's death Gwendolen Bragg became her father's closest companion, both as the centre of his personal home life and as hostess at the many friendly assemblies; great and small, that characterized the Institution. On her marriage in 1932 to Alban Caroe she gave up residence at the Institution for some eighteen months, but then returned and, with her husband, lived there for the rest of her father's life. Between Bragg and his daughter there existed a warm affection and comradeship which illuminated both their lives. 'We had grand times together', she wrote to a friend.

An essential part of Bragg's family life was the country cottage at Chidding-fold, Surrey, which he purchased in 1923. Hither he would retire for relaxation during holidays and at week-ends, and here .he would entertain those working at the Royal Institution in an atmosphere of intimacy which brought them very close to him. They, too, felt that they were part of the family.

Many important changes took place in the Institution under Bragg's rule. In 1929 an extensive reconstruction of the Royal Institution house, which lasted two years, began. It had become clear that the lecture theatre did not comply with the requirements for safety, especially against fire, demanded of a building to-day, and the complete reconstruction of the interior of the theatre and the provision of proper exits involved a number of alterations which were embodied in a comprehensive architectural scheme. Bragg took a leading part in the negotiations and the liberal contributions that were made to the expenses of the undertaking were largely the result of his persuasive charm. A further reconstruction took place in 1936, in which he was again active. Many notable additions were made to the equipment of the Davy-Faraday laboratory: in particular a giant X-ray generator, of a new type designed by Müller, was built and installed in the Institution. This possessed a rotating hollow target, with water cooling, and dealt with an input of fifty kilowatts. A smaller tube with an input of five kilowatts was also erected. The great intensity obtainable with these tubes very much accelerated work in general by permitting very short exposures and rendered feasible certain experiments which otherwise could not have been successfully carried out. Thus high dispersion was achieved by using much larger distances from object to plate than are practicable with weaker sources, and such experiments as those of Müller on the effect of compression on the lattice spacing of certain organic compounds were made possible.

In spite of the many calls upon his time which his name and fame entailed, and of advancing years, Bragg kept well abreast of physical thought and keenly welcomed all advances. Even when he\did not pretend to have mastered all the mathematical intricacies of the latest developments he was able to grasp very quickly the more important implications and to give them a new turn, which not only simplified them but set them in relation to the general advance of science. In 1928, for instance, he arranged for Schrödinger to deliver a course of lectures at the Institution on wave mechanics, then a very recent development. (Incidentally it is said that an enthusiastic, but unmathematical, yachtsman appeared at the first lecture, hoping to learn something about sailing in a rough sea). He gave an introductory lecture as preparation for the course, in which he returned again to the wave and corpuscle dichotomy. He referred back with pleasure and some pride to his old experiments with Kleeman and Madsen; which have already been mentioned, saying, 'I may say, I think, that in these experiments we were, though unwittingly, carrying out Einstein's suggestion that the corpuscular hypothesis deserved careful exploration. It is true, however, that I thought of the X-ray and γ-ray problems as distinct from that of light.' He ended on a typical thought, typically expressed, 'When the picture is finally clear there will no doubt be atoms in it, electrons, wave motions, energies, momenta and so on. But have we got them all rightly joined up? Perhaps wave motion belongs to more than the photon or to something else than the photon? We can only wait.' No doubt he was thinking of the experiments of Davisson and Germer and of G. P. Thomson, indicating the wave nature of the electron, of which accounts had just appeared.

On more than one occasion, while at the Royal Institution, Bragg showed his extraordinary power of taking up a new subject, adding something to it, and then laying it down again. A particularly good example is offered by his work on liquid crystals. In connexion with the Faraday Society's discussion on this subject, held at the Royal Institution in 1933, he became interested in the phenomena of smectic crystals, and he then showed, in an extremely simple and elegant way, that the static arrangement of a number of equidistant parallel layers, without rigidity, was a set of surfaces formed in successive layers around an ellipse, having as lines of discontinuity the ellipse itself and a hyperbola in a plane perpendicular to it, each curve passing through the focus of the other. In short, he explained the phenomena in terms of the cyclides of Dupin. He never seems to have concerned himself much with liquid crystals after his lecture on the subject at the Royal Institution in November 1933. The case serves to show how he retained his rare geometrical sense and could apply it at will.

Late in his work at the Institution Bragg carne to the conclusion that it was desirable to have some fundamental theoretical work proceeding at the same time as the experimental investigations and a systematic mathematical attack on details of the methane spectrum was made by H. A. Jahn, working sometimes in collaboration with W. H. J. Childs. This kind of work was far from anything that Bragg himself ever did, but it had his warm support and encouragement.

Towards the end of his life all his old enthusiasm was aroused by the so-called extra reflexions or diffuse spots, which can be observed with powerful X-ray illumination of single crystals. These had been adventitiously observed from time to time, but in 1939 G. D. Preston published a careful study of them in certain simple cases and explained them as due to small crystalline fragments in, imperfect alignment. The matter was immediately taken up at the Royal Institution, where the powerful tubes to which reference has already been made allowed the effects to be obtained with relatively short exposures. Mrs Lonsdale and H. Smith published a detailed study of the spots obtained with both organic and inorganic crystals, in which Bragg took the greatest interest. He was much intrigued by the effects, and was concerned about their explanation. The elaborate mathematical theories which had considered the thermal movements within the crystal as responsible for the reflexions did not appeal to him, and he evolved a simpler theory along the lines suggested by Preston, attributing the spots to a lack of regularity in the crystal structure. Discontinuities at the boundaries of the regular fragments or which the crystal is supposed to be composed lead to discontinuities in the phase relationship. Bragg wrote several short papers and notes on the subject, and, in fact, his last paper, written just before his death, and published posthumously, was on the secondary X-ray spectrum (extra reflexions) of sylvine.[5]

By 1930 Bragg had become not only one of the great figures of English science but also something of a national figure. In that year the Royal Society bestowed on him the Copley medal, its senior award. He had

---

[5] A simple account by Mrs Lonsdale of the extra reflexions may be found in *Engineering* p. 254,27 March 1942.

received the Rumford medal in 1916. He was an honorary doctor of some sixteen British and foreign universities, and a member of the leading foreign societies. In 1931 he received the Order of Merit. If any great scientific body had a ceremonial lecture to be delivered, Bragg was asked to act, and he generally consented if he found it possible. Whether he enjoyed delivering addresses it is hard to say. Beforehand he appeared to feel some diffidence and certainly the utterance that seemed to come so easily and so spontaneously was the fruit of more labour and thought than the audience often suspected. I think that once on his feet he did enjoy it, as a man enjoys doing anything in which he is supremely competent. Afterwards he received the compliments and congratulations, with an artless modest pleasure that became him well. He never lost his zest: if he always maintained the interest of his audience it was, perhaps, because he always maintained his own interest.

In 1935 Bragg, now seventy-three years of age, was elected to the Presidency of the Royal Society. His reputation, his fine presence, his dignity tempered with geniality, his wide knowledge and his ready sense of the right word made him a complete figure in the Chair. His kindness and ease of access commended him to the younger Fellows: his respect for tradition and his historical sense commanded the confidence of the older Fellows: his close connexion with Trinity College oiled many wheels. His kindliness led him to welcome certain ambiguous advances from learned bodies in Nazi Germany, and he did his best to further ostensible plans for an understanding between the two countries which, in his goodness of heart, he took at their face value.

The war found him burdened with a variety of heavy duties which he, never sparing himself when the good of the scientific community or of the country at large was in question, allowed to be increased. Soon he was not only President of the Royal Society and responsible for the many duties that attach to the head-ship of the Royal Institution, but Chairman of the Scientific Committee on Food Policy, Chairman of the Scientific Advisory Committee, and Chairman of the Science Committee of the British Council. He was also a member of the Advisory Council of the Department of Scientific and Industrial Research, and held a number of other appointments, none of them sinecures. Very few men of seventy-seven could have carried out in any fashion the variety of responsible tasks which he discharged with grace, dignity and. efficiency. His voice on the wireless service gave pleasure to hundreds of thousands, and he was keenly interested in the teaching of the elements of science to the boys of the Air Training Corps. He even wrote a little book The Story of Electromagnetism to help them in their studies. Naturally he did not feel inclined to embark on a campaign for reforming the whole organization of science in the war-time service of the country, but within the cumbersome, comp1icated and capricious frame set up by the Government through its peace-time servants he rendered great services. He laid down the Presidency of the Royal Society in 1940, and with it some other duties, but the demands on his services outside his duties at the Royal Institution were still many.

With his naturally strong constitution and strong sense of duty he carried on, bravely all through the worst period of the war, but his friends were distressed to see how exhausted he was from time to time. He would always make an effort when needed, but it cost him more and more. For some time his heart had been giving him trouble and he endeavoured to avoid physical exertion, while still using his mind like a young man. Reference has already been made to his enthusiastic interest in a new X-ray phenomenon, and as late as December 1941 he was writing in Nature about it. On Tuesday, 10 March 1942, the gallant veteran had to take to his bed: two days later he was dead. Bragg had an astonishing career. Up to the age of forty he never showed any desire to carry out original experiment. He then straightaway embarks upon a perfectly precise and important piece of work and within ~ few years his name is known wherever physics is seriously studied. He spends some years carrying out a careful series of experiments which can be interpreted to prove the corpuscular nature of X-rays, and he stresses this interpretation. He then himself conclusively demonstrates, by the work with which his name will always be, associated, the wave nature of X-rays. He starts life as an extremely shy and retiring youth, never, apparently, quite at home in Cambridge, and in his old age becomes a national figure, at ease in all surroundings, whose personal appeal is known all over England.

Yet there is nothing contradictory in his character. Bragg's nature was simple, straightforward and tenacious—incidentally, of course, he was a man of genius. It would never have occurred to him to embark upon research with the object of publishing a paper or papers to impress other people. He waited until he found something that seemed to him to ask for experiment and in the first case he did not come across this until he had trained both his critical faculties and his powers of exposition in the faithful discharge of his duties as a teacher. This long period of apprenticeship probably had a profound influence on his work. It helps to account for his great objectivity and for his power of formulating a problem simply and directly. Bragg's fundamental ideas, as he expounded them, were accessible to any undergraduate. His greatness was shown in .their originality and in the skill and perseverance with which he shaped them to well-recognized ends. He was a. very great experimenter who never wasted his time on the trivial or hid difficulties under the graceful veils of mathematical obscurity. His hypotheses were stated clearly as hypotheses, his experimental results as experimental results—the former he was ever willing to modify or to abandon as later discoveries might dictate, the latter he knew to constitute Knowledge won for all time. His work, like his personality, was simple yet profound, sincere and compelling.

Bragg preserved all through his fame many of the more admirable characteristics of youth: He was capable of a warm burst of enthusiasm at any new idea or achievement that appealed to him. He was apt to believe that all men were as sincere and diffident about their own achievements as he was about his own, which occasionally led him to lend support where it might, without disadvantage, have been withheld. But how much better this generosity of outlook became him than a coldly critical attitude would have done! Again, he never shrank from asking for information or advice: he was far too big a man to mind admitting gaps in his knowledge. His warm, simple, persuasive utterance, his personal, tone in lecturing which made each member of the audience think that the remark was intended for him, was also more reminiscent of the wise elder brother who was sharing with you the pleasure of a discovery, than that of the great sage who was instructing you. But he was a great sage.

Bragg was a man of very strong family feeling, who was never happier than with his children, and, later, his grandchildren. He took a particular pride in the achievements and career of his brilliant son. It was always a delight to his hearers to note the affection that came into his voice when, in lectures, he found occasion to deal with some one or other piece of work which had been carried out by 'my boy'. His unaffected pleasure and surprise at the news of some new high post or distinction awarded to him was very lovable. He was, for instance, obviously overjoyed when Sir Lawrence received his knighthood. The quiet affection that existed between him and his daughter was a source of pleasure to many outside his family circle.

There was nothing narrow about Bragg's interests. Reference has already been made to his love of games, and he was always ready to talk of life in the open air. He had a great affection for the sea and was always at his happiest on an ocean voyage, the first day of which seemed to wipe away his cares. He was a lover of simple music and enjoyed playing the old tunes on his flute, the sound of which, in the old days at the Royal Institution, could often be heard coming from his study at the evening hour. In his younger days he had been a very good draughtsman and painter in water colours, and he knew more than a little about .pictures. His range of reading was wide, both among the English classics and the moderns. With all this, he was an exceedingly good conversationalist and a first-rate impromptu speaker. Nature had given him a fine presence. In any company he was an unmistakable figure.

Religion was a strong influence in Bragg's life. He had no strong dogmatic views but he had a simple and genuine piety and was an enemy of unbelief, as he was of loose talk of every kind; He was not comfortable at any discussion, however serious and philosophical, that touched on the weaker side of man's nature. There were many things whose existence he preferred not to acknowledge. In religion, as in other matters, he was tolerant of the views of those who sincerely held views different from his own. Something of his own belief will be found in his address on 'Science and Faith'.

There was, we like to think, something peculiarly British about Bragg. His attitude towards physics was that characteristic of the great experimenters of our land, especially his strong pictorial sense. He was a lover of the traditions, especially those of the great institutions with which he was connected. His lack of pedantry, his gift for popular exposition, his strong feeling for the craftsman in factory and workshop are all characteristics which he shared with Faraday, with Tyndall, with J. J. Thomson. He was an ornament, not only of English science, but of English learning, a great teacher and a good man, whose death carne as a personal loss to all those who knew him. With him went an outstanding representative of a great period of English physics.

E. N. DA C. ANDRADE

BIBLIOGRAPHY

1891. The 'elastic medium' method of treating electrostatic theorems. Rep. Aust. Assoc. Adv. Sci. (Christchurch), 3, 57-71; Phil. Mag. 34, 18-35 (1892).

1892. The energy of the electromagnetic field. Trans. Roy. Soc. S. Aust. 15, 74-76.

1892. Presidential address to Section A. Rep. Aust. Assoc. Adv. Sci. (Hobart, Tasmania), 4. 31-47.

1895. The energy of the electromagnetic field. Rep. Aust. Assoc. Adv. Sci. (Brisbane, Queensland), 6.228-231.

1904. Some recent advances in the theory of the ionization of gases. (Presidential address to Section A.) Rep. Aust. Assoc. Adv. Sci. (Dunedin), 10, 47-77.

1904. On .the absorption of α-rays, and on the classification of the α-rays of radium. Trans. Roy. Soc. S. Aust. 28. 298-299 (abstract); Phil. Mag. 8, 719-725.

1904. (With R. KLEEMAN.) On the ionization curves of radium. Phil. Mag. 8, 726-738.

1905. Diea-Strahlen des Radiums. (Translation from English). J. Radioakt. 2. 4-18.

1905. (With R. D. KLEEMAN.) On the alpha particles of radium, and their loss of range in passing through various atoms and molecules. Trans. Roy. Soc. S. Aust. 29, 132-133 (abstract); Phil. Mag. 10, 318-340.

1905. On the α particles of radium. Phil. Mag. 10, 600-602.

1905. (With R. J). KLEEMAN.) On the recombination of ions in air and other gases. Trans. Roy. Soc. S. Aust. 29, 187-206; Phil. Mag. 11, 466-484.

1906. Die α-Strahlen des Radiums. (Translation from English.) Phys. Z. 7, 143-146.

1906. Uber die a-Strahlen des Radiums. (Translation from English.) Phys. Z. 7, 452-453.

1906. On the ionization of various gases by the α particles of radium. No. 1. Trans. Roy. Soc. S. Aust. 30,1-15; Phil. Mag. 11, 617-632.

1906. The α particles of uranium and thorium. Trans. Roy. Soc. S. Aust. 30, 16-32; Phil. Mag. 11, 754-768.

1906. On the ionization of various gases by the α-particles of radium. No.2. Trans. Roy. Soc. S. Aust. 30, 166-187; Phil. Mag. 13, 333-357 (1907); Proc. Phys. Soc. Lond. 20, 523-550 (1907).

1907. The influence of the velocity of the α particle upon the stopping power of the substance through which it passes. Rep. Aust. Assoc. Adv. Sci. (Adelaide), 11, 318 (abstract); Phil. Mag. 13, 507-516.

1907. A comparison of some forms of electric radiation. Trans. Roy. Soc. S. Aust. 31, 79-93.

1907. The nature of Röntgen rays. Trans. Roy. Soc. S. Aust. 31, 94-98.

1907. On the properties and natures of various electric radiations. Phil. Mag. 14, 429-449; Ann. Rep. Smithsonian Instn. 195-214; no. 1830.

1907. Uber die Zerstreuung der α-Strahlen. (Translation from Engl ish) Phys. Z. 8. 886-887.

1907. (With W. T. Cooke) The ionization curve of methane. Trans. Roy. Soc. S. Aust., 1 31.111-113; Phil. Mag. 14. 425-427.

1907. (With J. P. V. MADSEN.) The quality of the secondary ionization due to β-rays., Trans. Roy. Soc. S. Aust. 31, 300-304; Phil. Mag. 16, 692-697 (1908).

1908. (With J. P. V. MADSEN.) An experimental investigation of the nature of γ-rays. No.1. Trans. Roy. Soc. S. Aust. 32. 1-10; Phil. Mag.15. 663-675; Proc. Phyl. Soc. Lond. 21, 261-275.

1908. (With J. P. V. MADSEN.) An experimental investigation of the nature of γ-rays.No.2. Trans. Roy. Soc. S. Aust. 32, 35-54; Phil. Mag. 16, 918-939; Chem. News. 97, 162-165.

1908. The nature of 'γ- and X-rays. Nature, Lond. 77, 270-271; 78, 271, 293-294.

1908. The nature of X-rays. Nature, Lond. 78, 665.

1908. (With J. L. CLASSON.) On a want of symmetry shown by secondary X-rays. Trans. Roy. Soc. S. Aust. 32,300-310; Phil. Mag.-17, 855-864 (1909); PROC. Phys. Soc. Lond. 21, 735-745.

1909. The lessons of radio-activity. (Inaugural address as President of the Australasian Association for the Advancement of Science.) Rep. Aust. Assoc. Adv. Sci. (Brisbane, Queensland), 12, 1-30; Chem. News, 101, 101-103, 111-113, 134-137, 148-149 (1910).

1910. The secondary radiation produced by the beta rays of radium. Phys. Rev. 30. 638-640.

1910. The consequences of the corpuscular hypothesis of the 'γ- and X-rays, and the range of β-rays. Phil. Mag. 20, 385-416; (translation from Engl ish) Jb. Radioakt. 7, 348-386.

1911. Radioactivity as a kinetic theory of a fourth state of matter. (Lecture) Proc. Roy. Instn. 20, 1-10; Nature, Lond. 85, 491-494; Chem. News, 104, 110-113; Arch. Röntgen Ray, 15, 402-415.

1911. (With H. L. PORTER.) Energy transformations of X-rays: Proc. Roy. Soc. A, 85, 349-365.

1911. Corpuscular radiation. Rep. Brit. Assoc. Adv. Sci. (Portsmouth), 340-341; report in Engrg. 92, 418.

1911. The energy of the X-ray. (Lecture to Röntgen Society.) J. Röntgen Soc. 8, 16-20; report in Chem. News, 104, 291-292.

1911. Mode of ionization by X-rays. Phil. Mag. 22, 222-223.

1912. Studies in Radioactivity. Macmillan & Co. Ltd., London. Also translated into German (1913).

1912. On the direct or indirect nature of the ionization by X-rays. Phil. Mag. 28, 647-650.

1912. X-rays and crystals. Nature, Lond. 90, 219, 360-361, 572.

1912. Radiations old and new. (Evening Discourse) Rep. Brit. Assoc. Adv. Sci. (Dundee), 750-753; Nature, Lond. 90, 529-532, 557-560.

1913. The reflection of X-rays by crystals. Nature, Lond. 91, 477.

1913. Die Reflexion von Rontgenstrahlen an Kristallen. (Translation from English.). Phys. Z. 14, 472-473.

1913. (With W. L. BRAGG.) The reflection of X-rays by crystals. I. Proc. Roy. Soc; A. 88, 428-438.

1913. The reflection of X-rays by crystals. II. Proc. Roy. Soc. A, 89, 246-248.

1913. (With W. L. BRAGG.) The structure of the diamond. Nature, Lond. 91, 557.

1913. (With W. L. BRAGG.) The structure of the diamond. PROC. Roy. Soc. A, 89, 277-291.

1913. On the production of fluorescent Röntgen radiation. Phil. Mag. 25, 657-659.

1913. Crystals and X-rays. Rep. Brit. Assoc. Adv. Sci. (Birmingham), 386-387; report in Energ. 96, 422-423.

1914. The influence of the constituents of the crystal on the form of the spectrum in the X-ray spectrometer. Proc. Roy. Soc. A, 89, 430-438.

1914. The X-ray spectra given by crystals of sulphur and quartz. PROC. Roy. Soc. A, 89, 575-580.

1914. X-rays and crystalline structure. (Lecture) Proc. Roy. Instn. 21,198-207; Science, 40, 795-802.

1914. The intensity of reflexion of X-rays by crystals. Phil. Mag. 27, 881-899.

1914. An X-ray absorption band. Nature, Lond. 93, 31.

1914. Crystalline structures as revealed by X-rays. (Lecture to Manchester Literary and Philosophical Society.) Nature, Lond. 93, 124-126.

1914. (With S. E. PEIRCE.) The absorption coefficient of X-rays. Phil. Mag. 28, 626-630.

1915. X-rays and crystal structure. (Bakerian lecture) Phil. Trans. Roy. Soc. A, 215, 253-274.

1915. The new crystallography. Scientia, Bologna, 18, 378-385. (Suppl. in French, 240-249.)

1915. The relation between certain X-ray wave-lengths and their absorption coefficients. Phil. Mag. 29, 407-412.

1915. The distribution of electrons in atoms. Nature, Lond. 95, 344.

1915. The structure of magnetite and the spinels. Nature, Lond. 95, 561.

1915. The structure of the spinel group of crystals. Phil. Mag. 30, 305-315; 31, 88 (1916).

1915. (With W. L. BRAGG.) X-rays and crystal structure (British Association for the Advancement of Science, Manchester). Report in Engrg. 100, 305.

1915. (With W.L. BRAGG.) Die Reflexion von Rontgenstrahlen an Kristallen. Z. anorg. Chem. 90, 153-296. (Translations of papers in PROC. Roy. Soc. etc.; subsequently published by Leopold Voss, 1928.)

1915. (With W. L.BRAGG.) X-rays and crystal Structure. G. Bell & Sons.Ltd, London. 5 editions: also translated into Russian (1916) (1929), French (1921).

1916. The recent work on X-rays and crystals and its bearing on chemistry. (Lecture to Chemical Society.) J. Chem. Soc. (Trans.), 109, 252-269.

1916. X-rays and crystal structure, with special reference to certain metals. (May lecture to Institute of Metals.) J. Inst.. Metals, 16, 2-13.

1917. Physical research and the way of its application. Essay published with twelve others under the collective title Science and the Nation. Cambridge University Press.

1919. Transmitting and picking up sounds in water. (Report of lecture at British Scientific Products Exhibition.) Nature, Lond. 103, 393.

1919. Listening under water. (Tyndall lectures at the Royal Institution. 27 May and 3 June.) Report in Engrg. 107, 776-779.

1919. The examination of materials by X-rays. Trans. Faraday Soc. 15, (Part 2), 25-31.

1920. Analysis by X-rays. (3rd Silvanus P. Thompson Memorial lecture.) J. Röntgen Soc. 16, 127-133.

1920. The history of science. (Report of public lecture at University College, London. 7 October.) Nature, Lond. 106, 250.

1920. The World of Sound. G. Bell & Sons, London- 9 editions: also translated into Russian (1927), Polish (1933), Italian (1934).

1921. Electrons. (12th Kelvin lecture. 13 January.) .J. Instn. Elect. Engrs. 59, 132-137; Nature, Lond. 107, 79-82, 109-111; .J. Röntgen Soc. 17, 31-39; Engrg. 111, 120-122.

1921. Electrons and ether waves. (Robert Boyle lecture at Oxford. 12 May.) Humphrey Milford. Oxford University Press. Summary in Nature, Lond. 107, 374. Reprinted in Sci. Mon. 14, 153-160.

1920-1921. X-ray spectra. (Discussion) Proc. Phys. Soc. Lond.33, 1-9.

1921. Application of the ionization spectrometer to the determination of the structure of minute crystals. Proc. Phys. Soc. Lond. 33, 222-224.

1921. Sounds in Nature. (Discourse at 8th Annual Exhibition of Scientific Apparatus by Physical Society and Optical Society.)_report in Engrg. 111, 21.

1921. The intensity of X-ray reflection by diamond. Proc. Phys. Soc. Lond. 33, 304-310.

1922. The structure of organic crystals. (Presidential address to Physical Society) Proc. Phys. Soc. Lond. 34, 33-50; Proc. Roy. Instn. 23, 581-588; Nature, Lond. 110, 115-117.

1922. The crystal structure of ice. Proc. Phys. Soc. Lond. 34, 98-102.

1922. The significance of crystal--structure. (Lecture to the Chemical Society.) J. Chem. Soc. (Trans.) 121, 2766-2787.

1922. The significance of crystal structure. (Lecture at British Association for the Advancement of Science, Hull.) J. Soc. Chem. Ind. 41, (Review 5), 366-368.

1922. Nature's architecture. (Lecture.) Proc. Instn. Heat. Vent. Engrs. 21, 24-42; Proc. Univ. Durham Phil. Soc. 6, 117-135.

1922. Relation of the crystal-structure of some carbon compounds to those of graphite and diamond. Miner. Mag. 19, 316-318.

1922-1923. (With G. SHEARER and J. W. MELLOR.) Notes on the crystalline structure of some china clays examined by the X-ray powder method. Trans. Ceramic Soc. 22, 106-110.

1923. The crystalline structure of anthracene. Proc. Phys. Soc. Lond. 35, 167-169.

1923. New methods of crystal analysis and their bearing on pure and applied science, (6th Trueman Wood lecture.) J. R. Soc. Arts, 71, 267-277; Nature, Lond. 111, Suppl. (9 June) iii-x.

1923. The nature of the liquid state. Nature, Lond. 111, 428.

1923. Crystal structure of basic beryllium acetate. Nature, Lon 111, 532.

1923. (With G. T. MORGAN.) Crystal structure and chemical constitution of basic beryllium acetate and propionate. Proc. Roy. Soc. A, 104, 437-451.

1923. The electron theory of valency. (Faraday Society Conference. 13-14 Lengths of carbon chains in fatty acids and esters. Trans. Faraday Soc. 19, 478-479

1923. X-rays and crystal symmetry. Nature, Lond. 112, 618.

1923. X-ray measurements. (Introductory remarks at Physical Society discussion.) Proc. Phys. Soc. Lond. 35, 4D.

1923. Cohesion and molecular forces. (Opening address at discussion, British Association for the Advancement of Science, Liverpool.) Summary in J. Soc. Chem. Ind.42 (Chem. Ind. Reti. 1), 930-932; report in Nature, Lond. 112, 773; report in Engrg. 116, 359.

1923. Relation between the X-ray analysis of crystal structure and the conclusions of mathematical crystallography. (Reply to criticism by Dr R.W.G. Wyckoff.)J. Franklin Inst. 196, 675-677.

1923. (With W. T. ASTBURY.) Crystallography. Ann. Rep. Chem. Soc. 20, 230-260.

1924. Research work and its applications. (Address at the Sir John Cass Technical Institute.) Nature, Lond. 113, 311-312.

1924. Broadcast Appeal for British Science Guild. (29 January.) J. Brit. Sci. Guild (1921-1927), no.17, 29-30.

1924. The scope of modern chemistry (Address at the opening of new science schools, Durham University) Dutham Univ. J. 24,190-195.

1924. Recent research on crystalline structure (Lecture) Proc. Roy. Instn. 24,298-300 (abstract); report in Engrg. 117,183.

1924. (With W. T. ASTBURY.) Crystallography. Ann. Rep. Chem. Soc. 21, 220-237.

1924. The scattering of X-rays. (Lecture.) PROC. Roy. Instn. 24, 311-312 (abstract); report in Engrg. 117, 204-205.

1924. The crystalline structure of organic substances. (Lectures delivered at the Royal Institution.) Report in Engrg. 117, 211-212, 2-245, 274-275, 310-311.

1924. X-ray examination of metal films. Nature, Lond. 113, 639.

1924. X-ray examination of Oppau salt. Trans. Faraday Soc. 20, 59-60.

1924. The carbon atom in crystalline structure. (Address at Centenary Celebration of Franklin Institute) J. Franklin Inst. 198, 615-625; Sci. Mon. 19, 578-587; Nature, Lond.114, 862-865.

1924. The analysis of crystal structure by X-rays. (Presidential address to Section A, British Association, Toronto.) Rep. Brit. Assoc. Adv. Sci. 92, 34-52; Science, 60, 139-149; Engrg. 118, 349-350, 428-430; J. Soc. Chem. Ind. 43 (Chem. Ind. Reti. 2), 954-963; Pan-Amer. Geol. 42, 173-198.

1925. X-rays and crystal structure (Reprinted from the Handbook to the Exhibit of Pure Science, The State of Science in 1924, arranged by the Royal Society for the British Empire Exhibition.) Sci. Mon. 20, 115-121.

1925. The investigation of the properties of thin films by means of X-rays. (Lecture) PROC. Roy. Instn. 24,481-490; Nature, Lond. 115, 266-269.

1925. (With W. T. ASTBURY.) Crystallography. Ann. Rep. Chem. Soc. 22, 24Q-258.

1925. The modern demand for scientific knowledge. (Summary of address to British Science Guild. 21 April.) J. Brit. Sci. Guild (1921-1927), no. 20, pp. 18-21.

1925. Introductory remarks to discussion on International Units and Standards for X-ray work. (Physics section. International Congress of Radiology. 1 July.) Brit. J. Radiol. (Ront. Soc. section), 23 (1927), 64-66.

1925. The crystalline state. (The Romanes lecture.) Clarendon Press, Oxford.

1925. Concerning the Nature of Things. G. Bell & Sons. 8 editions. Also translated into Russian (1926), Czechoslovakian (1927), German (1930), Italian (1932), Polish (1933).

1925. Organic crystals. (English and French versions.) Inst. int. Chimie Solvay, Bruxellu, 2, 21-27.

1925. (With R.E. GIBBS.) The structure of α and β quartz. Proc. Roy. Soc. A, 109, 405-427.

1925. The structure of quartz. Trans. Soc. Glass Technol. 9, 272-282.

1925. The properties and structure of quartz. (Four lectures at Royal Institution.) Report in Engrg. 119, 169, 202-203. 236. 266. (Letter) Engrg. 119, 322.

1925. Modern developments of atomic and molecular theory. (Address at opening of new chemical and physical laboratories at United College of St Salvator and St Leonard, University of St Andrews. 4 December.) Summary in St Andrews Citizen, no.2909. 12 December.

1926. The work of the Davy-Faraday Research Laboratory. (Lecture) Proc. Roy. Instn. 25, 43-44 (abstract); report in Engrg. 121, 143.

1926. The invisible world. (Interview.) Review of Reviews, 73, 113-122.

1926. Atoms and electrons. Elect. Educator, part 15, 681-683.

1926. The influence of the Learned Societies on the development of England. (Presidential address, Birmingham and Midland Institute.) Published for Council of Birmingham and Midland Institute.

1926. The imperfect crystallization of common things. (Abstract of lectures given at Royal Institution.) Nature, Lond. 118, 120-122; report in Engrg. 123, 128-130.

1926. Old Trades and New Knowledge. G. Bell & Sons. 2 editions. Also translated into Russian (1928), Polish (1933).

1926. Science in industry (Presidential address to Institute of Physics.) Report in Engrg. 121, 700.

1926. The fine structure of animal and vegetable substances as revealed by X-rays. (Mather lecture.) J. Text. Inst. Manchester, 17, 1-10.

1926. Long chain molecules. (Lecture.) J. Soc. Dy. Col. Bradford, 42, 237-242; J. Soc. Chem. Ind. 45 (Chem. Ind. Rev. 4), 245-248.

1927. Tyndall's experiments on magne-crystallic action. (Lecture.) Proc. Roy. Instn. 25, 161-184; Nature, Lond. 119 (Suppl. 7 May), 61-72; Sci. Mon. 25, 65-79.

1927. The application of X-rays to the study of the crystalline structure of materials.

(14th Thomas Hawksley lecture.) Proc. Instn. Mech. Engrs. 2, 751-775.

1927. X-rays and the study of colloids. (British Association for the Advancement of Science, Leeds.) Report in Engrg. 124, 509-510.

1927. A note on the crystalline structure of certain aromatic compounds. Z. Kristallogr. 66, 22-32. (Sir William Bragg at this time became co-editor of the Z. Kristallogr.)

1927. X-ray analysis of metals. (Lecture at Royal Institution) Report in Engrg. 124, 811.

1927. (With W. L. BRAGG.) Stereoscopic photographs of crystal models, to illustrate the results of X-ray crystallography. 2 series (1927, 1930). Adam Hilger Ltd. London.

1928. The structure of an organic crystal. (Fison Memorial lecture, Guy's Hospital Medical School. 13 March.) Longmans, Green & Co. London. Also translated into Russian (1929).

1928. The enhancement principle in X-ray photographs. Nature, Lond. 121, 327-329.

1928. From Faraday's notebooks. (Report of three lectures at Royal Institution.) Nature, Lond. 121, 218, 254, 293.

1928. Photo-electricity. (Lecture) Proc. Roy. Instn. 25, 338-348; Sci. Mon. 27, 522-529.

1928. An Introduction to Crystal Analysis. G. Bell & Sons, London. Also translated into Russian.

1928. (With W. L. BRAGG.) Die Reflexion von Riintgenstrahlen an Kristallen. Leopold Voss, Leipzig. (Translations of early papers in Proc. Roy. Soc. etc.)

1928. Craftsmanship and science. (Presidential address.) Rep. Brit. Assoc. Adv. Sci. (Glasgow), 1-20; Nature, Lond. 122, 353-363; Science, 68, 213-223.

1928. Faraday's Diary. Spectator, 3 November, pp. 42-43.

1928. Speech at Annual Dinner of Institution of Electrical Engineers. J. Instn. Elect. Engrs. 66, 896.

1928. Tribute to the memory of H. A. Lorentz. Nature, Lond. 121,287.

1929. An instrument for measuring small amplitudes of vibrations. J. Sci. Instrum. 6, 196-197.

1929. Diamonds. (Four lectures at Royal Institution.) Report in Engrg. 127, 35-36.

1929. Further progress in crystal analysis. (Lecture.) Proc. Roy. Instn. 26, 6-8 (abstract) Chem. News, 139,55-56; report in Engrg. 127, 242.

1929. Organic compounds. (Introduction to a Conference of the Faraday Society.) Trans. Faraday Soc. 25, 346-347.

1929. Crystal structure of organic substance in its relation to medicine. (Huxley lecture, Charing Cross Hospital. 28 November.) Lancet, 217, 1293-1298.

1929. The early history of X-rays. (Report of lectures given at Royal Institution.) Nature, Lond. 123, 218, 253.

1930. Cellulose in the light of the X-rays. (Lecture) Proc. Roy. Instn. 26, 156-171; Nature, Lond. (Suppl.), 125, 315-322.

1930. New data on cellulose space lattice. Nature, Lond. 125,634.

1930. The meaning of the crystal. (Address at the Medal Meeting (21 May) of the Franklin Institute.) J. Franklin Inst. 210, 9-14; Science, 71, 547-550.

1930. Crystals: (Science Service Radio Talk, Columbia Broadcasting System.) Sci. Mon. 31, 338-340.

1930. Address at Graduation Exercises of Massachusetts Institute of Technology (Quoted by Pilgrim Trust lecturer, 1943, in Proc. Roy. Soc. A, 182, 1-2.)

1930. The contributions of Count Rumford and Michael Faraday to the modem museum of science. Science, 72, 19-23.

1931. The scattering of light. (Lecture) Proc. Roy. Instn. 26, 302-317.

1931. The Raman Effect (Lecture at Royal Institution.) Report in Engrg. 131, 291-292.

1931. Sizes of the atoms (Lecture at Royal Institution). Report in Engrg. 131, 350-351.

1931. X-ray investigation of the structure of liquids. (Lecture.) Proc. Roy. Instn. 26, 492-493; report in Engrg. 131, 706.

1931. X-rays in the factory. Times, 21 August; Brit. J. Radiol. 4, 513-516.

1931. Faraday's first successful experiment on diamagnetism. Nature, Lond. 127, 337.

1931. Faraday relics. Nature, Lond. 127, 486.

1931. Faraday's Diary. Rev. Mod. Phys. (Faraday-Henry Memorial number), 3, 448-463.

1931. Michael Faraday. (8th National Broadcast lecture.) B.B.C., London; Sci. Mon. 33, 481-499.

1931. Faraday Commemorative Oration. (Queen's Hall. 21 September.) J. Inst. Elect. Engrs. 69, 1363-1367; Proc. Roy. Instn. 27, 40-56.

1931. X-rays and the new range of vision. (8th Hurter and Driffield Memorial lecture.) Photogr. J. 55, 143-148.

1932. Response to toast of 'Our Guests' at annual dinner, English Association. Bull. English Assoc. 75, 8-11.

1932. Physical laboratories and social service. (Address at opening of new Physics Building, University of Leeds.) Nature, Lond. 129, 495-497.

1933. The crystals of the living body. (Lecture.) Proc. Roy. Instn. 27, 606-624; Nature, Lond. 132, 11-13, 50-53.

1933. Focal conic structures. Trans. Faraday Soc. 29, 1056-1060.

1933. The Universe of Light. G. Bell & Sons, London. 4 editions. Also translated into Danish. (1933), Dutch (1933), Polish (1933), Italian (1934), German (1934), Russian (1935), Swedish (1937).

1933. (Co-editor with W. L. BRAGG.) The Crystalline state (Vol. I written by W. L. Bragg) G. Bell & Sons. 2 editions. Also translated into Russian (1938).

1934. Liquid crystals. (Lecture) Proc. Roy. Instn. 28, 57-92; Nature, Lond. 133, 445-456.

1934. Molecule planning. (3rd Spiers Memorial lecture.) Trans. Faraday Soc. 30, 665-673.

1934. Refrigeration. (British Science Guild Research and Development lecture.) Ann. Rep. British Science Guild (1933-1934), pp. 26-~9; report in Nature, Lond.133, 715.

1934. Structure of the azide group. Nature, Lond. 134, 138.

1935. X-rays and the coarse structure of materials. (15th Mackenzie Davidson Memorial lecture.) Brit. J. Radiol. 8, 144-154.

1935. School science after school. (Presidential address, Science Masters' Association.) Ann. Rep. for 1935, Science Masters' Assoc. pp. 8-16.

1935. Chemistry and the body politic. (7th S. M. Gluckstein Memorial lecture.) Published as separate monograph by the Institute of Chemistry, London.

1935. The theoretical strength of materials and their practical weakness. (Lecture.) Proc. Roy. Instn. 28, 490-511; Sci. Mon. 41, 111-120; Trans. Soc. Engrs.(Inc.), 26, 27-43.

1935. X-ray crystal analysis. Nature, Lond. 135, 69Q-692.

1935. Opening survey: International Conference on Physics, London, 1934. Vol. II. Solid State of Matter, p. 1~. Phys. Soc. London. .

1935. The molecular structure of dielectrics. (Kelvin lecture.) J. Instn. Elect. Engrs. 77, 737-748.

1935. Address given as Special Visitor at Assembly of Faculties of University College, London. Ann. Rep. Univ. Col. Lond. (1935-1936), pp. 113-121.

1935. (Hon. co-editor with M. v. LAUE.) Internationale Tabellen zur Bestimmung von Kristallstrukturen. 2 vols. Gebrüder Bomtraeger, Berlin. G. Bell & Sons, London, etc.

1935. Aeolian Tones. (Lecture at Royal Institution.) Report in Engrg. 139, 553.

1936. The progress of physical science. (Sir Halley Stewart Trust lecture. 24 October.

1935. Published in book form with five other lectures under the title Scientific Progress, pp. 39-77. George Allen & Unwin, London.

1936. The electric properties of crystals. I and II. (Two lectures.) Proc. Roy. Instn 29, 225-230, 290-295.

1936. Presidential address at Anniversary Meeting. Proc. Roy. Soc. A, 157, 697-727 B, 121, 396-426.

1936. Progress in the technique of crystal analysis. Nature, Lond. 138, 953-954.

1936. Bells. (Lecture at Royal Institution.) Report in Engrg. 142. 509.

1937. The development of crystal analysis. (Trueman Wood Memorial lecture.) J. R. Soc. Arts, 85, 228-241.

1937. Recent crystallography. (Lecture.) Proc. Roy. Instn. 29, 484-495; Nature, Lond. 139, 865-866, 9.11-913.

1937. Presidential address. International Association for Testing Materials Congress. 19 April. Report in Engrg. 143, 460.

1937. The crystal and the engineer. (43rd James Forrest lecture.)J. Inst. Civ. Engrs. 6, 181-202; Engrg.-1-43, 499-501 (Summary).

1937. The application of X-ray methods to industry. Rep. Brit. Assoc. Adv. Sci. (Nottingham), p. 333.

1937. Tribute to the memory of Lord Rutherford. Nature, Lond. 140, 752.

1937-1938. (With W. L. BRAGG.) The discovery of X-ray diffraction. Curr. Sci. 7, Suppl. Special Number on 'Laue diagrams' (1937), pp. 9-13.

1937. Presidential address at Anniversary Meeting. Proc. Roy. Soc. A, 163, 455--482; B, 124, 369-396.

1937. The grain-like structure of solids. (From Presidential address.) Nature, Lond. 140, 954-956.

1938. Infra-microscopic magnitudes. (From Presidential address.) Sci. Mon. 46. 146-149.

1938. Science in schools. (Address given at Conference of National Union of Teachers, 1937.)G. Bell & Sons, London.

1938. Clay. (Lecture.) Proc. Roy. Instn. 30, 39~7; Letter to Nature, Lond. 41, 649.

1938. Social relations of science. Nature, Lond. 141, 725.

1938. Ice. (Lecture.) Proc. Roy. Instn. 30, 283-301.

1938. Nature's Architecture. (Address at Oil Industries Club Luncheon.) Published for Oil Industries Club.

1938. Opening address as Hon. President, Conference on Rubber Technology. Proc. Rubber Tech. Conf. p. 1-4.

1938. Crystallography and the engineer. (lst Annual Lecture to Manchester Association of Engineers.) Trans. Manchr. Assoc. Engrs. (1937-1938), pp. 349-358.

1938. Combination tones in sound and light. (Lecture.) Proc. Roy. Instn. 30, 424-433; Nature, Lond. 143, 542-545.

1938. The molecular basis of the strength of materials. (7th Andrew Laing lecture.) Trans. N.E.. Coast Instn. Engrs. Shipb. 55, 33-53.

1938. Early links between science and engineering. Jr. Instn. Engrs. J. 48,463-478.

1938. Presidential address at Anniversary Meeting. Proc. Roy. Soc. A, 169, 1-24; B, 126, 263-286; Science,-88, 579-583. .

1938. (With SIR W. MOBERLY and LORD KENNET.) Moral Rearmament. S.C.M. Press, London.

1938. Address of President of the Royal Society, introducing Pilgrim Trust lecturer, Dr I. Langmuir. Science, 88, 611-612.

1939. Liquid films. (Lecture.) PROC. Roy. Instn. 30,687-696.

1939. History in the archives of the Royal Society. (Pilgrim Trust lecture to National Academy of Sciences.) Science, 89,445-453; Nature, Lond. 144, 21-28.

1939. Presidential address at Anniversary Meeting.. Proc. Roy. Soc. A, 173, 286-312; B, 128,1-27.

1939. X-ray analysis and the structure of matter. (From Presidential address.) Nature, Lond. 144, 961-963.

1940. Science and the worshipper. (In response to Sir Richard Tute.) Hibbert J. 38, 289-295.

1940. History of the vacuum flask. (Two lectures at Royal Institution.) Nature, Lond. 145,408-410; report in Engrg. 149, 42-43.

1940. Tribute to the memory of Sir J. J: Thomson. Nature, Lond. 146, 352.

1940. The extra spots of the Laue photograph. Nature, Lond..146, 509-511.

1940. Physics in war-time. (Lecture.) PROC. Roy. Instn. 31, 70-79.

1940. Presidential address at/Anniversary Meeting. Proc. Roy. Soc. A, 177, 1-26; B; 129, 413-438.

1940. Science and national we1fare. (From Presidential address.) Nature, Lond. 146, 731-732.

1940. Introduction to The Nation's Larder and the Housewife's Part Therein. G. Bell & Sons, London.

1941. The Story of Electromagnetism. G. Bell & Sons, London. The diffuse spots in X-ray photographs. Proc. Roy. Soc. A, 179, 51-60.

1941. The diffuse spots in-X-ray crystal photographs. Proc. Roy. Soc. A, 179, 94-101.

1941. Diffuse spots in X-ray crystal photographs. Nature, Lond. 148, 112.

1941. Diffuse spots in X-ray photographs. Nature, Lond. 1.48, 780.

1941. Science and faith. (Riddell Memorial lecture, University of Durham. 7 March.) Oxford University Press. Report in Nature, Lond. 148, 181-183.

1942. Science and the community. Endeavour, 1, 4-5.

1942. The secondary X-ray spectrum of sylvine; theory and experiment. Proc. Phys. Soc. Lond. 54, 354-361.

1942. (With twelve OTHERS.) Science lifts the veil. A series of broadcast talks on the conquest of the sub-visible universe. Published for the British Council. Longmans, Green & Co. London.

K. LONSDALE

William Lawrence Bragg (1890 – 1971)

Sir David Phillips

Sec. Royal Society

WALKING along the Backs in Cambridge one day in the autumn of 1912 William Lawrence Bragg had an idea that led immediately to a dramatic advance in physics and has since transformed chemistry, mineralogy, metallurgy and, most recently, biology. He realized that the observations of X-ray diffraction by a crystal, which had been reported by von Laue and his asSociates earlier in that year, can be interpreted very simply as arising from reflexion of the X-rays by planes of atoms in the crystal and hence that the X-ray observations provide evidence from which the arrangement of atoms in the crystal may be determined. A few weeks of intensive work on simple inorganic compounds were enough to demonstrate the correctness of these ideas but the development of the method, at first in asSociation with his father and later as the leader or guiding influence of a host of workers, was the labour of a lifetime. When he died on 1 July 1971, X-ray crystallography had revealed the arrangement of atoms in matter of all kinds from the simplest of salts to the macromolecules of the living cell. The story of his life is very largely the story of that achievement and the circumstances that led to his unique part in it.

FAMILY BACKGROUND

The Bragg family (a)* had its roots in Cumberland where the men farmed or went to sea. In the 1820s one of them, John Bragg, married a Workington girl, Lucy Brown, who had settled near Belfast and they had four children, William, Robert John, James Brown and Mary McCleary. When William the eldest was twelve, in about 1840, his father "as lost at sea between Cumberland and Belfast. Some time later the family moved to Birkenhead where William was apprenticed to a chemist, James went into an office and, in 1846, Robert John went to sea as an indentured apprentice in the *Nereides* of Workington, 530 tons. Robert completed his apprenticeship in October 1851 and soon afterwards sailed again for India in the *Nereides*, now as second mate. The outward voyage to Calcutta was uneventful but the return had barely begun, with Robert now chief mate, when the ship was wrecked in the Hoogli River. Only Robert, four crew members and the pilot were saved.

After this disaster he went back to sea again but only until the late 1850s when he bought Stoneraise Place, near Wigton in Cumberland, and settled down to farm there in the parish of Westward within sight of the Solway Firth. In 1861, he married Mary, daughter of the Vicar, the Rev. Robert Wood and his wife Ruth (Hayton). Mary, who was 28, was remembered as a gracious figure with a natural bent for mathematics. On 2 July 1862, when Robert Bragg was returning from a visit to London to see the Great Exhibition of that year, their eldest son, William Henry , was born. Two more sons, John and James, came later.

There are few records of W. H. Bragg's early childhood, but he remembered his mother teaching him to read and then, at the age of five, going to the Westward village school. But in 1869, when he was only seven, his mother died at the early age of 36. It was Uncle William, long accustomed to thinking of himself as head of the family and now established as a chemist in Market Harborough, who carne to the rescue. He had just helped to re-establish the old Grammar School in Market Harborough, and the young W. H. Bragg was promptly moved there, to live with his grandmother and uncles (James ran the grocer's shop next door).

Summer holidays were spent in Cumberland and in 1875, at his father's insistence, W. H. Bragg was sent to school at King William's College in the Isle of Man. In 1889 he was awarded an exhibition at Trinity College,

* The labels (a) etc. refer to the General References and (A) etc. and (1) etc. to the two lists (books and papers respectively) in the Bibliography.

Cambridge, but he stayed at school a further year before going up for the long-vacation term in July 1881. At about the same time his father retired to Ramsey, in the Isle of Man, and Market Harborough, not so far from Cambridge and dominated by Uncle William, was clearly the young man's home.

At Cambridge W.H.B.* read mathematics and in 1884 he graduated third in the class with first class honours, Third Wrangler. The following year, encouraged by J.J. Thomson, he applied for the position of Professor of Mathematics and Physics in the University of South Australia. The electors were Horace Lamb (the retiring Professor), J. J. Thomson and the Agent General, Sir Arthur Blyth, but before deciding to send out so young a man they consulted the Government Astronomer of South Australia, Charles Todd, who happened to be in London at the time. He had no objection and the 23-year-old Third Wrangler was appointed to start as soon as possible.

Up to this time W.H.B. had done very little physics, but the electors assumed that he could learn enough as he went along. On the six week voyage to Australia he read Electricity and magnetism by Deschanel but on arrival in Adelaide he found that the physics classes were small and elementary and was well able to keep ahead of them. One of the principal difficulties was a shortage of apparatus so he went to work with a firm of instrument makers in the town and helped to equip the laboratories himself. His head mechanic, Rogers, was a key figure who worked with him throughout his period in Australia, initially making apparatus for classes and later the apparatus designed by W.H.B. for his researches.

On the day after his arrival in Adelaide W.H.B. went for supper with the Todd family. Charles Todd, born in England in 1826, and his wife Alice, had been in Australia since 1855 when he was sent out from Greenwich as Government Astronomer with the particular task of installing an electric telegraph system in South Australia. Alice, who carne from a Cambridge family named Bell, was 19 and already pregnant when they arrived in Adelaide, a town which had been founded only 19 years earlier by Charles Sturt and William Light and named after the Consort of William IV. By 1872, after years of surveying the bush, the overland telegraph line from Adelaide to Port Darwin in the north was completed and direct communication with Europe by the submarine cable from Darwin was made possible. The final connection was made near an oasis in central Australia,

---

* From this point ii is convenient to distinguish W. H. Bragg from his son, the subject of this memoir, by referring to him as W.H.B.

known since as Alice Springs after the Astronomer's wife. Postmaster General and Superintendent of Telegraphs, South Australia, Todd was elected F .R.S. in 1889 and appointed K.C.M.G. in 1893.

The Todds lived at the Observatory in West Terrace, Adelaide, a spacious house surrounded by paddocks and the buildings housing the offices, the transit telescope and other equipment. With six children, Elizabeth, Charles, Headley, Maude, Gwendoline and Loma, they were at the centre of Adelaide Society and they soon made the young professor a regular member of their lively circle. A happy working relationship between the two scientists was quickly established and a closer connection followed. On 1 June 1889, W.H.B. and Gwendoline Todd were married. She was just on 19, lively, Sociable, without much formal education but a gifted artist who had been a star pupil at the Design School. They set up house, with a cook and a housemaid not easily managed by the young wife, in Lefevre Terrace, North Adelaide and on 31 March 1890 their elder son, William Lawrence, was born. A second son, Robert Charles, was born a year later and a daughter, Gwendolen Mary , in 1907.

ADELAIDE, 1890-1909

W. L. Bragg's earliest memories (b) went back to the period following the birth of his brother, Bob, when his mother was convalescing after a difficult delivery .He remembered the pleasures of listening to her stories when she was still in bed but also the indignities of being wheeled out in the pram together with his baby brother and of wearing the blue tunics with red belts and straw hats favoured by his artistic parent. Quite early a nursemaid named Charlotte Schlegel was recruited who stayed with the family for nearly 30 years. She carne from the part of Denmark that was lost to Prussia in the war of 1867 which she remembered. Having a fierce and repressive temperament she sternly discouraged any games in which the participants could conceivably get dirty. Writing years after Bragg noted sadly 'I do not think my brother Bob was much affected by her—even as a child he had considerable calm self-confidence—but I was very impressionable and unsure of myself and I am certain that Charlotte was very much the wrong person for me'. Even so there was clearly plenty of fun in the backyards and tree house in Lefevre Street with Eric Gill, later well known as an artist, who lived nearby, the principal playmate.

Even more memorable were the regular Sunday visits to the Observatory, a veritable paradise for small boys with its rambling construction and exciting outbuildings. In the storerooms treasures of all kinds-souvenirs of the Todd's journeys, discarded scientific equipment, old letters and so on-provided the basis for endless play and later for the design and construction of electrical gadgets.

Holidays were spent at the seaside, either on the shore of the Gulf of St Vincent, a horse-tram ride away, or at Port Elliott on the ocean coast. Here the Bragg and Todd families combined to take a large part of a boarding house. Gwendoline Bragg sketched and coached W .H.B. also, so that in time their styles were said to be indistinguishable. W .H.B. also played games of all sorts but Bragg remembered best his happy relationship with his Aunt Loma, then still in her teens, who 'looked after me, invented games for me, read Grimm's Fairy Tales to me and altogether constituted herself my guardian'.

Bragg started school when he was five, attending a convent school, to which he walked, on the far side of North Adelaide. Then there was a setback. Riding his tricycle one afternoon, he was overturned by his brother and shattered his left elbow. The damage was so serious that the doctor thought nothing could be done but allow the joint to set rigid. But the family intervened. Uncle Charlie (Todd) was himself a doctor and devised an heroic treatment. Every few days Bragg was anaesthetized so that his arm could be flexed to prevent stiffening while a new joint was formed. The treatment worked but he was left with a slightly short and crooked left arm and, 40 years later, in Manchester, had to have an operation to relieve pressure on a nerve that was giving rise to paralysis in his left hand.

This incident involved Bragg in his first encounter with X-rays. While working hard to develop his subjects at the University, W.H.B. took a keen interest in all the latest developments in science as news of them reached him from the other side of the world. In 1895 the greatest excitement was created by Röntgen's discovery of X-rays and W.H.B. eagerly set out to repeat the experiments in Adelaide. The apparatus was ready just in time for examination of Bragg's elbow. The scene was described later by Stanley Addison, W.H.B.'s laboratory assistant (a). 'In the area below the laboratory proper the Professor had set up a primitive X-ray apparatus based on the discovery which Dr Röntgen of Würzburg, Germany, had announced to the world a few weeks earlier. Although Professor Bragg' s machine [had been] hurriedly brought into being, [it] nevertheless worked, despite its crudity, and was put into motion. Fascinated, the Doctor watched as the big induction coil buzzed loudly, electric sparks crackled and the vacuum tube emitted a weird green glow. Later on there appeared on the screen an X-ray photograph which distinctly showed the extent and location of the injury to the boy's arm. This was the first recorded surgical use of the Röntgen Ray, as it was then called, in Australia'. But W.H.B.'s son remembered (b): 'I was scared stiff by the fizzing sparks and smell of ozone and could only be persuaded to submit to the exposure after my much calmer small brother Bob had his radiograph taken to set me an example.'

An account of Röntgen's experiments, with demonstrations, was one of the university extension lectures W.H.B. gave at that time (1896) as he developed his skills and reputation as a popular lecturer. Another of his major interests was the school system in South Australia since he recognized early on that the quality of his students at the university depended in large part on how well they had been taught at school. He gave an address on this subject at the University Commemoration in 1888 and in 1897 the interest led to a momentous decision. He would take a year's leave—it was 12 years after his appointment—and study the educational system in England. But the major inducement, no doubt, was the opportunity this would provide of seeing the uncles again and introducing his family to them.

The family was divided for the journey. W.H.B. and his wife went ahead to make a tour of Egypt and Italy while the two boys followed with Charlotte and Aunt Lizzie, who was now married to a Cambridge solicitor, Charles Squires, and had been visiting her parents. The two parties were reunited at Marseilles and went on together to Market Harborough. Here the boys stayed most of the time while their father visited schools, enquired into the educational system and renewed his acquaintance with leading scientists. But they also visited W.H.B.'s cousins, William Addison and Fanny (now Kemp-Smith) and their families and they stayed in Cambridge near to the Squires family, so that Bragg and his brother Bob got to know their English cousins. And they saw the sights of London. Among many vivid memories Bragg remembered that his father started a series of bed-time stories during these holidays: 'they were always the same-about the properties of atoms'. He also remembered reading a large volume on the voyages of Captain Cook.

On returning to Adelaide the family stayed at first at the Observatory saddened no doubt by the death of Grandmother Todd during their absence while a new house was being built for them. It was called Catherwood House, after Uncle William's house in Market Harborough, and looked out across the park lands towards the race-course. It had a large garden in which

Bragg had a plot of his own and began his life-long interest in gardening. He also went to a new school, a preparatory school named Queen's on the far side of North Adelaide, which was reached by horse-tram-with a walk at each end of the journey. The headmaster named Hood did not believe in sparing the rod and kept a supply of canes ready to discourage spelling mistakes and other misdemeanours. Looking back, Bragg decided 'I was a misfit at school, being so very immature in some ways and so precocious in others'. He remembered (b) being in the same room as a very senior class doing Euclid. 'From what I overheard I realized what it was all about. Somehow Hood must have caught on to what was happening for he pulled me, a very small boy, out of my class and made me explain the theorems to the large boys while he crowed with delight.'

When about 11 years old, he was sent to St Peter's College, the leading Church of England School in South Australia. There were between 300 and 400 dayboys and some 70 boarders under a headmaster named Girdlestone whose main distinction was a passion for good English. Bragg took English language and literature, French, Latin, Greek, scripture, mathematics and chemistry, all to an equal level. The only available subjects that he did not take were German and physics—and he always regretted learning no German. His closest friend was Bob Chapman, son of the Professor of Applied Mathematics at the University, and together they revelled in mathematics, with their fathers' help and encouragement. But again there were difficulties which Bragg attributed later to his being always in a class of older boys and too proud to play with those of his own age and stature. But 'it was a kindly school, because the boys treated me as an amusing freak instead of the teasing and bullying which might so easily have been my lot'. He was in the sixth form at 14 and at 15 his father decided that he should leave and enter the university.

This was in 1905 and the previous year had been a turning point for W.H.B. Throughout his career in Adelaide he had maintained a lively interest in the latest scientific developments and had struggled to repeat the key experiments as they were reported by the leading research workers of the day and to explain them to the interested public. His latest efforts in this direction had been particularly fruitful. In the years following his return from England he had collaborated closely with his father-in-law in emulating Marconi's experiments on wireless telegraphy which had achieved a dramatic success in 1896-97 and must have been much discussed during the visit to Europe. Together they set up a transmitter in the grounds of the

Observatory and a receiving station seven miles away at Henley Beach. The successful outcome, described in a popular lecture in 1899, aroused great interest in the isolated community of South Australia and the scientists were local heroes. Bragg remembered this as his first real contact with practical science: 'Bob and I took a great interest in these experiments, especially because it meant a picnic on Sunday afternoons, when my grandfather and father drove to Henley Beach with us in the official Post Office Wagonette to see the signals coming in.' But this was not really original research and W.H.B. had no experience of that until 1904 when he was 41 years old and had been professor for 17 years.

On 7 January 1904 he was to give the Presidential address to the mathematics-physics section of the Australian AsSociation for the Advancement of Science meeting in Dunedin, New Zealand. This was not his first lecture as President of the Section, he had been President at the Hobart meeting in 1892, but now he was to speak on Rutherford's home ground and, stimulated perhaps by the lively interest of his son, he chose to talk on 'Some recent advances in the theory of the ionization of gases'. Reading the published work of Rutherford, Mme Curie and others, he had noticed that the absorption of α- and β-rays had been assumed to be analogous to the exponential decrease in the intensity of a wave traversing an absorbing medium and that this had led to some absurd conclusions. At the same time Mme Curie had described experiments which implied that all of the α-particles emitted by radium travelled about the same distance into the surrounding air. W .H.B. realized that the α-particles must behave 'like bullets fired into a block of wood' and that they must pass through the air atoms hat they meet, losing some of their energy with each encounter. He concluded: 'It cannot be correct to say that the amount of the radiation which penetrates a distance x is proportional to the expression exp(-ax): it must rather be proper to say that (1) the number of α-particles penetrating a given distance does not alter much with distance until a certain critical value is passed, when there is a rapid fall; (2) the energy of the α-particles penetrating a given distance gradually decreases as the distance is increased and dies out at the same critical value. These statements are the expression of what we should expect if ionization, consuming energy, were alone responsible for absorption of the radiation.'

On his return to Adelaide he obtained some radium, through the generosity of a Mr. Barr-Smith, and with the assistance of R. Kleeman set out to test his ideas. They were brilliantly confirmed—α-particles of 'four different ranges were shot out from the radium preparation'—and he was able to write to E. Rutherford and J.J. Thomson with the news. Papers describing the results were published in the Philosophical Magazine on 8 December 1904. In them he described the absorption of β-rays with characteristic imagery: 'the general effect will be that of a stream whose borders become ill-defined, which weakens as it goes, and is surrounded by a haze of scattered electrons. At a certain distance from the source all definition is gone and the force of the stream is spent.'

Rutherford replied promptly and generously and thereafter kept closely in touch through the regular exchange of long letters. Proposed by Lamb, with the support of Rutherford and J.J. Thomson, W.H. Bragg was elected F.R.S. on 2 May 1907.

This sudden flowering of his father's research coincided with Bragg's career as an undergraduate at Adelaide University. He read mathematics with subsidiary courses in chemistry and physics, graduating with first class honours in mathematics in 1908 at the age of 18. Most of his tuition was from his father—he even had a desk in his father's office—and from Chapman, the Professor of Applied Mathematics. At his father's suggestion he also took a course in English and was particularly pleased at winning the University prize for the best English essay 'from under the noses of the professionals'. Through it all there were detailed discussions of the latest researches with W .H.B. trying out his ideas and his papers on his son.

The most important of these ideas was broached one day just as they were boarding the horse-tram for the Observatory (b). W.H.B. had turned to the consideration of γ- and X-rays and, building on his earlier studies of the ionizing properties of α- and β-rays, he questioned the commonly held view, proposed by G. G. Stokes, that X-rays are formless pulses of electromagnetic radiation caused by the electrons in the X-ray tube hitting the anticathode. Instead, arguing from the similarities between X- and γ-rays, he suggested that many of the properties of these rays are easier to explain if the rays are supposed to consist mainly of neutral pairs of material particles (d). In a paper 'On the properties and natures of various electric radiations' published in 1907, he compared the known properties of the various rays (α-, β-, X- and γ-rays and ultraviolet light) and discussed 'the possibility that γ- and X-rays may be of a material nature'. He noted that earlier corpuscular theories of γ- and X-rays had been discounted because 'it was always felt that the difficulty of accounting for the great penetration of these radiations was insuperable' and he argued that this difficulty 'was quite exaggerated and even imaginary' since neutral pairs, consisting perhaps of one α or positive particle [the charge of an α particle was not yet known] and one β or negative particle' would be expected to have weak ionizing power and hence great penetration. On the other hand, if X-rays were electromagnetic pulses which spread as they travelled why did they not ionize most of the atoms as they passed through a gas, instead of the relatively few actually observed, and how could a spreading pulse concentrate enough energy on an atom to ionize it.

But many physicists at this time believed that X-rays had been shown unequivocally to be electromagnetic pulses by the experiment of C.G. Barkla. These experiments were inspired by J.J. Thomson's theory of X-ray scattering which showed that X-rays, regarded as unpolarized electromagnetic waves, should be scattered through an angle θ with intensity proportional to $(1 + \cos^2\theta)$. In single-scattering and double-scattering experiments Barkla showed that the scattering of relatively soft X-rays is consistent with this formula and that the X-rays can be polarized. Naturally a controversy developed, with Barkla the chief proponent of the ether-pulse theory. It was conducted mainly in Nature and the *Philosophical Magazine* and it stimulated both W .H.B. and Barkla to engage in further experiments that were concentrated initially on more detailed studies of the angular distribution of secondary X- and γ-radiation. W.H.B. emphasized the asymmetric scattering and emission of y-rays (with higher intensity at $\theta = 0°$ than at 180°) which he claimed was consistent with the neutral pair theory and 'fatal to the ether-pulse theory of the γ-rays'. Barkla in reply claimed close agreement between his X-ray experiments and the predictions of the electromagnetic pulse theory but had to admit some difficulty with the harder X-rays and y-rays: 'My argument has not been concerned with γ rays but with the type of radiation with which I am experimentally most familiar-X rays of ordinary penetrating power.'

In retrospect it is easy to see the connection between the properties of X- and γ-rays that concerned W.H.B. and the quantum properties of light but W.H.B. had not read Einstein's paper on the photoelectric effect and neither he nor Barkla saw the connection. But both made important new discoveries whose true significance was recognized later. Thus Barkla and C. A. Sadler found that secondary X-rays in general consist of two distinct types: (1) the ordinary Thomson-scattered X-rays; and (2) entirely new homogeneous X-rays, the hardness of which is *characteristic* of the emitting element. W.H.B. and J.P.V. Madsen found that y-rays excite f3-rays of the same velocity from a number of different scatterers and they proved that the velocity of the f3-rays depends only on the hardness of the primary γ-rays.

Bragg followed the arguments closely, acting as a sounding board for his father's ideas: he remembered living 'in an inspiting scientific atmosphere'. Then, in 1908 at the height of the debate W .H.B. accepted the Chair of Physics at Leeds University. They left in the *Waratah* in January 1909 and arrived in Plymouth in March. The *Waratah* was lost with all hands on her next voyage.

It was more than 50 years before Bragg saw Australia again but his memory of it remained vivid. He remembered particularly the long summer holidays, in the hills or by the sea to escape the heat in the cit)-; collecting shells on the coastal reefs (including a new Sepia that was called in his honour *Sepia Braggi*, *Verco*); galloping bareback along the sands and into the sea: and learning from his mother, together with the rest of the family, how to draw and paint. England under snow in March 1909 must have seemed very different.

CAMBRIDGE, 1909-14

The family were left in lodgings in Plymouth while W.H.B. and his wife went into Leeds to report for duty and find a house. Initially they rented a furnished house and through the spring and early summer lived rather miserably there, surrounded by the unfamiliar grime of an industrial city. Bragg afterwards regretted not having done something more constructive in this period but eventually, following his father's example, he went up to Trinity College, Cambridge, for the long-vacation term. His brother Bob went to Oundle School, before following him to Trinity in 1912.

Bragg began by reading mathematics at Cambridge. His tutor was the Rev. E.W. Bames, afterwards well known as a radical Bishop of Birmingham, and he attended lectures by A. N. Whitehead, G. H. Hardy and Professor A. R. Forsyth and was coached by R. A. Herman, all of Trinity .Having been unable to take the Scholarship Examination before going up he sat it in the spring of 1910—while in bed suffering from a serious attack of pneumonia—and was awarded a major scholarship in mathematics, the Master of the College commenting on the brilliant imagination shown in his essays. In the Part I examinations he gained first class honours in mathematics, and then, strongly urged by his father, he transferred to physics for Part II of the degree course. C. T. R. Wilson, who ran the practical class at that time and lectured on optics, made the strongest impression and left him with a love of physical optics which never deserted him. But he also remembered dull lectures from Searle on heat, stimulating fireworks from J. J. Thomson (who staunchly favoured the wave-theory of X-rays) and the excitement of J. H. Jeans's lectures on statistical m echanics and the emerging quantum theory. One of his letters home (b) recounts in detail Jeans's discussion of the black-body radiation of a cavity, showing that the observed energy distribution cannot be derived from Maxwell's equations but without mentioning Planck's quantum hypothesis of 1900. He wrote 'I got an awful lot from a Dane who had seen me asking Jeans questions, and after the lecture carne up to me and talked over the whole thing. He was awfully sound on it, and most interesting, his name was Bohr or something that sounds like it.' Bohr soon went on to Rutherford in Manchester but he remained one of Bragg's friends for life.

Bragg sat the final examination in Part II Physics in 1911, gaining first class honours, and he spent his third year mainly doing 'a crude research into the velocity of ions in various gases, suggested by J. J. Thomson' (c). It was not an encouraging experience since, contrary to popular belief, the facilities for research in the Cavendish at that time were extremely primitive. The large number of research workers attracted there by J. J. Thomson's reputation quite over- whelmed the meagre resources. Bragg remembered for example: 'that there was only one foot bellows between the forty of us for our glass blowing which we had to carry out for ourselves, and it was very hard to get hold of it. I managed to sneak it once from the room of a young lady researcher when she was temporarily absent, and passing her room somewhat later I saw her bowed over her desk in floods of tears. I did not give the foot pump back' (226).

But life in Cambridge, or more particularly in Trinity, had proved to be rewarding and provided compensations. A. L. Goodhart remembered dining regularly in Trinity with Bragg, E. D. Adrian, F. W. Aston and G. P. Thomson, probably during 1913, and listening fascinated to the chaffing of the scientists: Braggs were good at experiments, Thomsons were not!

The beginnings in 1909 were not so happy, perhaps not surprisingly for a young man away from home for the first time, but Bragg's colonial background led to a rewarding activity. On 6 November 1909, following the example of one of his Squires cousins, he enlisted as a Trooper in King Edward's Horse. This Cambridge unit of the Special Reserve had been called originally 'The King's Colonials' and was composed of men who had come from, or had some close connection with, the colonies. They were mounted infantry and trained in the tactics developed during the Boer War, concentrating on marksmanship, riding and the care of horses. Joined by his brother in 1912, Bragg trained during the year and at summer camps for four years until his discharge in November 1913. The company seems to have produced no close friends but it must have helped in the first experience of a new culture to have this link with the old.

It was College life that provided the closest friends. Chief among these was Cecil Hopkinson who carne from a family of engineers and was the youngest brother of Bertram Hopkinson the Professor of Engineering in Cambridge at that time. Hopkinson introduced Bragg to skiing, sailing, shooting and climbing and 'dragged him into adventures which he thoroughly enjoyed once he was launched into them'. Sailing remained a particular pleasure, though an early cruise off the south coast of Ireland gave rise to another dangerous bout of pneumonia. The only available hospital was the infirmary of Skibbereen Workhouse, where Bragg was looked after by nuns and spent much time in long conversations with the local schoolmaster, Jeremy O'Regan, who remained a lifelong friend. He went to convalesce with the Townshends at Castle Townshend, a summer resort of hunting people. This and other adventures cemented his friendship with Cecil Hopkinson whom he regarded as a major formative influence.

At Trinity Bragg also became close friends with a group of contemporaries who shared his interest in intellectual exploration. They formed a discussion group, not only sitting up late discussing the world in the universal manner of undergraduates but also reading formal papers to each other on a wide variety of subjects. H. Townsend was a mathematician, C. S. S. Higham a historian, H. W. St C. Tisdall a classicist and B.S. Gossling a physicist. Bragg remembered giving a joint paper with Townsend on Minkowski's interpretation of relativity and he remembered also a paper by Gossling on the theory of crystal structures with particular emphasis on the latest ideas of Pope and Barlow. This was Bragg's first introduction to crystallography and to the work of W. J. Pope, Professor of Chemistry in the University, 1908-39, which were soon to be of critical importance.

Throughout this undergraduate period, while enjoying new friendships and experiences, Bragg remained closely in touch with his family as they settled down in Leeds. After the initial shock, his mother had come to terms with the industrial north. Securely established in 'Rosehurst', a commodious stone home in Grosvenor Road, and with a cottage near Bolton Abbey in Wharfedale for weekends and holidays, she entered wholeheartedly into the

Social life of Leeds and its surroundings. According to her son, she was gregarious by nature and clever at making people enjoy themselves and these qualities found full expression in a ceaseless round of visiting, entertaining and welfare work.

For W.H.B. the move to Leeds had been less successful. Reproaching himself for his wife's initial unhappiness and engaged in organizing a large department and giving more formal lectures and courses than those he had given in Adelaide, he found little time or inspiration for new research. The controversy with Barkla was still unresolved, however, and W.H.B. continued to think and to write about the nature of the X- and γ-rays. Early in 1910 he began a correspondence with A. Sommerfeld of Munich (d). The occasion was the publication of papers by Sommerfeld and by J. Stark of Aachen in which they advocated rival views of the origin and nature of X-rays. Stark, apparently the first to be stimulated by Einstein, had proposed that X-ray quanta are produced when a beam of electrons collides with electrons in a metal plate and he described the individual collisions on the assumption that momentum is conserved, using the quantum value for that of the X-ray. Sommerfeld, a strong advocate of the view that X-rays are classical electromagnetic radiation, challenged Stark's theory, claiming that the experimental evidence, including the asymmetric emission of the X-rays, could be explained on the pulse theory W.H.B.s letter to Sommerfeld (7 February 1910) admitted the difficulty he had in accounting for the polarization of X-rays with his neutral-pair theory but emphasized his belief that the difficulties that arise from the spreading of waves are even greater. Sommerfeld's reply conceded the 'weakness of his position regarding the production of the secondary rays' but claimed that 'concerning the emission of the primary rays, we are by contrast on familiar ground'. A year later Sommerfeld sent W.H.B. a reprint of the article in which he showed that if an electron moving at nearly the speed of light is brought to rest in a distance of atomic dimensions nearly all of the resultant radiation (Bremsstrahlung) is emitted into a narrow region between two concentric cones around the direction of motion of the electron. Thus, according to Sommerfeld, the radiation has 'the character of a projectile and in its energy localization is no longer appreciably different from a corpuscular radiation' (f). W.H.B. replied immediately (17 May 1911) 'Your hollow cone is most interesting and the ring structure of the y-ray. But this does not meet the real difficulty to my mind. How do you propose to get the energy back again from this ever spreading ring to a single electron? In other words, how are you going to account for the production of a β-ray by a γ-ray?' But he went on: 'I am very far from averse to the reconcilement of a corpuscular and a wave theory: I think that some day it must come. But at present it seems to me that it is right to think of the X- or the γ-ray as a self contained quantum which does not alter in form or any other way as it moves along... My chief point is that it does not spread: and it seems that spreading is the inevitable accompaniment of the electromagnetic theory.' Finally he drew attention to C.T.R. Wilson's recently published 'pictures of the fog formed instantly after the passage of ionizing rays through a gas' noting that: 'there is no visible general fog due to the direct action of the X-rays: nor is there any corresponding effect in the γ-ray picture. All that is seen is the track of the [secondary] β-ray like a fine hair right across the chamber.'

Despite these difficulties, Sommerfeld believed at this time that only a demonstration of the diffraction of X-rays was needed to exclude every corpuscular theory of X-rays and early in 1912 he thought he had found such evidence in the diffuse broadening of the image of a wedge-shaped slit illuminated by X-rays. But he had also appointed a young assistant, Walter Friedrich, to investigate whether polarization and directional emission could be found for characteristic radiation. Meanwhile, W.H.B. had been confronted by further evidence that his neutral-pair theory was inadequate to account for all the properties of X-rays: Otto Stuhlmann and R.D. Klee discovered independently that ultraviolet light ejects β-rays asymmetrically from a thin metal plate. Thus a relationship had been established between X-rays and light, an accepted electromagnetic radiation, using the very phenomenon that he had used most often in support of his own theory. Nevertheless, while accepting more and more openly that something new was needed to embrace both the corpuscular and wave concepts, he continued to argue the case for a corpuscular theory, sustained by Whiddington's observations, clearly consistent with such a theory, that X-rays cannot excite the characteristic rays of any substance unless they have themselves been excited by cathode rays of energy exceeding a certain limit. In his book Studies in radioactivity' which was published in 1912, he wrote: 'it still seems to me that the neutral pair theory correctly pictures the chief processes of the X-ray, which the old form of the spreading pulse, even the modified Thomson's pulse, are unable to do. But I should now add that we ought to search for a possible scheme of greater comprehensiveness, under which the light wave and the corpuscular X-ray may appear as the extreme presentments of some general effect'.

It was another ten years before A. Compton showed that X-rays are scattered, by matter in two ways: for the soft X-rays studied by Barkla, Thomson scattering predominates; for y-rays and hard X-rays the asymmetric Compton scattering is the more important. And it was longer still before it was accepted that both radiation and matter have particle-like and wave-like properties. Meanwhile, the balance of the argument was pushed dramatically to the side of the wave theory as the result of experiments conducted by W. Friedrich and P. Knipping at the suggestion of Max von Laue in Sommerfeld's Institute for Theoretical Physics in Munich.

These experiments and the ideas underlying them have often been described. (Ewald (e), who played a major part in the story, and Forman (f) give most detail in somewhat contradictory efforts to recapture the scientific climate of the time.) In brief, Ewald had worked out the theory governing the interaction of electromagnetic radiation with a simple orthorhombic lattice of dipoles and he sought to discuss his results with Laue. Without going deeply into the theory, Laue was struck by the possibility that a crystal irradiated with X-rays might give diffraction effects, if X-rays were indeed electromagnetic radiation. He therefore persuaded Friedrich and Knipping to try an experiment. Friedrich and Knipping first irradiated a copper sulphate crystal and subsequently zinc sulphide (ZnS) and various other crystals of cubic symmetry and observed that the X-rays were scattered by the crystals in discrete directions close to the direction of the incident X-ray beam. The pattern of spots made by the scattered X-rays on a photographic plate depended upon the orientation of the crystal and upon its symmetry so that, for example, cubic crystals irradiated along a four-fold axis gave a characteristic four-fold pattern of spots, though the intensities and absolute positions of the spots varied from substance to substance.

An account of these experiments, with an introduction by Laue on the theory of diffraction by a three-dimensional lattice, was presented at a meeting of the Bavarian Academy on 8 June 1912 followed on 6 July 1912 by Laue's attempt to explain in detail the effects observed with ZnS (e).

Unfortunately, after his faultless derivation of the basic conditions for diffraction from a three-dimensional lattice, Laue went wrong in his consideration of the experimental results with ZnS. First, in order to estimate the size of the unit cell of the ZnS crystal structure (which was, of course, unknown at the time and which he needed to derive the wavelengths of the X-rays) Laue assumed that ZnS has a primitive cubic structure with one ZnS molecule per unit cell whereas in fact, as we shall see, the structure is face-centred cubic with four ZnS molecules per cell. Secondly, and more disastrously, he was preoccupied with the idea that the observed effects were asSociated with characteristic X-rays arising in the crystal. Consequently he sought to explain the results in terms of a limited number of X-ray wavelengths and showed that, if the conditions for diffraction need be fulfilled only approximately, the observed spots could be explained in terms of only five different wavelengths. He noticed that some additional spots should be expected to appear, but nevertheless he was strongly persuaded by his success in explaining the observed spots that the phenomenon could be explained by diffraction and hence that it established the wave nature of X-rays.

The papers of Laue and his colleagues appeared in late August 1912 and aroused intense interest, but the conclusions were not immediately accepted, even by convinced advocates of the wave theory .Thus Barkla, writing to Rutherford on 29 October 1912, remarked 'I have had a copy of Laue's paper for some little time and certainly am sceptical of any interference interpretation of the results. A number of features do not point that way... this in no way affects my absolute confidence of the truth of the wave theory of X-rays' (f). The confusion and the lack of understanding of Laue's theory is well illustrated by the fact that Laue himself, in March 1913 (e), was arguing that the effects could not be due to the interaction of a crystal with a continuous range of X-ray wavelengths since this would lead (by analogy with the properties of a simple diffraction grating) to the photographic plate being blackened everywhere. If he had followed up his initial discussion with Ewald, who shortly showed his mastery of the theory , this story would have been different.

Bragg and his father were of course deeply interested in this new evidence on the nature of X-rays and they had the details in time to discuss them during the family holiday, which was spent in 1912 as guests of Leeds friends at Cloughton on the Yorkshire coast. Naturally the first thought was to explain the results in terms of corpuscles and on returning to Leeds Bragg set up an experiment to test whether the spots on the photographs might be due to neutral particles shooting down the avenues between rows of atoms in the crystal structure. The same idea had occurred to Stark (1912) in Aachen and it led to the first mention of Bragg in a published paper. In a letter to Nature, written on 18 October 1912, W.H.B. (g) noted 'a fact which my son pointed out to me, viz. that all the directions of the secondary pencils in this position of the crystal are "avenues" between the crystal atoms'.

Back in Cambridge, Bragg continued to think about Laue's papers and here several bits of knowledge carne together to provide the answer. First he remembered that J.J. Thomson had lectured on Stokes's theory of X-rays as very short pulses of electromagnetic radiation, picturing an electron moving with its asSociated lines of force and X-rays as the whip crack which would run along these lines when the motion of the electron is arrested suddenly at the anticathode of the X-ray tube. Next he recalled C. T. R. Wilson's lectures on the nature of white light which had shown (following Schuster)* that white light could be regarded either as a combination of light of all wavelengths, each wavelength being diffracted at the appropriate angle by a diffraction grating, or as a succession of formless pulses which the lines of the grating converted into a train of waves. Then, drawing on the ideas about crystal lattices that his friend Gossling had described at one of their discussion meetings and remembering that the Laue spots became increasingly elliptical as the photographic plate was moved further from the crystal, he had the idea that the formless X-ray pulses could be regarded as reflected by sheets of atoms in the crystals. The pulses reflected from successive equidistant sheets then would form a wave train, just as in Wilson's treatment of the diffraction grating. Since the path difference between the waves of the reflected train is $2d \cdot \sin\theta$, where θ is the glancing angle at which the radiation falls on the planes and d is their spacing, it followed immediately that the wavelengths (λ) of the different orders of reflexion would be given by

$$n\lambda = 2d \sin\theta ,$$

where n is an integer.

The critical test was to see whether these ideas explained the observations from ZnS, including the absence of some spots predicted by Laue's analysis. Here Bragg inverted the argument and used the fact that the X-ray pulses can be regarded as equivalent to a 'white-light' spectrum extending over a characteristic range of wavelengths and with maximum energy at a certain wavelength. The intensities of the Laue spots ought, therefore) to fall in a regular series depending upon which part of the

---

* In the preface to the second edition of his textbook (h), Schuster notes .the treatment of white light and of interference problems has been made more consistent-and I hope clearer by introducing the theory of impulses at an earlier stage'.

spectrum was responsible for each of them. Examination showed that this did not work.

At this juncture Gossling's talk on crystal structures again carne to the rescue since his discussion of Pope's ideas (i) had embraced not only the simple cubic lattice but also the face-centred lattice. Remembering this Bragg tried to explain the ZnS pattern on the assumption that the structure is face-centred cubic and everything fell into place. Thus he showed that the Laue pictures were made by a continuous range of X-ray wavelengths, a kind of 'white' radiation, and that X-ray diffraction. could be used to get information about the crystal structure. This was the start of the X-ray analysis of crystals.

The work was described (1) at a meeting of the Cambridge Philosophical Society on 11 November 1912 and briefly reported in *Nature* of 5 December, though not before W.H.B. (j) had remarked in another letter (28 November) 'my son has given a theory which makes it possible to calculate the positions of the spots for all dispositions of crystal and photographic plate'.

The paper, which appeared in January 1913, was called 'The diffraction of short electromagnetic waves by, a crystal' because Bragg was still unwilling to relinquish his father's views that the X-rays were particles; he thought they might possibly be particles accompanied by waves. W.H.B., however, did not cling to his ideas but, noting that 'the problem then becomes, it seems to me, not to decide between two theories of X-rays but to find one theory which possesses the capacities of both' (J) he went on vigorously to exploit the new possibilities. At this stage they began to lead in two directions, towards the analysis of crystal structures and studies of the nature of X-rays.

Naturally enough, Pope was very pleased at the support for his theories provided by this work on ZnS and he suggested to Bragg that studies of the alkali halides NaCl, KCl, KBr and KI might be even more rewarding, presenting him with suitable crystals. 'The Laue pictures which they gave were simpler than those of zinc blende and led to a complete solution of their structure. These were the first crystals to be analysed by X-rays' (G). But before these results of Bragg's individual work were published (S) there had been other developments.

During the discussion following the presentation of Bragg's epoch-making first paper (1) C. T. R. Wilson suggested to Bragg that crystals with very distinct cleavage planes such as mica might show strong specular reflection of the X-rays. Bragg tried the experiment and well remembered J.

J. Thomson's excitement on seeing the still-wet photographic plate with a mirror reflection of X-rays on it (G). This observation was published in Nature on 12 December (2) in a letter dated 8 December, the first of Bragg's papers actually to appear in print; and it aroused great interest, not least on the part of W.H.B. who saw at once the possibility of using the effect to study the nature of reflected X-rays by methods with which he was long familiar. On 23 January 1913 (k), in a letter dated 17 January, W.H.B. reported in seven lines his success in measuring the ionization produced by X-rays reflected from mica. This urgency illustrates the high excitement of the time and the competitiveness: the following week Moseley and Darwin (l), in a letter dated 21 January, published a rather more detailed record of similar experiments in which they acknowledged the stimulus of Bragg's original observation of reflection (2) even though they seem to have had similar ideas independently (m).

The apparatus used by W.H.B. in this experiment was rapidly developed in Leeds into the X-ray spectrometer and with this instrument he examined in detail the reflection of X-rays from a number of crystal faces, including those of rock salt. This provided the next great discovery. 'In addition to the "white" X-irradiation of all wavelengths which Bragg had called the X-ray pulses, W.H.B. found that each metal used in the X-ray tube as source of radiation gave a characteristic X-ray spectrum of definite wavelengths, just as elements give spectra in the optical region.' This work was presented at the Royal Society on 7 April1913 in a joint paper (4). Bragg later (M) disclaimed having played more than a general part in the design of the spectrometer or in its use, though his knowledge of the crystals that were studied and which he had been analysing in Cambridge must have been important. Such was the start of X-ray spectroscopy. Moseley and Darwin in similar experiments had missed the characteristic spectra, apparently through the use of too-fine collimating slits, but prompted by details communicated privately by W.H.B. they went on immediately to improve the measurements, discover the fine structure of the spectra and start the classical survey which led Moseley to establish the atomic numbers of the elements.

The development of the X-ray spectrometer by W.H.B. and Jenkinson, his instrument maker in Leeds, highlighted the inadequacies of the Cavendish Laboratory where Bragg had great difficulty in getting on with his experiments. Years later he remembered (M) 'When I achieved the first X-ray reflections I worked the Rumkorff coil too hard in my excitement and burnt out the platinum contact. Lincoln, the mechanic, was very annoyed as a contact cost ten shillings [a week's wages at the time], and refused to provide me with another for a month. I could never have exploited my ideas about X-ray diffraction under such conditions'. Furthermore, the X-ray spectrometer promised a far more powerful way of analysing crystal structures than the laborious and indirect method of the Laue photograph. Accordingly it was at this stage, in the early summer of 1913, that Bragg and his father began to work together during the vacations in Leeds. But the Cambridge picture was not entirely black. Pope was always helpful and, although the Professor of Mineralogy had given strict orders that no minerals should ever leave the collections, Bragg got his specimens from Hutchinson, who was then a lecturer and later became Professor (155), and he was able to take them to Leeds with him.

The next papers were published at about the same time in a somewhat strange order. In the first of them W .H.B. (n) derived the wavelengths of various radiations and correlated them with Barkla's characterististic radiations, making use of the structure of rock salt which had been worked out by his son, but not yet published. This paper was followed immediately by Bragg's detailed account (5) of NaCl and related structures in a paper described by Ewald (e) as 'the great break-through to actual crystal structure determination and to the absolute measurement of X-ray wavelengths'. The analysis depended mainly on Laue photographs taken in Cambridge, supported by some measurements with the spectrometer, and led to a conclusion that was to disturb chemists for many years that 'in sodium chloride the sodium atom has six neighbouring chlorine atoms equally close with which it might pair off to form a molecule of NaCl'.

Finally, in this group of consecutive papers, carne a paper jointly by Bragg and his father (7) on the structure of the diamond. Bragg later (234) attributed the credit for this analysis mainly to W.H.B., who succeeded with the spectrometer where he had himself failed with Laue photographs, but as Ewald (e) points out, this paper again employed all of the arguments developed in the preceding paper. Ewald goes on to note: 'Diamond was the first example of a structure in which the effective scattering centres did not coincide with the points of a single (Bravais type) lattice. Whereas in the structures of rock salt, zinc blende and fluorite the absence of molecules in the accepted sense created an element of bewilderment, the beautiful confirmation of the tetravalency of carbon on purely optical principles made this structure and the method by which it was obtained immediately acceptable to physicists and chemists alike'.

The summer of 1913 in Leeds was described by Bragg (G) in glowing terms. 'The X-ray spectrometer opened up a new world. It proved to be a far more powerful method of analysing crystal structure than the Laue photographs which I had used. One could examine the various faces of a crystal in succession, and by noting the angles at which and the intensity with which they reflected the X-rays, one could deduce the way in which the atoms were arranged in sheets parallel to these faces. The intersections of these sheets pinned down the positions of the atoms in space. ...It was like discovering an alluvial gold field with nuggets lying around waiting to be picked up. ...It was a glorious time when we worked far into every night with new worlds unfolding before us in the silent laboratory.'

During this period W.H.B. was more interested in X-rays than he was in crystals and he left the crystal structures to Bragg whose next great paper (10) described a refined analysis of NaCl, using spectrometer measurements of the reflected intensities and went on to describe the structures of zinc blende (ZnS), fluorspar (CaF_2), iron pyrites (FeS_2) and calcite ($CaCO_3$). This paper represents further remarkable progress, in particular showing how the intensities of the reflexion must be measured and evaluated in a complete structure analysis and in demonstrating the possibility of solving structures in which atomic positions are not fixed by the symmetry but have to be found by a detailed analysis of the X-ray intensities. Iron pyrites and calcite were the first structures involving undefined atomic parameters-one in each structure.

Apart from the structure of copper (11) these were the last to be completed for publication before the outbreak of war in August 1914 brought this period to an end. But even before that happened there were signs of difficulty in the unique working relationship between father and son. As the importance of the work was recognized W.H.B., the established scientist, naturally was consulted or asked to talk about it an:d, however hard he tried, he did not quite avoid leaving the impression that his was the guiding part: what was no more than fair looked like parental generosity. Thus W.H.B. gave his son credit, though without mentioning him by name, in the earliest Nature letters (g, j) but they were still W.H.B.'s letters and Bragg's first papers came later and were over-shadowed by the first joint paper .

Then in 1913 W.H.B. described the work at the annual meeting of the British AsSociation and was invited to the Second Solvay Conference on Physics, 27-31 October 1913. The subject was 'The structure of matter' and Laue, Pope and Barlow were there as well as W.H.B., each of them relying on Bragg's work for major parts of their presentations. They all acknowledged the importance of his contribution and the members of the Conference, including Sommerfeld, Laue, Einstein, Lorentz and Rutherford, sent him a postcard congratulating him on 'advancing the course of natural science'. But there is no doubt that a cloud remained that overshadowed his future relations with W.H.B. and was remembered 60 years later with pain mixed with gratitude for his father's part in making possible the rapid development of the work.

It was in the discussion of the Solvay Conference that Bragg, who was known to family and friends as Willy, was first referred to as W. Lawrence Bragg, to distinguish him more clearly from W.H.B.; and he used this style in his subsequent publications.

Back in Cambridge during 1914 Bragg struggled to extend his methods of analysis to more complex crystals. By this time he had a spectrometer of his father's design but work remained difficult at the Cavendish ('They keep the wretched liquid-air machines going most of the day, which makes the leaf jump all over the shop') and he did not find it easy to get on with writing the book that he and W.H.B. had planned. In a typically undated letter he wrote: 'I find it impossible to do my experiments and write the book at the same time, the book requires one to be absolutely on the spot. I nearly faint when I think of the article for the Jahrbuch. All this kind of thing does make ones brain boil so. Its a curse this continual writing'. But he kept up and extended his contacts with classical crystallographers. Barlow wrote frequently and in March Bragg spent a weekend with W. J. Sollas, the Professor of Mineralogy in Oxford, during which he had a 'great' talk with Barker and visited Moseley who was then near the end of experiments on X-ray spectroscopy that established the idea of atomic number.

In the summer of 1914 Bragg was elected to a lectureship and Fellowship at Trinity College and he occupied a set of rooms there with his close friend Cecil Hopkinson. His brother Bob was also in Trinity, in his second year as an undergraduate, and he had his first research student, E. V. Appleton. At this stage he was working on aragonite, a structure involving several undefined parameters, and writing to his father on 19 July 1914 he said: '1 have been writing up the aragonite but am a bit puzzled about the structure. I would really like to do a little more work on it. I wish I could go over it with you, I don't know quite how infallible I am.' As a postcript he added: 'Could you tell me about our programme in Germany? Where is the meeting and when?' There was no meeting: on 24 August 1914 he was commissioned as a 2nd Lieutenant in the Leicestershire Royal Horse Artillery.

WORLD WAR I, 1914-18

On the strength of his service with King Edward's Horse and knowledge of mathematics, Bragg was posted to a Territorial battery of the Leicestershire R.H.A. and spent the first year of the war training at Diss in Norfolk. Describing this experience he wrote (L) 'I was very much out of my element as my knowledge of horses was not at all extensive, and my fellow officers and men were Leicestershire hunting enthusiasts'. But he managed and wrote daily letters home describing his new life and yearning to continue his research. Meanwhile, despite the war, recognition of the importance of the new crystallography was growing. In February 1915 'X-rays and crystal structure' (A), describing the results obtained so far, was published with a preface in which W.H.B. tried hard to put the record straight: 'I am anxious to make one point clear, viz. that my son is responsible for the "reflection" idea which has made it possible to advance, as well as for much the greater portion of the work of unravelling crystal structure to which the advance has led.' In May the Barnard Medal of Columbia University was awarded to them both jointly. Throughout the early part of 1915 there was much family discussion about W.H.B.'s move to University College London, and all the time there were exchanges about crystal structures.

Then in August 1915 everything changed. The French Army had been experimenting with a method of locating enemy guns from the sound of their firing and the War Office decided that the British Army should follow suit. Bragg was selected to do the work helped by H. Robinson, a member of Rutherford's staff in Manchester, and they left for France at the beginning of September.

Bragg has described the general principle of sound-ranging (L) as follows: 'A series of listening posts or microphones are situated in known positions along a base behind the front line. The time differences between the arrival of the report at the posts are measured. Suppose the sound to reach post 1 at time T_1, post 2 at time T_2 and so forth. Then if one draws a circle on the map around post 2 with radius $V(T_2 - T_1)$, where V is the velocity of sound, and similar circles for the other posts, a great circle which passes through post 1 and touches the other circles represents the form of the report wave, with the gun at its centre.'

The chief requirements for an effective system were to identify the sound due to a particular gun and to record the time intervals precisely. After a study of the French equipment in the front line in the Vosges, Bragg chose the recording equipment developed by Lucien Bull of the Institut Marey in Paris. This was the most elegant and accurate of the recorders but it was complex and required photographic development of a cine film on which the displacements of galvonometer wires were recorded (L). The more difficult problem was to find a microphone that distinguished between the report of the gun and the shock wave asSociated with the shell and this was not solved until late in 1916. Bragg (240) has described how the solution was found. 'We were living in tarred felt huts in bitterly cold weather at the time and we noticed that whereas the shell wave was a deafening crack, the faint gun wave blew jets of very cold air through the readily available holes in the sides of our hut. Now I had in my unit a certain corporal Tucker, who in peace time was a lecturer at Imperial College and who had made experiments on the cooling of heated fine platinum wires by currents of air. The joint brainwave carne to us, I think mainly to him, that we could use this effect. We sent to England for a supply of the thin wire. We scrounged some ammunition boxes, bored a hole in each, and stretched the wire across the hole. We incorporated this in one arm of a Wheatstone Bridge, with a sufficient current to heat it to a dull red. ...The idea was that the high-pitched noises which were so troublesome would have such rapid fluctuations that they would hardly displace the shell of warm air around the wire, whereas the low-pitched gun wave would blow a blast of air through the hole, sweep the warm air away, cool the wire, and so reduce its resistance, upset the bridge, and make the galvanometer record a current.'

The Tucker microphone worked 'like a charm' and this and other developments (L), all devised at the Front, made sound ranging a powerful and trusted method.

In the meantime, however, Bragg's personal life was deeply affected during the autumn of 1915 by the death of his brother Bob at Gallipoli, where Moseley was also killed, and by the eventually fatal wounding of his friend Cecil Hopkinson. News of a different kind, the award of the 1915 Nobel Prize for Physics to Bragg and his father, carne when he was setting up the first sound-ranging station near the front line south of Ypres. He wrote home on 17 November: 'Just got Dad's letter and yours with the cheery news in it.' The village cure with whom he was billeted produced a bottle of Lachryma Christi.

From the beginning of 1917 to the end of the war Bragg supervised the successful application of sound ranging and a great expansion of the number of units. Many young scientists, including Andrade, were involved. In the process Bragg was awarded the O.B.E. and M.C., was mentioned in dispatches three times, and rose to the rank of Major. This service also brought him new friends. One of them was Harold Hemming, who was in charge of a complementary operation, flash-spotting, and another of lasting importance was R. W. James (232). Bragg and James had been in the same Part II class at Cambridge, where G. R. Crowe, then a precocious lab. boy who had recognized James's more conventional ability had predicted that: 'Mr. James would get a first and Mr. Bragg might get a first. James had subsequently had an adventurous time on Shackleton.s expedition to the Antarctic and then returning home he joined Bragg's unit near Ypres. Later when Bragg moved to G.H.Q.. James helped him set up a school for sound rangers where they worked together a great deal and no doubt discussed the future development of crystallography.

As the war drew to an end, their thoughts turned towards jobs and letters home were full of the possibilities. Thus Bragg wrote to W.H.B. on 21 October 1918: 'I got your letter yesterday about the G.E.C. work. I had never contemplated anything else than a University career ...training men in the University to take up applied science afterwards. In December he wrote about James, who was thinking of applying for the Professorship in Cape Town: 'I know that I would feel absolutely happy and confident in taking on any University Chair, such as Leeds even, with a fellow like James to back me up. But in January 1919 this was followed by 'I am just a bit doubtful of my powers of tackling the Birmingham University job right now as I know so little physics. I've had no experience in lecturing.

After appropriate celebrations of the end of the war (Major Bragg and Captain G. P. Thomson made their first appearances at the Royal Society Dining Club on 30 November 1918) Bragg returned to Cambridge early in 1919 to take up his duties at Trinity College and demonstrate in G.F.C. Searle's (193) practical class at the Cavendish laboratory. With his brother, Hopkinson and Tisdall all dead it must have been a sad return but he found Cambridge a lively place with the Social life dominated by young demobilized officers Blackett amongst them. He had just enough time to fall in love with Alice Hopkinson. Cecil's first cousin who was then reading History at Newnham College before accepting the Langworthy Professorship of Physics at Manchester, in succession to Rutherford.

MANCHESTER. 1919-37

At this time the Vice-Chancellor of Manchester University was Sir Henry Miers. F.R.S. who had previously been Professor of Mineralogy in Oxford (1895-1908). In Manchester the University had created a special Chair of Crystallography for him and he was deeply interested in the development of the subject that had been made possible by Bragg's work. His influence in the appointment is clear. Lecturing on 'the old and the new mineralogy' (o) he had said: 'In my opinion, the importance of the study of crystals has now become so great, not only for the identification of substances by crystal measurements but also on account of the new knowledge which modern crystal study is contributing to problems belonging to different sciences, that there is a real need for a department of pure crystallographic research, one in which such studies can be carried out quite independently of elementary teaching or of immediate applications, and without being tied to mineralogy. I venture to hope that it will not be long before some such department is founded either in connection with one of our Universities or elsewhere. Through Bragg's appointment and the support and encouragement he gave to him Miers was able to realize much of this dream. He also became a firm family friend, godfather to Bragg's younger son David (1926), and his diary records many contacts.

Despite Miers's support, Bragg's early days in Manchester were not easy. Most of the other professors were relatively old—Horace Lamb the Professor of Mathematics had preceded W.H.B. in Adelaide—and they were used to dealing with Rutherford who had made the Physics Department world-famous. Bragg took over at a difficult moment. During the war most of the staff had been away on war work and the teaching had been continued mainly by E. J. Evans and N. Tunstall. When Bragg joined the Department in the autumn of 1919 he brought with him R.W. James and E.C.S. Dickson. H. Robinson (with D.C.H. Florance) had returned a little earlier so that he had at the beginning a nucleus of sound-rangers to support him. He lived with another, a classics lecturer named Drew, in an establishment arranged by his mother with his formidable old nanny, Charlotte, as housekeeper.

The first priority was to organize the teaching and Bragg at the age of 29 was very conscious that he had had essentially no previous experience that was relevant to the elementary teaching—even as an undergraduate. The beginnings were disastrous. Many of the undergraduates were returning ex-servicemen and they had no mercy on the novices. Tunstall remembered that there were 'rowdy, boisterous goings on in the lecture room particularly when medicos were being lectured to. One could hear this not only on the same floor but in the laboratory under the large lecture theatre and there was visible evidence in the fact that panels of the benches were kicked into matchwood during the lecture periods taken by Bragg, James and Dickson'. In one dramatic episode a student set off a firework under the reading desk and Bragg boxed his ears. To make matters worse anonymous letters began to arrive, addressed to the Vice-Chancellor and others, in which Bragg and his young colleagues were accused of incompetence with evidence quoted that was clearly based on a detailed knowledge of events in the department. Bragg was brought close to the edge of breakdown but recovered when the letters began to attack his father and Rutherford and when his research began to flourish again. But he was deeply scarred and it took a year or two for him to gain confident control. Miers gave what support he could and noted laconically in his diary at the end of 1924: 'there was a plague of anonymous letters at the University against certain members of the Professorial staff. But these ceased with the disappearance of one of the Junior staff to another post (with his wife).'

Research had a more promising beginning and in the autumn of 1919 Bragg was writing optimistically, but somewhat guardedly, to his father: 'My own apparatus is nearly set and James and I are eager to get going. I have one or two ideas I am anxious to try right away,' The heart of the apparatus was the X-ray spectrometer made by Jenkinson but it was used initially with a 'very inadequate gas X-ray bulb with a palladium target' (p). It wasn't until the following year, when the General Electric Company at Schenectady gave them a Coolidge tube, that rapid progress became possible. In the meantime Bragg returned to the study of zinc oxide that he had begun in 1914 (15) and this work, which included comparisons with related structures, combined with reconsideration of the pre-war analyses in terms of the new ideas about atomic structure and chemical bonding that had emerged during the war led him to the idea that interatomic distances in ionic compounds obey an additive law as if ions had characteristic sizes. He assigned sizes to the common ions and showed that the sums of their radii agreed quite closely with measured interatomic distances (16, 23). This important idea provided the subject for his first Friday Evening Discourse at the Royal Institution, on

28 May 1920, but unfortunately in working it out he had been too ambitious in attempting to embrace the sizes of individual atoms in compound ions (such as CO_3^{2-}) and in simple ions (such as Cl^-) in the same system. Accordingly, as he was fond of saying later, his scheme was inside out; all the negative ions were too small and the positive ions too large, though the sums of their radii were correct. J. A. Wasastjerna and v. M. Goldschmidt soon corrected Bragg's values (38) and the idea of atomic radii has played an important part in crystal structure analysis ever since.

Established for the first time in his own laboratory Bragg also had to consider the relationship between his own and his father's research programme. W.H.B. was now at University College London with a growing research group and Bragg continued the letter quoted above: 'I have been wondering what you were intending to go on with. I do hope you will never keep from doing any bit of work, Dad, because you think that may be the line I am going on. ...If we did happen to do the same thing its all to the family credit, isn't it? and I am sure I would never be the loser if people weren't quite sure which of us did a piece of work.' This philosophy was to prove difficult to live by, and they arrived later at a tacit agreement to work on different aspects of the subject, but at this stage he went on: 'I wish you would go on with some of those experiments on the temperature coefficient of the strength of reflection, you did get such interesting results on the ones you started and I don't think anyone else has done that. I have had one or two brain-waves which I am keen to develop but I want to work at ordinary temperatures and get good numerical results.'

This foreshadowed the research programme in which Bragg, James and Bosanquet (18-20, 24) set themselves the task of making X-ray analysis a quantitative science (240). James (1962) has described how they set themselves to make 'a series of measurements of the absolute intensity of reflexion of X-rays from rocksalt, a crystal whose structure ,vas definitely known, with no uncertain parameters'. The idea was, firstly, to test the applicability of the formulae governing X-ray reflexion which had been derived by C. G. Darwin in 1914. Secondly, they hoped to measure the atomic scattering factors, the ratio of the X-ray amplitude scattered by an atom to that scattered by a simple classical electron under the same conditions, since this would provide evidence of the distribution of electrons in the atoms and fundamental data that would be needed in the analysis of more complex crystal structures. The experimental work was very demanding but they succeeded in showing that the rock-salt crystals reflected X-rays very nearly according to Darwin's formula for an imperfect crystal-a 'mosaic' crystal in Ewald's graphic description-according to which the reflected X-ray intensity is proportional to the square of the structure amplitude (F, the effective number of electrons in the unit cell contributing to that reflexion). Furthermore, they were able to derive experimental atomic scattering factors for sodium and chlorine which were 'in fair agreement with what was to be expected from what was known at the time of the electron distribution in those atoms.'

It turned out later that the choice of rock-salt had been fortunate, since the crystals conformed closely to the mosaic model, but the work did much to establish the methods needed for the quantitative analysis of crystal structures. Bragg's unique contribution here was to see the value of making experimental measurements of the absolute intensities of the X-ray reflexions which showed directly the effective number of electrons contributing to each reflexion. The work on rock-salt a1so stimulated work on the theoretical derivation of the atomic scattering factors that were needed to calculate the intensities of reflexions corresponding to any model structure for comparison with the observed values. In 1925, D. R. Hartree, who was then at Cambridge but soon afterwards became a close colleague as Professor of Applied Mathematics in Manchester, calculated promising atomic scattering factors (f-curves) based on the Bohr-orbit model of the atom and, stimulated by the need for better f-curves in crystal-structure analysis, he went on to devise the method of the self-consistent field. Recognizing the importance of this work at any early stage, Bragg (33) laid down the criterion that has guided all subsequent structure analysts: 'The structure which leads to the best agreement between observed and calculated values [of the X-ray intensities] is chosen as the closest approximation to the truth.'

The painstaking work of measuring individual X-ray reflexions in detail and studying their dependence upon temperature and crystal perfection was not Bragg's forte. In later years, comparing himself with his father he said that W.H.B. was the better physicist: 'His points always lay on smooth curves; mine didn't,' So from 1922 onwards he left this side of the work to James and concentrated on the analysis of structures.

In the meantime, however, there had been happy domestic changes. Bragg was elected F.R.S. on 12 May 1921 and among the letters of congratulation was one from Alice Hopkinson, then in the final year of the History Tripos, whom he had not seen since leaving Cambridge. They were married in December and moved into a house in Didsbury. Bragg had misgivings about bringing a lively 22-year-old wife to grimy Manchester and introducing her to the sober Society of middle-aged professors but she had, after all, been brought up in Manchester, where her father had been a much-loved physician, and she knew the place well and was able to help her rather reserved husband in his developing contacts with the city as well as the university, while he introduced her to the wider scientific world in which he was becoming a leading figure. Earlier in 1921 Bragg (but not W.H.B.) attended the Third Solvay Confetence on Physics, the subject was 'Atoms and .electrons', and in the autumn of 1922 he had the pleasure of taking his young wife with him to Sweden where he delivered his Nobel lecture (25) and they established friendly relations with Arrhenius, Westgren and other Swedish scientists. Then, in the summer of 1924 when their first child was a few months old, they made the first of many visits to North America where Bragg lectured at Ann Arbor and they attended the meeting of the British AsSociation in Toronto. In all respects it was an eventful and fruitful marriage which remained a romance to the end. After Stephen Lawrence, born in 1923, there were three other children, David William born in 1926, Margaret Alice in 1931 and Patience Mary in 1935.

Bragg's contacts outside the university had begun early in his career in Manchester Miers's diary for April 1920 notes that he discussed cotton industry research with the science professors and visited the Tootal Broadhurst Mill. At this time the leading figure in that firm, was Kenneth Lee who was a friend of the Hopkinsons and became a close friend of the Braggs and godfather to their daughter Patience. Bragg (243) has described the contact that this friendship gave him with the work of Dr R.S. Willows at the Shirley Institute which certainly coloured his views of scientific research in industry. Important contacts with other firms, especially Metropolitan-Vickers, carne later but another Manchester activity must be mentioned here. Bragg's first lecture to the Manchester Literary and Philosophical Society, on 'Sound ranging', was given soon after his arrival in the city (14) and, thereafter, he lectured regularly to the Society. He was President in 1927-28 (49) and Dalton Medallist in 1942.

In 1924 Bragg turned again to the study of aragonite which he had begun in Cambridge just before the war. This analysis required the determination of nine variable parameters and was much the most complex yet attempted (28). The solution depended upon careful absolute measurements of the X-ray intensities made with the spectrometer and Coolidge tube and, for the first time, Bragg discussed the symmetry of the crystals in terms of formal space-group theory.

Miers was deeply interested in this work and his influence, both direct and through his publications, can be discerned in Bragg's next achievement. Having a consuming interest in how best to use physical observations of all kinds to establish the arrangement of atoms in crystals, he turned briefly to the inverse problem of deriving the physical properties of crystals from their atomic structures. He showed that the refractive indices of calcite and aragonite could, with simple assumptions, be calculated essentially from the arrangement of carbonate ions in them (29) and in a second paper (30) he made some progress in generalizing the result and showing how consideration of the refractive indices may help in the analysis of crystal structures. This work aroused considerable attention and Bragg interested Sydney Chapman in it. Chapman had been a contemporary at Cambridge, though not a close friend, and had succeeded Lamb as Professor of Mathematics. In taking a further step they were prompted, perhaps, by a letter from W.H.B. (12 iii 22) discussing the structure of ruby, in which he noted the question 'of why the rhomb takes the particular shape it does', and they showed how the rhombohedral angle of crystals of the calcite type can be predicted quite precisely (31). But Bragg then left this field to Chapman, who moved away from Manchester shortly afterwards. Although neglected for some time these were important early stages in crystal physics.

The study of aragonite also underlined a problem that struck at the heart of Bragg' s quantitative method of structure analysis: the intensities of the reflexions appeared to be more nearly proportional to the structure amplitudes than to their squares (28), as W.H.B. had also observed in his work on diamond and on calcite. This was a consequence of the state of perfection of the crystals, as the theories of Darwin and Ewald had shown, but the mosaic model had worked so Well for rock-salt that the theory of 'perfect' crystals was temporarily overlooked. Ewald was aware of the problem, however, and he wrote to W.H.B. about experimental measurements for comparison with his theory. W.H.B.'s response was to send some data but also to write an illuminating letter to Bragg (21 i 25) in which he reported Ewald's suggestion that new measurements on iron pyrites would be useful, and continued: 'I feel this is your province entirely and am thinking of writing to tell him so.' Later in the same letter W .H.B. promised to send an organic paper but in a discussion of mercuric chloride remarks: 'This also is your line rather than mine,' By this time, then, a division of work between W .H.B., who had moved to the Royal Institution in 1923, and Bragg seems to have been understood between them: W .H.B. worked on organic structures (and silica) while Bragg concentrated on inorganic compounds and the physics of crystals and diffraction.

The result of Ewald's interest was a study conference that he organized at Holzhausen in Bavaria in 1925 which was attended by the leading exponents of theory and experiment, including Bragg, James and Darwin. Bragg used to tell how they put forward their champion, Darwin, to present his theories of crystal reflexion only to find that he had forgotten how to derive them. However, the result of the meeting (43) was a much clearer understanding of the role of crystal perfection in determining the intensities of X-ray reflexions and this led, in turn, to increased confidence and success in the quantitative analysis of crystal structures.

From 1925 Bragg concentrated on an intensive programme of research into the structures of silicate minerals. The initial objective was to develop a technique for the analysis of crystals in which the atomic positions were defined by a large number of parameters and the silicates provided excellent material for this purpose owing to their complexity and the ease with which well-formed natural crystals could be obtained (67). The main source of supply was still Hutchinson in Cambridge. The first of these structures to be described were the olivines (40) and this success was soon followed by the analysis of beryl, $Be_3Al_2Si_6O_{18}$ by Bragg in collaboration with J. West (42), who played a large part in the silicate analyses and the training of those who worked on them. This was a remarkable study in a number of ways. The high symmetry of the crystals, which they described by means of the recently published diagrams of Astbury and Yardley, made the analysis very straight for Ward: Bragg later (59) noted that 'when West and I had determined the space group, I remember well that we found all the atomic positions in about a quarter of an hour, and all subsequent work only altered our first estimates slightly'. This also depended, of course, on their developing knowledge of atomic sizes, which Bragg had discussed again at the Solvay Conference on Chemistry in April1925 (45), and the expectation that these structures would be defined mainly by the close packing of the oxygen atoms with the other, smaller, atoms tucked in between them (240). Careful measurements were made of absolute intensities and, referring to the earlier difficulties over crystal perfection, they noted that 'accurate allowance for extinction appears to be the key to the analysis of complex structures'. Finally, for the first time in this paper Bragg and West reported their use of a

Fourier synthesis of an electron density distribution. Based upon WHB's Bakerian lecture of 1915 and subsequent work by Duane, Compton, Havighurst and others, this synthesis was in only one dimension but it pointed the way ahead and Bragg clearly saw what would follow: 'The Duane method cannot be applied until the signs of the coefficient F (in the Fourier series) are fixed by preliminary analysis, for the observed intensities only give the squares of these quantities. Probably the most convenient procedure will be to combine the trial-and-error method of assuming structures and calculating the spectra to be expected from them, with the Fourier-analysis method, the latter being used to make the final adjustments of atomic position and to indicate the accuracy of the results.'

A discussion of one-dimensional Fourier syntheses as used by various workers, including J. M. Cork in his studies of isomorphous replacement in the alums which were carried out during a visit to Manchester in 1926-27, formed a large part of Bragg's contribution to the 5th Solvay Physics Conference on 'Electrons and photons' in October 1927 (54).

Over the next few years the structures of many silicates were determined by the standard method, of which Bragg and West (55) gave a definitive account in one of his favourite papers, entitled 'A technique for the X-ray examination of crystal structures with many parameters'. Many means 20 or 30. In addition to Bragg himself and West, the workers most involved were W. H. Taylor, a research student who was appointed to a lectureship in 1928, and two of the many visitors from abroad, W. Zachariasen (1927-29) and B. E. Warren (1929).

Bragg introduced Warren to crystal-structure analysis during a visit to M.I.T., where he spent a term lecturing in the spring of 1928. Before leaving Manchester he had begun a study of diopside, $CaMg(SiO_3)_2$, and had encouraged West to make a much more complete set of intensity measurements for this mineral than had been achieved in any previous study. This was made possible by the crystals, provided by Hutchinson (by this time professor in Cambridge), which had been cut perpendicular to the principal axes and permitted the measurement of complete zones of reflexions. Bragg took these data to M.I.T. with him and worked there with Warren on the analysis of the structure which proved to be of key importance (56). Throughout his work on the silicates there was doubt about the extent to which .the oxygen atoms were associated more closely with the silicon than with the other atoms, but it was clear in diopside that tetrahedral SiO_4 units were joined by their corners to form long chains running through the structure. Systematic analysis of the other structures showed that the essential properties of the various types of silicates were largely determined by the SiO4 tetrahedra which could join together sharing corners or edges, but not faces, to form three-dimensional structures, plates or chains.

The classification of 8ilicate structures on this basis was initiated by Machatschki (236), a visitor in 1928-29, and it was elaborated by Bragg in his review of the whole programme in 1930 (67). Even at this stage, however, Bragg was reluctant to recognize the importance of SiO4 units as complex ions and preferred to consider the individual atoms separately-as had been found appropriate in rocksalt. He was still deeply attached to the views expressed in his Royal Institution Discourse on 20 May 1927 when he had said (48): 'Some of the very earliest structures which were analysed caused us to revise our ideas of what was meant by the "molecule" of the chemist. In sodium chloride there appear to be no molecules of NaCl. The equality in numbers of Na and Cl atoms is arrived at by a chess-board pattern of these atoms; it is a result of geometry and not of a pairing-off of the atoms. This is, of course, not universally true, for this absence of the molecule in solids is in general only found in inorganic compounds. It would appear, however, that the silicates are of this non-molecular type and that in seeking to assign formulae to them, and to the hypothetical acids of silicon on which they are based, it should be borne in mind that they are really extended patterns. The relative numbers of their constituent atoms are characteristic of the extended pattern, and essentially a result of their solid state, so that it is doubtful whether a grouping of the atoms into molecules has in this case a meaning.' *

In contrast with this view, Linus Pauling (r), who was a visitor in 1929-30, had proposed a method of describing the structures in terms of SiO4 and other polyhedral units and he went on to propound a set of principles governing the assembly of ionic compounds which depended heavily on the evidence accumulated in Manchester, as Bragg rather sadly remarked (67), and subsumed the less-well-developed rules that had been evolved there (James (e)).

* It was the reporting of this Discourse that provoked H. E. Armstrong's well-known letter to *Nature* (q), which included the sentence: 'It were time that chemists took charge of chemistry once more and protected neophytes against the worship of false gods: at least taught them to ask for something more than chess-board evidence.' In his later years, Bragg enjoyed showing a slide of this letter in his lectures—usually a little out of context.

Nevertheless, no subsequent analysis can diminish Bragg's achievement in guiding the studies of these complex mineral structures to a fruitful conclusion. His was clearly the guiding hand and his colleagues of those days have written of the excitement of working with him as more and more complex arrangements of atoms yielded to their attack. Silicate chemistry was shown to be inherently a chemistry of the solid state, intelligible only in terms of the three-dimensional structures, and Bragg never tired of using the story of its explanation to illustrate his conviction that the analysis of increasingly complicated structures can lead to the discovery of unimagined new principles.

After 1930 Bragg left further research on mineral structures to others, though he remained closely interested and spent the spring of 1934 at Cornell University delivering the Baker lectures and writing The atomic structure of minerals (E). He was more concerned with crystal physics and the physical methods of structure analysis and he had, by this time, already defined the method that was to dominate crystal-structure determination for the next 20 years. This was another benefit from the work on diopside. Given the complete two-dimensional data that had been measured for the structure analysis he was able to explore the possibility of using Fourier series in two dimensions to calculate the electron density projected on the three faces of the unit cell (60). This was possible because, following the procedure outlined in the paper on beryl (42), the phases (or in this case the signs) of the Fourier coefficients could be calculated from the model of the structure proposed by himself and Warren (56); that is to say, the crystallographic phase problem, which arises because only the structure amplitudes and not their phases are given directly by X-ray observations, was already solved. Unlike most professors, he did not ask someone else to do the calculations for him. Armed only with a slide rule and mathematical tables, he did them himself and so produced the first two-dimensional electron-density projections. The results were very pretty and, although they did not lead to much new structural information since the positioning of the atoms in the original structure analysis had been closely correct, they did show the possibility of identifying atoms (or ions) from their electron counts and they suggested the method of the future. In discussing the results, Bragg posed the question whether the phases could be derived without knowing the complete structure: would partial information do? Investigation showed that the answer could be yes. In the projection on (010), for example, the Ca and Mg atoms overlap giving the equivalent of an atom of atomic number 32.

Furthermore, this 'atom' always gives its maximum contribution whereas the other atoms, although they total 74 electrons, tend to cancel out. Consequently only one reflexion differs in phase from that given by the Ca/Mg alone. Bragg was, therefore, able to point the way ahead more precisely than before: 'a preliminary analysis of the crystal which gives approximate positions of the heavy atoms suffixes to fix the signs of the coefficients F. The Fourier series may thus be formed and the positions of all the atoms accurately read off on the projections.' In a further paper with West (65) the defects of the image that arise from errors and terminations of the series were defined and possible remedies discussed.

Subsequently, Bragg was sorry not to have published his work on Fourier syntheses jointly with his father, since it derived directly from W.H.B.'s Bakerian lecture and the two of them had discussed it in detail. He did not use the method himself to determine new structures but he encouraged members of his Laboratory to explore it further. In 1929 Zachariasen used two-dimensional projections in his refined determination of the structure of sodium chlorate, in which the oxygen positions were read from the maps; in 1930 West studied potassium dihydrogen phosphate in the same way (incidentally demonstrating the presence of regular PO_4 tetrahedra) and in 1932 Parker and Whitehouse reexamined iron pyrites. Characteristically Bragg was content to be thanked for his advice and help at the ends of their papers. He had demonstrated the potential of the method and left it to others to develop in detail: it has been an essential feature of X-ray crystallography ever since.

However, he did not leave it without trying to impress upon others the physical basis of the new method. He began to lecture on X-ray optics (58) and he had the idea of showing directly that the production of an X-ray image was an optical process. Remembering again C.T.R. Wilson's lectures on optics, he knew that a microscope image could be regarded as the superposition of individual waves (or fringes) and that Fourier synthesis was merely a way of adding these waves mathematically instead of experimentally through a lens. He therefore tried to see whether he could produce an image of diopside by superposing fringes on a piece of photographic paper.

Again he carried out the painstaking business himself, with the help of his laboratory superintendent, William Kay. His fringes were the out-of-focus images of a set of parallel rods: each set had to be given the right spacing, orientation, displacement (phase) and exposure to represent the corresponding Fourier coefficient. In spite of the odds against carrying out all these operations without mistakes for about 25 terms, the results were very convincing (61).

Although this analogue method was not taken up generally, and satisfactory numerical methods were devised instead, it is notable as the first of Bragg's instructive developments of optical methods and as a clear illustration of his over-riding interest in the physics of his subject.

At about the same time the possibility of his returning to Cambridge forced Bragg to consider his hopes for the future, and on 3 June 1929 he wrote to Rutherford (c) as follows:

'I have been thinking very hard about the possible post at Cambridge which you mentioned to me some time ago... 'To come to Cambridge would in itself be delightful and anything which would bring me there has all the attractions of Cambridge to recommend it. Here on the other hand are a few of the "cons". I have got used to the running of a big physics laboratory, and although I often grumble at the administrative work and would like to be freed from much of it, it is very pleasant to have a constant supply of men eager to do research. I have about twenty or twenty five doing research here and do much of my work through them. Then again I take a great interest in the teaching of physics in general and do not want to drop that. I do not want to label myself a crystallographer as against a physicist and think indeed that though my research is concerned with crystals it is the physical side of it which attracts me. I might at any time wish to switch over into a more purely physical line, and want to feel quite free to do that.

'Could you tell me what the post at Cambridge might involve? I realice of course that it is all very much in the air at present. Would it carry with it a laboratory in which I could house a large group of research men? What funds for research would be available? What students would come other than men from abroad coming to research under me on crystal problems? Could I take part in the teaching of physics at the Cavendish?

'I could perhaps explain my views best by describing what would seem to me a very attractive post. This would be an additional chair of experimental physics with especial charge of the physics of the solid state. You talked to me about the future of the Cavendish, and the possibility that in time to come the work might be divided as the ground to be covered in experimental physics is so great. Is there any possibility of this being done in the next year or two and of your devolving part of the responsibility? If I might put it quite frankly, I feel that such a post would offer unlimited scope for one's energy, whereas the subject of crystallography by itself is very limited. The parts of it which are growing and important are really pure physics, and it would best be developed as part of a big Physics School, not as a subject on its own.'

Nothing carne of this negotiation and J. D. Bernal, who had moved from W.H.B.'s group at the Royal Institution t0 Cambridge in 1927, remained in charge of crystallography there. But it seemed to focus Bragg's views and may have suggested to him that there would be a danger of losing contact with mainstream physics if he concentrated too much on analysing structures. Disappointment at the outcome may also have contributed to the crisis that overtook him in 1930. In addition to the normal strains of his office, which at this time included the planning of a new building, the introduction of a new series of lectures for industrial physicists and writing a book (C), other factors also contributed. There was the excitement but also the tension involved in bringing the silicate work to its climax, especially when understanding the results drew him into the unfamiliar, fast-developing and competitive field of chemical bonding; there was the continuing conflict between his simple enjoyment of family life and his obsessive preoccupation with research at critical moments; there was the distant and guarded relationship with his father, * and the recollection of his lack of rapport with his mother, who had died in 1929 after a long illness; and there was the worsening economic situation which was more evident in Manchester than in many other places.

A real chance to move brought matters to a head. He was offered the Professorship of Physics at Imperial College, London, but, after anxious thought and unhelpful discussion with his father, he refused the appointment, largely through strong feelings of loyalty to the university that had given him so much. His friend G. P. Thomson was appointed instead. It was at this point, fearful that he had lost his last opportunity to move his family to more attractive surroundings, that he broke down.

Bragg's colleagues appear to have been largely unaware of the tensions that produced this crisis and, with the staunch support of his family, Bragg recovered his balance quite quickly. In this he was greatly helped by

* On 17 November 1926, for example, he wrote to W.H.B.: 'I am too lucky in too many ways to have any reason for feeling low at all. I think we ought to talk more about the lines we are doing and I would like to co-operate in the sense that each of us specialized in some particular line which supplemented the work of the other.

spending the spring of 1931 on leave in Sommerfeld's laboratory in Munich, where he sought to broaden his command of the latest developments in physics, and by the birth in June 1931 of his third child and first daughter, Margaret Alice.

Back in Manchester at Easter, however, he changed his style of working, involving himself less closely in the day-to-day experiments and analysis, and looked for new and more physical lines of research while encouraging the continuation of mineral-structure studies by Taylor and others. With his reputation renewed by the silicate work, he also became somewhat more involved in general scientific affairs which took him outside the university. In 1931 he was awarded the Hughes Medal of the Royal Society and from 1931 to 1933 he served on its Council. He was elected a member of the Dining Club in 1934. During this period, in the summer of 1932, he and his wife made a trip to Russia, together with R. H. Fowler, Dirac, Gurney and the Kapitzas who arranged the trip. This may have made Manchester seem more attractive but, however that may be, everything brightened up in 1933 when he was able to move his family from Didsbury to Alderley Edge, to a house overlooking the Cheshire plain, within sight of the Welsh hills and with a beautiful garden. At the same time research became exciting again and Bragg faced the world with renewed vigour. In 1934 he not only spent a term lecturing at Cornell, he also made his first contribution to broadcasting by giving a course of six lectures on 'Light' and, at the end of the year, delivered the Royal Institution's Christmas lectures to a juvenile auditory on the subject of 'Electricity' (D).

Bragg's new research subject was a development of one of his established interests. A short excursion into electron diffraction (76, 79) did not promise much but his attention was attracted by another of Pauling's papers, this time on 'Rotational motions of molecules in crystals', and he made the new idea, that there is a much greater freedom of movement in the solid state than had been suspected, the theme of his lecture at the Centenary Meeting of the British AsSociation in 1931 (77). Although he did not include an account of it in his lecture, this topic was directly related to new work in Manchester that was growing out of the long-standing studies of metals and alloys.

Bragg's first research student in Manchester was A. J. Bradley and, when he had completed his doctorate, Bragg asked him to explore the use of the powder diffraction method, which had been shown by Hull in America and Debye and Scherrer in Switzerland to be valuable in studies of elements and other simple structures. This approach had also been taken up by Westgren in Sweden, whose laboratory had become one of the main centres for the developing study of metals and alloys, so in 1926 Bragg sent Bradley to spend a year with Westgren and Phragmén in Stockholm to learn the trade. This proved extraordinarily successful. Bradley was in his element and was soon solving problems that others had considered impossible. On his return to Manchester he became the leading exponent of powder methods, solving problems of surprising complexity and attaining an accuracy in measurements that has hardly been surpassed in present times (s).

Bradley's most outstanding contribution was to determine the structure of 'γ-brass, Cu_5Zn_8.' which has 52 atoms in a cubic unit cell and had defied Westgren and Phragmén whose single crystal Laue and rotation photographs and powder diagrams Bradley used in his analysis. The structure of γ-brass—an otherwise quite useless alloy—played a considerable part in the development of modern solid-state physics: the y-structure was found to exist because its Brillouin zone is more nearly spherical than those of the simpler body-centred and face-centred structures. Bradley was no theoretician, his genius was to provide the evidence, but Bragg saw the importance of his work and discussed it with a succession of lecturers and visitors in the department whose contributions to the development of the theory are universally recognized. They included N. F. Mott (1929-30), W. Hume-Rothery(1932-33), H. A. Bethe (1933-3+) and R. E. Peierls (1933-35). The part that Bragg played in this work is not now fully appreciated; he published no original papers on the subject and he would have been the last to claim that he had made any significant contributions nevertheless, his role as a catalyst cannot be denied and it is entirely appropriate that the term 'Bragg reflexion' now occurs naturally in solid-state physics and is used by people who have never carried out any X-ray diffraction work themselves.

During the period 1928-32 Bradley was paid by Metropolitan-Vickers, though he continued to work in the university, and it was natural, therefore, for Charles Sykes to consult him when he encountered a problem with iron-aluminium alloys at the Company's Research Laboratories in Trafford Park. Sykes was carrying out an industrial investigation into the properties of these alloys and, during the work, he found that their electrical resistivity varied as a function of aluminium content in what appeared to be a completely haphazard manner. Bragg was immediately drawn into the discussions and suggested, as it turned out correctly, that superlattice formation might be the cause of this behaviour. This started a long and fruitful association with Sykes and it stimulated Bragg's interest in order-disorder phenomena.

Bradley investigated the iron-aluminium system and found that the resistivity effects were due to the difference between an ordered structure of the alloy when cooled slowly and a disordered structure when it "as cooled rapidly. The former had the lower resistance. Bradley then gave a colloquium in the department at which he described his results and Bragg put forward some general qualitative ideas about the nature of the order-disorder change. This occasion was attended by E. J. Williams, the Lecturer in Mathematical Physics, who was fascinated and overnight drafted a thermodynamic theory of the phenomena which he showed to Bragg the following day. There is no doubt that the detailed theory which developed from this beginning was very largely due to Williams, as Bragg made clear in his Royal Institution Discourse of 17 March 1933 (81), but Bragg took it up enthusiastically and used it as the basis of his Bakerian lecture in June 1934 which was published, most unusually, in their joint names (83). At this stage they discovered that theoretical treatments on similar lines had been published earlier by other workers and they discussed these, and a new approach to the problem by Bethe (who, together with Peierls, had contributed importantly to the discussions in Manchester), in a second paper (86). Williams rounded off the series of papers with an independent publication in which he acknowledged help from Bragg and from Peierls, who subsequently made further developments to Bethe's approach.

Although the Bragg-Williams contribution was not as original as it first seemed, there is no doubt that it attracted great attention and stimulated much further work all over the world. In Manchester experimental studies were continued in the university in collaboration with Sykes at the research laboratory of Metropolitan-Vickers and they were summarized in joint papers in 1937 (96) and 1940 (111, 113, 114). Meanwhile Bragg encouraged Bradley to devote himself to the study of alloy systems in the hope that X-ray methods, which made possible the unequivocal recognition of individual phases, would clarify the interpretation and use of complex phase diagrams. The work developed naturally into a classification of alloy phases in binary, ternary and even quaternary systems but this did not generally fire Bragg's imagination, though he was excited by the work on magnetic alloys (106), and he left this field in the main for Bradley to develop.

Single-crystal structure analysis continued to be an important part of the work in Manchester during the 1930s but Bragg did not involve himself closely in any particular study after the work on [3-alumina which was completed in 1930 (75). But he continued to encourage this branch of research and it is interesting to note that he was present at the Washington meeting of the American Physical Society where Patterson presented his first account of the Fourier syntheses formed with the observables F^2 as coefficients which are now known universally as Patterson functions. Patterson, who noted (e) that 'Bragg [was] in the audience to ask the right questions', had used West's data on KH_2PO_4 in his first tests of the method but Bragg told him about the work by C. A. Beevers and H. Lipson at Liverpool on copper sulphate pentahydrate, which they had had solved by Fourier methods but not yet published, and suggested that he might use their data in more detailed tests. Bragg also described the new numerical methods of Beevers and Lipson for summing Fourier series which Patterson then used, together with the copper sulphate data, to calculate the two-dimensional synthesis that illustrates his classic paper (t). First Beevers and then Lipson subsequently moved to Manchester where they continued their development of the famous 'strips' and worked on a number of structures, Lipson for the most part with Bradley. Their work on a method for computing Fourier syntheses of electron density from the structure factors probably stimulated Bragg's invention of a graphical method to solve the inverse problem, the calculation of structure factors from atomic positions (90), which was elaborated in a joint paper with Lipson (91).

By the spring of 1937 there must have been an end-of-term feeling in Bragg's Department in Manchester: W. H. Taylor had left in 1934; West (for a professorship in Rangoon) and Peierls in 1935; Williams in 1936; and James (who now realized his earlier ambition to become professor at Cape Town) in April 1937. In May Bragg was invited to succeed Sir Joseph Petavel as Director of the National Physical Laboratory. He accepted and; arranging for Bradley and Lipson to move with him, took up his new duties on 1 November. He was succeeded at Manchester by Blackett.

In his final departmental report at Manchester, Bragg wrote: 'It is with great regret that I leave a department in which I have spent eighteen happy years. I came to it directly after the war, with no previous teaching experience. Manchester University has taught me all I know about the running of a department, and the fascinating and intricate life of a modem university. I shall always remember with gratitude the kindness of my colleagues and the inspiring atmosphere of this University.' He had certainly served it well, not least by making Manchester the centre of his subject visitors came from all over the world and they found him a charming and helpful person, with none of the pomposity they might have expected in so eminent a man. Many have remarked upon the personal kindness of Bragg and his wife and their hospitality at Alderley Edge but Sir Charles Sykes may have provided the most illuminating snapshot: 'when we got down to the serious discussion he would adjourn to the billiard room which contained a full-size billiard table. On all my visits to this room, I never saw any balls on the billiard table; it was covered with reprints of papers and, as the argument developed, Bragg would get up from his chair, wander round the table, pick out the appropriate reprint, and we would then examine it in terms of the ideas we were discussing at the table.'

THE NATIONAL PHYSICAL LABORATORY 1937-38

The National Physical Laboratory was at that time administered by the Department of Scientific and Industrial Research whose Secretary, Sir Frank Smith, was much concerned in Bragg's appointment. W.H.B., who was now President of the Royal Society and had strong views on the importance of applied science, very much approved of the move.

But Bragg found the work disappointing. Although some of the research was flourishing, many things were being done that had long ceased to be useful. In taking Bradley (who was then Royal Society Warren Research Fellow) and Lipson with him he hoped to set an example in research. They worked in the Metallurgy Division, whose Superintendent was C. H. Desch, but most of the short time they were there was spent in writing up for publication work done in Manchester. Bragg's only original paper published from the N.P.L. was a short note with Lipson on high-dispersion X-ray photographs of metals (101). One of his few innovations was to attempt the introduction of a series of lectures from prominent scientists: he found that there was no proper lecture room for them; there had been one but it was so little used that it was converted into an extension of the library .But there were plenty of committees and formal occasions, which he did not enjoy.

Rutherford died on 17 October 1947, before Bragg had left Manchester, and there was immediate speculation about who should succeed him as Cavendish Professor of Experimental Physics in the University of Cambridge. Bragg was clearly a possibility but W.H.B., who was himself ill at ease in Cambridge and opposed the move, wrote advising his son's wife not to be disappointed at not being able to go to the Cavendish having just moved to the N.P.L. Sir Frank Smith, who was an elector to the chair, took a different view and with the rest of the Board (The Vice-Chancellor, H. E. Dean; R. H. Fowler; C. G. Darwin; H. Thirkill; W. J. Pope; W. Wilson; O. W. Richardson; and G. I. Taylor) decided to offer Bragg the appointment which was announced in March 1938. Election to a Professorial Fellowship at Trinity College followed shortly afterwards. There was undoubtedly some consternation, especially among nuclear physicists, but Darwin is reported to have remarked that 'nuclear physics is a passing phase' and *Nature* commented: 'The Cavendish laboratory is now so large that no one man can control it all closely and Bragg's tact and gift of leadership form the best possible assurance of the happy cooperation of its many groups of research workers.' The next 15 years showed that this was closer to the mark than most editorial comments.

CAMBRIDGE DURING WORLD WAR II, 1938-45

Bragg moved to Cambridge with his wife and four children in October 1938 where, at the age of 48, he succeeded Rutherford in a major appointment for the second time. Again it was not easy since many members of the laboratory had hoped that Rutherford would be succeeded by another nuclear physicist and Bragg not only had different interests, his style of management was quite different also. P. I. Dee and N. Feather were the principal nuclear physicists remaining in Cambridge and the other research interests at the Cavendish were the work of the Mond Low-temperature Laboratory under J. D. Cockcroft, the ionospheric research under E. V. Appleton, and the crystallography, which had been directed hitherto by Bemal. During the previous ten years Bemal had inspired a very lively group which had been particularly successful in starting studies of biologically important molecules and macromolecules, but he had been appointed to succeed Blackett at Birkbeck College in the round of musical chairs that followed Bragg's leaving Manchester. Bragg brought Bradley, with Lipson to assist him, from the N.P.L. to succeed Bemal in charge of the Crystallographic Laboratory and, in the next year, he strengthened metal physics by attracting E. Orowan from Birmingham.

Another change was quickly needed. Appleton, who had been Bragg's first student, resigned the J acksonian chair in order to succeed Sir Frank Smith as Secretary of the D.S.I.R. and Cockcroft was appointed in his place. In this first year Bragg worked closely with Cockcroft, who had played a large part in designing the new Austin wing of the laboratory and was now supervising its construction—on the understanding that it would be used by the Services in the event of war (138). As war grew increasingly probable, however, ideas for reorganization of the laboratory were set on one side and, by September 1939, most of the basic research work was suspended as the members of staff left for war service or turned their attention to related research.

Nevertheless, Bragg's first year back in Cambridge was scientifically fruitful and, in particular, his first weeks there were marked, most significantly as it turned out, by a dramatic revival of his interest in the analysis of more and more complicated crystal structures. M. F. Perutz, who was the only member of Bemal's group of biological crystallographers left behind in Cambridge, has described what happened (u): 'I waited from day to day, hoping for Bragg to come round the Crystallographic Laboratory to find out what was going on there. After about six weeks of this I plucked up courage and called on him in Rutherford's Victorian office in Free School Lane. When I showed him my X-ray pictures of haemoglobin his face lit up. He realized at once the challenge of extending X-ray analysis to the giant molecules of the living cell. Within less than three months he obtained a grant from the Rockefeller Foundation and appointed me his research assistant. Bragg's action saved my scientific career and enabled me to bring my parents to Britain' (as refugees from Hitler's invasions of Austria and Czechoslovakia).

This meeting with Perutz introduced Bragg to the problem that was to be his main research interest for the rest of his career and his first paper related directly to protein-structure analysis followed almost immediately (107). Here he discussed the problems involved in determining complex structures directly from their Patterson functions and, referring to recently published claims that the Patterson of insulin could be interpreted, he pleaded 'for a due sense of proportion'. This note shows him already thinking about proteins but its main interest now is as an introduction to the two letters that followed immediately afterwards from Bernal and J. M. Robertson. In his letter Bernal paid attention to the then uncertain chemical constitution of proteins, a problem ignored by Bragg, and referred to the difficulty of locating a zinc atom with 28 electrons in a molecule with 20 000 electrons. Robertson, however, while agreeing with Bragg's comments about Patterson diagrams, went on to suggest how the structure of insulin might be determined: 'the molecule does, however, contain a few zinc atoms, and if these could be replaced by mercury, as has been suggested, a very profitable study might ensue'. It was to be a long time before essentially this method was used to solve the structure of a protein crystal and, in the meantime, Bragg played a major part in keeping alive the hope that success would be achieved at last despite the scepticism of most crystallographers including many of those in his own laboratory.

Meanwhile, in a new burst of creativity, there was time for one more advance before the war intervened to stop most of the research. This was a development of Bragg's earlier attempt to devise an optical method for making Fourier syntheses of electron-density maps (61). Now he had the idea of using Young's fringes instead of the out-of-focus images of a set of rods to represent the Fourier components. Each set of fringes could be produced by a pair of holes with the right orientation and spacing. He suddenly realized that he had rediscovered the reciprocal lattice. The holes had to have areas proportional to the structure amplitudes but there was no easy way of simulating the phases. Nevertheless, the method could be used to synthesize the familiar (010) projection of diopside for which all but one of the phase angles were zero. Crowe, the precocious lab boy of his undergraduate days, built the necessary apparatus-Bragg called it the X-ray microscope-and Chapman, the instrument maker of the crystallography section, drilled the plate simulating the reciprocal lattice. Its diffraction pattern showed clearly the arrangement of atoms in the crystal (108). Bragg was delighted with this result and demonstrated it at a Royal Society Conversazione.

Further developments of these ideas continued intermittently during the war and it is convenient to describe them without interruption here. The first (119) in 1942 was a relatively simple extension of the method just described in which a photographic method was used to produce the reciprocal-lattice plate. This took advantage of a photographic process developed by the British Scientific Instrument Research Association, whose director, A. J. Philpot, was a fellow member of war-time committees, and it made possible the synthesis of a Patterson projection of horse haemoglobin from measurements by Perutz. The possibility of synthesizing centrosymmetrical electron-density projections, for which the phase angles are all 0° or 180°, by using half-wave plates of mica was also described. Having obtained images in this way, Bragg next turned his attention to the opposite problem, how to simulate optically the production of a single crystal X-ray diffraction pattern-in two dimensions. His solution to this problem, first demonstrated at a Royal Institution Discourse in 1942 (128, 136), was to make a set of images of the contents of a unit cell by using a multiple pin-hole camera, in which the holes were arranged on the points of the crystal lattice-the so-called 'fly's eye'. Crowe again carried out the work and it was a success which led to practical results: C.W. Bunn used the device in helping to solve the structure for penicillin.

Finally, Bragg realized that the fly's eye was unnecessary: one image of a unit cell would give rise to a continuous diffraction pattern from which the reciprocal lattice picked out the appropriate intensities. In a joint paper with Lipson, he therefore proposed the use of a simple optical diffractometer and illustrated its value in studies of alloys (123).

These optical devices were taken up and further developed by several people, especially by Lipson and his colleagues. They proved very important educationally in emphasizing the Fourier-transform theory that underlies X-ray diffraction phenomena and they have been used intensively in the analysis of electron micrographs of periodic structures.

But in September 1939 Bragg's main preoccupation was to make what contribution he could to the war effort and he took pains to see that his staff and ex students were efficiently deployed. His problem then was to carry on the teaching of physics in Cambridge, suitably adapted to meet the demands of the war. Queen Mary College and Bedford College, London, were evacuated to Cambridge and their physics classes and teaching staff were combined with those of the Cavendish. The Queen Mary College professor was an old friend and colleague, H. Robinson, and he stayed with the Braggs. Most remarkably, Searle was brought back from retirement and he took charge of the practical classes until the end of the war (193). Much of the teaching was concentrated on short two-year Honours courses with special classes in electronics for potential radar personnel.

In addition to the X-ray optics already described, Bragg continued the metals research with a good deal of emphasis on practical problems as it had been in Manchester. Unhappily, however, Bradley was not good at the day-to-day running of a laboratory and he went into a sad decline, finally having to give up charge of the Crystallography Group. Lipson took over responsibility for running the section. Apart from the continued work with Bradley and Sykes on alloys and order-disorder phenomena that has already been mentioned (111,113,114) Bragg's most important contribution to metals research in this period was the invention of his remarkably useful bubble model of a metal structure (118). Inspired by his discussions about the strength of metals with Orowan, this simple model illuminated the behaviour of domain boundaries in plastic deformation and, developed further after the

War, did much to popularize the theory of dislocations (145, 156). It illustrates very well Bragg's uncanny gift for visualizing atomic arrangements and expressing their essence in simple models, and the fact that Feynman, most unusually, quotes one of Bragg's papers (145) verbatim in his Lectures on physics clearly expresses his appreciation of the model and of Bragg's gift for exposition.

Bragg played no part in the war research that was conducted in the Cavendish, by Halban, Kowarski and others, but he did contribute to services research of two main kinds. As early as 1937 he was consulted about the equipment and tactics of the Sound Ranging Section in the Army and he continued throughout the war to advise on its development. The centre of research and teaching was on Salisbury Plain and there he renewed contact with World War I friends, including Hemming who had been in charge of the complementary method of flash spotting. This latter method was no longer useful, since gun flashes had been eliminated, but sound ranging was. Bragg found it in much the state in which he had left it in 1918, without even an effective radio communications system though with some not-very-useful accretions (L). Refined under the pressures of war it again proved valuable and, in addition to being employed in essentially the old way in the main land engagements of the war, the same principles were used in plotting the trajectories of the V2 rockets.

Secondly, Bragg was consulted by the Admiralty on the development of Asdic (sonar). This method of underwater detection by the use of sound waves had been developed to some extent during World War I by a research group led by W. H. Bragg at Parkeston Quay (a) but it carne into use only in the World War II. Bragg regularly visited the Admiralty Research Station at Fairlie on the Clyde for discussions of the problems encountered in the further development and use of the system. Writing about it later he remarked modestly: 'I find it hard to estimate how much I helped. Only the people on the spot could appreciate the practical difficulties and such help as an outsider could give came from a knowledge of the man to consult about this or that special point... Quite apart from direct help, I think the researcher liked talking about their problems to someone who understood and could appreciate their work.' He continued to serve as an advisor on this work for about 15 years.

Bragg also served on committees set up by the Ministry of Supply to keep its scientific activities under review, as Chairman of the General Physics Committee and member of the Metallurgy Committee, and this enabled him to keep the Ministry closely aware of the work still going on in Cambridge. Other members of these committees included A. V. Hill, Andrade and A. J. Philpot. From the end of 1942 he was also a member of the Advisory Council of the D.S.I.R.

Early in 1941 Appleton asked Bragg to serve a term of six months in Canada as Scientific Liaison Officer between Canada and the U .K., in succession to R. H. Fowler. He sailed to Halifax in March accompanied by C. G. Darwin who was on his way to Washington, D.C., and was attached to the team of scientists working under the leadership of C. J. Mackenzie at the National Research Council Laboratories in Ottawa. Bragg and Mackenzie had met briefly at a sound-ranging course on the Vimy Ridge front in World War I. In Ottawa they collaborated closely and Mackenzie has described Bragg's performance, which was based upon his close relations with the scientific community in Canada and the United States and his standing in the U.K., as that of 'a superb liaison officer for the exchange of secret information and arranging useful and congenial meetings between distinguished allied scientists'. Bragg's report to Appleton on 12 August 1941 advocated the policy of keeping the liaison office small and encouraging experts in each subject to travel backwards and forwards between the two countries that was largely followed. After visits to Vancouver and other centres he flew home in September in a bomber on its way to active service and was succeeded in Ottawa by G. P. Thomson.

A second wartime journey abroad, of a more cultural but possibly equally hazardous nature, took place in 1943 when he visited Sweden at the invitation of the British Council to talk to Anglo-Swedish Societies and re-establish contacts with Swedish scientists. Bragg flew to Stockholm on 16 April and gave some 14 lectures, mainly on his research interests in X-ray optics, proteins, metals and minerals, in six Swedish centres before returning home on 12 May. He met many old friends-Westgren was especially remembered-and his report, which ranged over the availability of scientific journals and the importance of further exchanges, concluded: 'I cannot exaggerate the warmth of my welcome.' With the tide of war changing and thoughts turning more confidently to its end, Bragg clearly made a valuable contribution to Anglo-Swedish relations.

Bragg performed at least one other public function during the war which was of great importance. From October 1939 to September 1943 he served as President of the Institute of Physics and worked hard to maintain the activities of the Institute at as normal a level as possible and to initiate constructive discussions about the likely needs of the postwar world (117). Throughout his time in Manchester he had fostered the activities of the Institute in that area and he had also taken a special interest in the application of X-ray methods to industrial problems (78). At the beginning of his Presidency plans were made to hold a Conference of the Institute on 'X-ray Analysis in Industry' but they were frustrated first by the outbreak of war and then by the events of 1940. It was then decided to publish the papers that had been prepared (116) and a Conference to discuss them was at last held in Cambridge on 10-11 April 1942. The report in Nature noted that 'some anxiety was felt by those responsible for the arrangements, lest preoccupation with war work would prevent many from attending, but the decision to proceed was made because the X-ray tool is being widely used for problems directly connected with the war .The large attendance at the Conference (some 280 participated) and the generally expressed appreciation of this opportunity for discussion have shown that this anxiety was unnecessary.' Bragg gave an historical review and mentioned proteins and the 'fine' structure of deformed metal as problems on the threshold of solution.

This meeting was such a success that its members decided to set up an organization under the aegis of the Institute of Physics to arrange similar conferences from time to time. As a result, a discussion meeting on the determination of equilibrium diagrams by X-ray methods was held in September at the Royal Institution, with Bragg in the chair, and a second full conference on X-ray analysis in industry was held in Cambridge on 9-10 April 1943, again with Bragg as chairman. This led to the establishment of the X-ray Analysis Group of the Institute of Physics to arrange meetings and perform other functions connected with X-ray research for a membership drawn from both university departments and industry. The committee of the new group, which met in July 1943, was made up largely of Bragg's asSociates and he was elected chairman with Lipson as secretary.

The X-ray Analysis Group (XRAG) adopted the pattern of meetings that was set in 1942 and meetings were held regularly in the spring and autumn of each year until the end of the war and, with few exceptions, this has continued to the present time. Bragg remained chairman until April 1947 and was thereafter a vice-chairman until his death.

At the end of the war the XRAG, led by Bragg, played a critical part in the organization of crystallographic research internationally. At the committee meeting held in July 1945, following suggestions by Ewald at the

Oxford Meeting a year earlier, there was discussion of the need for a new journal to succeed

he then-defunct *Zeitschrift für Kristallographie*. Bragg consulted Wyart, Mauguin, Ewald and others and it was agreed that advantage should be taken of the summer meeting of XRAG in 1946 to hold international discussions of the problem. This XRAG meeting was held with Bragg as chairman on 9-11 July 1946 at the Royal Institution. The subject was 'X-ray analysis during the war years' and it was a memorable occasion which provided the first opportunity after the war for crystallographers of all nationalities to re-establish contacts. In addition to about 250 U.K. participants some 75 visitors from 15 different countries around the world were present, happily including Laue, and many moving accounts were given of research during the hostilities.

On the following two days formal meetings were held at Brown's Hotel of a Provisional International Crystallographic Committee to explore the question of publishing an international journal of crystallography and to consider other questions of crystallographic interest. Bragg opened the meeting which agreed, after much discussion of detail, that a new journal was needed and that it should be called Structural crystallography. Bragg then noted that it would be necessary for some organization to assume formal responsibility for the journal and suggested that one possibility would be to form an International Union within the existing framework of Unions, either as a Union of Crystallography or as a Commission of the Union of Physics or of the Union of Chemistry. Unanimous agreement in favour of a separate International Union of Crystallography was quickly reached and Bragg was asked to explore the possibility further.

At this time the General Secretary of the International Council of Scientific Unions was Professor F. J. M. Stratton of Cambridge and Bragg was able quickly to arrange a meeting at which Stratton gave his opinion that a new Union would be acceptable. In this way the International Union of Crystallography was conceived. Bragg was not able to attend its first formal meeting at Cambridge Mass. in 1948 but he was there elected its first President. He also served as a founder member of the Editorial Board of the new journal, which was actually given the fittingly general and international title 'Acta Crystallographica' at the request of the Russians, and he played a large part in raising the money that was needed to launch it.

This account of the International Union has taken the story beyond the end of the war but some brief details of Bragg's personal life in those years remain to be added.

Bragg's knighthood was announced in the New Year Honours list of 1941 and W.H.B. wrote to Lorna Todd in Adelaide on 5 January: 'Isn't that fine? ... He will have to be Sir Lawrence: we can't have confusion worse than ever. I am so very glad for his sake. In spite of all care, people mix us up and are apt to give me a first credit on occasions when he should have it: I think he does not worry about that at all now, and will never anyhow have cause to do so now. I think I am more relieved about that than he is' (a).

W.H. Bragg died on 12 March 1942, still in his post at the Royal Institution where Bragg had spent part of the day with him. Since 1938, when Bragg was appointed (non-resident) Professor of Natural Philosophy in the Institution in succession to Rutherford, they had met more often and their relationship seems to have grown more easy. Writing about their father 20 years later on the centenary of his birth (223), Bragg and his sister described him in terms which show clearly the qualities that endeared him to so many and reveal to some extent how difficulties arose in family relationships.

From the summer of 1938, Lady Bragg was heavily involved in the work of the Women's Voluntary Service (W.V.S.), initially at the head office in London and subsequently as head of the Service in Cambridge. Bragg took great pride in his wife's work which made her a well-known public figure in Cambridge and led to her election to the Council and, in 1946, to her becoming Mayor. With their four children, the two girls still at school at the end of the war, they had a busy time.

CAMBRIDGE AFTER THE WAR, 1946-53

Bragg's cogitations during the war about the future need for physicists (117) and the organizations that would be needed to provide them (130) prepared him to meet some of .the problems that faced him in Cambridge at the end of 1945. He was clear, at least, that the Cavendish Laboratory would no longer be dominated by any single research group under one dominating figure since he believed that 'the ideal research unit is one of six to twelve scientists and a few assistants, together with one or more first-class mechanics and a workshop in which the general run of apparatus can be constructed'. With this model in mind he waited, for the most part, for people to emerge with ideas that engaged his interest.

But the teaching as well as the research had to be reorganized and in both departments he relied heavily on J. A. Ratcliffe who returned early from his war service and became Bragg's most trusted helper, especially in the handling of University Committees at which Bragg did not excel. Ratcliffe's intervention, 'We think, don't we Professor...', seems to have been heard at awkward moments in more than one meeting. To some extent the changes within the laboratory were forced by the veterans returning from the war whose ideas were very different from the old ones: senior members of staff expected an office, a secretary and a telephone whereas before the war they had none of these. Bragg met these wishes and, as Ratcliffe recalls, 'modernised the very antiquated notepaper, introduced a departmental secretary to help him run the laboratory and opened up the Austin wing. There was a complete transition to a new style laboratory and the place worked in a completely different way from the old one.' A. B. Pippard, one of his successors as Cavendish Professor, wrote long afterwards in 1972: 'Bragg performed a notably excellent job in decentralizing the work of the Cavendish, and thus effectively breaking away from what would have ultimately become the dead hand of the Rutherford tradition. His decision to give each research section as near as possible autonomy, consistent only with very general central principles and of course financial control, has played a significant part in the subsequent developments. Ever since then, the Cavendish has been notable among Cambridge departments for the democratic way in which it conducts its business. There has been no suspicion, I believe, of essential decisions being taken by the head of the department without consultation.'

The development of autonomous sections was made difficult in 1945 by the need to replace staff and fill vacant positions. A.J. Bradley had suffered a serious breakdown and his appointment in charge of the Crystallography Laboratory could not be renewed. To replace him Bragg appointed W. H. Taylor, his most valued and productive asSociate and successor in the silicate work, who had been Head of the Physics Department at the Manchester College of Science and Technology where Lipson succeeded him. From the end of the war Cockcroft was expected to resign the Jacksonian Chair in order to direct government research in atomic energy but his new appointment was delayed and it was not until 1947 that O. R. Frisch was appointed in his place to take charge of the of the laboratory's continuing effort in nuclear physics. At the same time there were delays in finding a new Plummer Professor of Mathematical Physics before Hartree, another old Manchester colleague, was appointed. However, in November 1946, Bragg was encouraged by the award of a Royal Medal of the Royal Society.

The rapid build up of work after the war and the new range of interests are illustrated broadly by the Departmental reports (printed in the University Reporter) and by the public lectures which Bragg gave about the general work of the laboratory (149). In 1948 the major groups were 'nuclear, radio and low-temperature physics, crystallography, metal physics and mathematical physics, with some minor groupings'. Of these Bragg was, of course, most directly interested in the crystallography and metal physics, with the asSociated electron microscopy under V. E. Cosslett, but his interest had also been engaged in radio physics, which was directed by J. A. Ratcliffe and embraced M. Ryle's developing radio astronomy. Bragg recognized in this latter work a further application of the principles of physical optics to set alongside X-ray crystallography and he supported it vigorously. After he had left Cambridge in 1953 Ryle wrote to him: 'The fact that you were so enthusiastic about our early work on the sun really made me feel that it was worthwhile. The same enthusiasm has made such a tremendous difference ever since-and it has always been a most happy thing to come to you with some new result.'

Nuclear phyS1CS, under Frisch and E.S. Shire, remained the largest group in the laboratory but Bragg wrote of it without the same evident enthusiasm. Noting the heavy investment in equipment and the need for technical officers to run it, he must have recalled his thoughts about research institutes (117): 'The strikingly successful places of this kind are those which may be regarded not as a body of men but as a body of equipment. Such a place has a nucleus of permanent staff and accumulates traditions of technique peculiarly its own, but its main service is as a place open to all for short periods of intense work and its main population a changing one'.

Echoing his earlier letter to Rutherford, Bragg described his own central interests as follows (153): 'The department which we call crystallography would perhaps be better described as the department for discovery of the structure of the solid state. ...Mainly by X-rays we seek to discover the way the atoms are arranged in crystals and in other forms of solids. The scope of the work is very considerable. At one end we are investigating such substances as minerals and alloys in the inorganic field; other researchers are examining complex organic compounds...; finally at the other extreme we have a little group which is financed by the Medical Research Council under the direction of Perutz, which is engaged in a gallant attempt to work out the structure of the highly complex molecules which build up living matter, the proteins. ...

'This section of the laboratory is closely linked to metal physics under Orowan. His students are particularly studying the mechanical properties of metals and relating them to their structure. The effects of cold work on a metal, recrystallisation, the yield point and plastic flow, brittle fracture, distortion under rolling or drawing and so on, are being investigated as physical phenomena. A satisfactory theory of the strength of a metal has yet to be formed.' Bragg's contributions to research in the early days after the war were concerned mainly with these problems in metal physics, especially through his further development of the bubble-raft model (145, 156) which he discussed now in terms of dislocations. He also promoted the development of X-ray microbeam methods by Taylor's group, with a view to studying directly the variations in grain size in cold-worked metals, and he encouraged J. N. Kellar and P. B. Hirsch to build a big rotating-anode X-ray tube to produce a high-intensity microbeam. With this instrument they produced encouraging pictures of aluminium but work elsewhere then suggested that electron microscopy might provide a better approach. Happily an essential shift in technique in this direction was possible because of Bragg's earlier encouragement of electron microscopy: in 1946 he recruited Cosslett who had proposed a programme of work on electron microscopy. But in 1948 there was an upset. As the new rotating-anode tube carne into operation Bragg's interest was becoming more and more focused on proteins and he stunned the metals group by suggesting that it should be diverted to this work. Taylor helped to preserve a balance and, in the end, both lines flourished (v).

Perutz had engaged Bragg's interest in proteins in 1938 and this was maintained throughout the war, although Perutz was prevented by internment and subsequent war work from doing much research until he returned to the Cavendish in January 1944. He was joined in January 1946 by J. C. Kendrew. By early 1947 Bragg was seeking some way of ensuring long-term support for the group and on 21 May 1947, with Keilin's encouragement, he wrote a long letter to Sir Edward Mellanby, the Secretary of the Medical Research Council, asking for help. In this letter Bragg gave an outline of the proposed research and its difficulties and described its promise with reference to his earlier experience: 'We thought it a great triumph to analyse quite simple inorganic salts by X-ray methods in the early days, and a complex organic molecule then seemed almost as far beyond our reach as the proteins might seem now. Yet a patient accumulation of clues, and improved techniques, have made it possible to enter the organic field. If the structure of a few molecules of a new type can be analysed, a rich harvest is then reaped, because the structures of many others will then be clear by analogy .I foresee the same happening in the protein field. ...'

After discussions at the Athenaeum Club, Mellanby agreed that a case could be made to the M.R.C. and, after further talks and correspondence during the summer of 1947, he wrote to Bragg on 20 October: 'Rather to my surprise, your project for the establishment by the M.R.C. of a Research Unit at the Cavendish Laboratory, on molecular structure of biological systems, was adopted by the Council at the meeting on Friday October 1ith, although I had put it forward only for a preliminary run.' Such was the birth of the Medical Research Council's most famous Research Unit, now the M.R.C. Laboratory of Molecular Biology .The original application was for a grant of 12550 rising to 12650 per annum to support Perutz, Kendrew and two research assistants for five years.

The work that kept Bragg's enthusiasm for protein research alight was the attempt by Perutz, which had continued on and off through the war years, to derive structural information directly from the diffraction patterns of haemoglobin crystals and, most especially, from a complete three-dimensional Patterson synthesis of horse haemoglobin. On the assumption that the polypeptide chains were arranged in some kind of regular fold (which alone seemed to offer any hope of solution) Perutz devised a model of haemoglobin in which the molecules were shaped like 'pill-boxes' with the chains folded to give prominent repeat distances of about 5 and 10 A.

At this stage Bragg became deeply interested and, together with Perutz and Kendrew, turned his attention to the possible forms of the folded polypeptide chain. Various models had already been discussed, especially by Astbury, but Bragg was attracted to the idea propounded by Huggins that the most likely structure was a helix because it placed each amino-acid residue in the same kind of position in the chain. There were various observations to take into account, in particular Astbury's studies of IX-keratin indicated a repeat distance of 5.1Å, closely similar to one of the distances observed by Perutz whose data also suggested that the number of amino-acid residues in such a repeat was 3.3. Furthermore, it was regarded as very probable that the chain was held in a folded condition by hydrogen bonds between NH and CO and that these bonds were nearly parallel to the axis of the chain.

With these features to guide them Bragg, Kendrew and Perutz (161) tried various forms of helical chain. They allowed free rotation about all the single bonds in the chain and various symmetries in which there were 2, 3 or 4 aminoacid residues per turn of helix but they failed to find any structure that was especially convincing. Sadly they concluded: 'In X-ray analysis in general, when a crystal structure has been successfully analysed and a model of it is built, it presents so neat a solution of the requirements of packing and interplay of atomic forces that it carries conviction as to its essential correctness. In the present case the models to which we have been led have no obvious advantages over their alternatives.'

Pauling and Corey showed within a year by their description of the IX-helix that Bragg and his colleagues had missed an important feature of protein structure: lacking the necessary chemical insight, they had not realized that the peptide units would be planar-thus reducing the number of possible bends in the chain to one per residue. They had also adhered too firmly to the apparent keratin repeat distances of 5.1Å (which turned out to arise from a higher level of structure) and they had given insufficient consideration to the possibility of non-integral helices.

Otherwise their four-fold helix, which had planar peptide units and the correct hydrogen-bonding pattern, might have been refined to the IX-helix with its 3.6 residues per turn and 5.4A repeat. But the main shortcoming was clearly in the chemistry and Bragg never forgave himself. Years later he wrote (233): 'I have always regarded this paper as the most ill-planned and abortive in which I have ever been involved.' It was especially aggravating to have asked the right question only to have Pauling provide the answer.

Discouraging though this was, the protein work was continued with increasing intensity and Bragg's influence on the next stage can be seen especially in the use of absolute measurements of the X-ray intensities which derived directly from the Manchester methods (55). Following a lead of Crick's, Bragg and Perutz showed by careful analysis of projection data that the reflexions were too weak to be consistent with a model in which the polypeptide chains were straight and parallel throughout the molecule and this finding raised the possibility that less regular models would have to be considered (171). Analysis of the changes in absolute intensities when a salt solution was substituted for water as the medium permeating a haemoglobin crystal next revealed the approximate outer shape of the molecule (172), a result that was clarified further by examination of different: crystal forms (173, 189). Then, in a fourth paper in 1952, Bragg and Perutz (175) at last succeeded in determining the signs (phases) of some protein; reflexions. Using the fact that the cell dimensions of haemoglobin crystals change as their liquid content varies, they were able to plot the variation of the molecular transform along the c*-axis of the crystals. Since the projection on the c*-axis had a centre of symmetry the transform was either positive or negative and passed through zero at intervals along the axis. A key question was to decide the minimum wavelength of these variations and Bragg (176) illuminated the problem with a characteristic example, showing that the transform of any random function, such as the times of arrival of the Cambridge trains at Liverpool Street Station on Sundays between 8 a.m. and midnight, contained a set of loops which change sign only at certain minimum intervals. These intervals were determined by the width of the function, 2 x 16 hours for the trains (when a centre of symmetry is added) or 38A for the haemoglobin molecule. Bragg's principle of minimum wavelength was used to determine the signs of the 001 reflexions and hence the electron density of the molecule projected on the c*-axis.

This method of plotting the molecular transform was extended by Perutz and it proved especially valuable in providing a check on the working of the much more powerful method that emerged soon afterwards. Bragg's (233) account of its origin illustrates well his enthusiastic involvement in the work: 'I remember going to Perutz in great excitement one day because I had heard from Professor Roughton that an American worker had succeeded in attracting a mercury complex to haemoglobin in stoichiometric proportions, only to have Perutz tell me very coldly that he had given this information to Professor Roughton.' The possibilities opened up by this discovery were explored very quickly and Perutz (u) remembered that in July 1953 he was able to show Bragg an X-ray photograph from a haemoglobin crystal which had two atoms of mercury attached to each molecule of haemoglobin. At this moment they both realized that the phase problem was solved, at least in principle, and that the way was at last open to unravelling the structure of proteins by X-ray analysis. The signs of the *h0l* reflexions of haemoglobin were quickly determined by Green, Ingram and Perutz and compared with the other evidence (175, 188): 'Everything checked and double-checked perfectly; it was a thrilling time' (231). This work gave directly an image of the haemoglobin structure projected down the b-axis of the crystals (190) and although this told very little about the structure of the protein it paved the way for a detailed investigation in three dimensions. But at this stage

Bragg left Cambridge for the Royal Institution and continued his collaboration with Perutz and Kendrew at a distance.

The period 1951-53 during which these critical advances were made in the protein work witnessed also the first great triumph of the M.R.C. Unit. Bragg played no direct part in the study of DNA; indeed at one stage he actively discouraged Crick and Watson from working on it in an attempt to avoid competition with the M.R.C. Unit at King's College, London, but Watson (w) has given a colourful and irreverent account of his growing appreciation of its importance, his encouragement at a critical stage and his quick comprehension of the result. Watson also noted Bragg's concern that the chemistry underlying the final model should be checked by A. R. Todd, Pope's successor as Professor of Chemistry and the leading expert on the chemistry of nucleic acids. When Todd approved he was more than willing to promote rapid publication. As Watson saw it: The solution to the structure was bringing genuine happiness to Bragg. That the result carne out of the Cavendish and not Pasadena was obviously a factor. More important was the unexpectedly marvellous nature of the answer, and the fact that the X-ray method he had developed forty years before was at the heart of a profound insight into the nature of life itself.'

In the report for 1952-3, his final year as Cavendish Professor, Bragg described these dramatic advances in work on .The molecular structure of biological systems' very briefly together with progress in the other sections of the laboratory. By this time there were seven sections to be listed and their relative sizes were indicated by the distribution of research students between them: nuclear physics 30, radio waves 17; low temperature physics 10; crystallography 16; electron microscopy 4; meteorological physics 4; fluid dynamics 4. In addition there were a number of theoretical physicists housed in the laboratory and working with the various experimental groups. The descriptions of research in progress began with nuclear physics, still the largest group, and recorded the decision to install a linear accelerator and the work that had been done to implement it. But the longest and most obviously enthusiastic section described the work on radio waves which was divided between the physics of the ionosphere under Ratcliffe and radio astronomy under Ryle. On the last topic the report concluded: The new knowledge of the Universe which it is yielding is proving to be of intense interest and, as the Cambridge unit under Mr Ryle has already established a leading position, the opportunity to develop this new science should be exploited vigorously.'

The report on crystallography noted particularly that 'The analysis of crystal structure by X-rays is a typical borderline subject' and went on to record the collaboration with chemistry, metallurgy and mineralogy. Bragg's own papers, which included a description of a simple device for calculating structure factors (174), were included in the list of the M.R.C. Unit and it is remarkable that, although he undoubtedly played a significant part in advising and helping the crystallographers, he scrupulously avoided (as did Taylor) sharing the authorship of papers unless he had made a major contribution to the experiments or theoretical developments recorded in them.

At his Farewell Dinner on 18 December 1953 (v) Bragg spoke with pride of the resurgence of the laboratory after the war, attributing the success to advances made by one member of the staff after another. Characteristically he emphasized that the atmosphere of 'affairs of state' was not one that he found easy to breathe and he thanked Ratcliffe particularly for his help with these matters and E.H.K. Dibden, whom he had appointed General Secretary of the laboratory in 1948, for his skilled administration.

From Dibden's point of view Bragg was, in fact, a good administrator because he knew what needed doing and believed in delegation. Ratcliffe has provided a more comprehensive summary which embraces this view: 'A Cavendish Professor plays at least four parts. He must be a scientist, run the laboratory, uphold the interests of the department in the University, and act as an Elder Statesman of Science outside. Bragg was pre-eminently the active scientist, and he ran the laboratory extremely well. I do not think he played the part that some others have done in the University itself, and I am not sure that his part as Elder Statesman was quite as large as theirs would have been. I found him extremely helpful and kindly, and above all things a real gentleman in every way. He was quite open and straightforward and ready to help anyone who had the good of the laboratory at heart. I think there was an extremely good feeling in the laboratory during his time and all liked him.'

THE ROYAL INSTITUTION, 1953-66

From the time of his appointment to the non-residential Professorship of Natural Philosophy at the Royal Institution in 1938 Bragg had played an increasing part in the affairs of the Royal Institution. At first with his father as Resident Professor and then, after his father's death in March 1942, with Sir Henry Dale (1942-46) and E.K. Rideal (1946-49) as Resident Professors for short periods he lectured regularly in every year except 1941 when he was away in Canada. In this way he remained in touch with the staff after the family connection had been broken and he followed the fortunes of the Institution with close interest as it sought a new role in the difficult post-war period.

When Rideal arrived at the R.I. it was suffering from the inevitable neglect of the wartime period but he established a lively research group, with grants from industry and Government, worked hard at the programme of Friday Evening Discourses and, with his wife's help, struggled to maintain the tradition of weekly dinner parties at a time of shortage and rationing. It seems to have been because of the difficulties associated with this regular entertaining that Rideal suggested unexpectedly in 1949 that E. N. da C. Andrade should take over the Resident Professorship and responsibility for the lectures and entertaining while he remained in charge of the research. When this proved unacceptable to the President, Lord Brabazon, and the Managers, Rideal resigned and Andrade was appointed in his place.

Andrade took up his appointment in January 1950 and began his attempt to refashion the Institution. Unhappily, but perhaps not surprisingly in view of Andrade's temperament and the traditions of the place, this led to trouble. The root of the difficulty was that the Resident Professor, although enjoying the resounding titles of Fullerian Professor of Chemistry, Superintendent of the House and Director of the Davy-Faraday Research Laboratory, was specifically not the Director of the Royal Institution and much of the responsibility for day-to-day affairs and the staff of the Institution remained with the President, and other honorary officers and the committees of managers and visitors elected by the members. It was widely held at the time that the R.I. had much of the character of a club, though one with scientific objectives, and in Brabazon's opinion the position of Director of a club was unthinkable. Andrade, however, understood that the terms of his appointment gave him powers within the Institution analogous to those of a managing director and this led to friction and discord especially with the honorary secretary, Professor A. O. Rankine. Despite strenuous efforts by many people to find a solution the situation deteriorated rapidly until in March 1952 at a meeting of the members a vote of confidence in Andrade was lost by a substantial majority. Andrade then resigned, though arbitration of his claim for compensation and the litigation that followed were not completed until March 1953.

This unfortunate affair deeply divided the members of the Royal Institution and strong feelings were expressed on both sides. Inevitably Bragg was consulted and, although he tried to stand aside as a servant of the Institution (he was still Professor of Natural Philosophy giving his annual lecture), there was no concealing his concern for the future of the Institution or his disapproval of Andrade's approach. Early in 1950, following the accelerated retirement of W. J. Green who had been the principal lecture assistant in his father's day, he commented in *Nature* 'We shall miss him greatly, for he has come to be a part of the Institution he has served', and later, in a letter to the President, he recalled the difficulties he had encountered when Andrade was in command of one of his Sound- Ranging Sections in France. By mid-1952 he was believed by Sir Henry Dale and other leading figures in the Royal Society to be an advocate, with Brabazon, of a policy to get rid of Andrade and then put the organization of the Institution on a proper basis-a policy that seemed to them grossly unfair and improper on their interpretation of Andrade's letter of appointment. Bragg was faced, therefore, by a difficult and embarrassing decision when the Managers of the Royal Institution in April 1953 offered him the vacant post of Resident Professor. There can be no doubt that he saw it as his duty to revive the fortunes of the Institution but equally he realized that the task was a difficult one and that his motives for intervening at all in the recent troubles would be called into question. Adrian, one of his oldest friends who was now President of the Royal Society and Master of Trinity College, Cambridge, reluctantly acknowledged that he would have to accept the post .because no-one else would' and he did so, taking up the duties of the Fullerian Professorship immediately and the residential duties on 1 January 1954.

Thus, for the third time Bragg accepted a challenging appointment at a difficult time and against a background of disapproval. But despite the difficulties it was not all gloom. There was a chance to develop a new role for the Institution and, perhaps most important, a chance to continue with the protein research for a few more years just when success seemed imminent and retirement from Cambridge in the normal way, he was already 63, might have deprived him of an active part in it.

From the outset Bragg asserted that his appointment required him to work closely with the Honorary Officers (Brabazon continued as President but the Secretary and Treasurer were new), the Managers, Visitors and members of the Institution and he set out, in the main with their grateful help, to rebuild the reputation of the Institution. His aim above all was to avoid further public discord. Although he knew that the administrative structure would have to be changed he reconciled himself to a patient process of persuasion in order to prepare the ground for his successor and, as it turned out, he was made Director of the R.I. in 1965 when he was about to retire.

In 1953 the immediate problem was financial. The costs of keeping up the premises and running the traditional activities of the R.I. could no longer be covered by the subscriptions of the members and the endowment income and for some years the Institution had been drawing on its reserves. Bragg argued at once that new sources of support would have to be found and that the Institution should seek to provide a public service of some kind, in addition to research, that would justify an appeal for funds. At this time the main activities, apart from research, were the Friday Evening Discourses, occasional Afternoon Lectures on a variety of subjects, and the famous Christmas Lectures adapted to a juvenile auditory. Of these, the Afternoon Lectures no longer attracted large audiences, even though free tickets were issued to undergraduates, and the Discourses were largely reserved for the members and their friends. It was the Christmas Lectures that pointed the way ahead: Bragg offered courses of lectures for London schoolchildren, initially at the sixth form level.

This initiative was based securely on the long experience of the R.I. in presenting science to essentially lay audiences by the lavish use of experimental demonstrations. The idea was to show schoolchildren the experiments they would otherwise only read about, and it received an enthusiastic response from the schools and the Education Authorities. Bragg gave the first course of three lectures on Electricity during the session 1954-55 and it was repeated four times. The lecture theatre holds 500 so that 2000 tickets were issued, but this by no means satisfied demand. Later in the year Bragg gave a further single lecture on 'Famous experimenters in the Royal Institution' for sixth formers that was repeated four times and this eventually set the pattern. The Advisory Committee, which included representative science teachers and members of the Education Authorities, recommended that the demand would best be met by single lectures each repeated four times. They were given usually on the Tuesdays and Wednesdays of consecutive weeks and by 1965 a regular pattern had developed with 20000 schoolchildren of various ages attending the lectures every year (199, 235).

Armed with this evidence of the Royal Institution's concern for and contribution to the training of future scientists, Bragg turned to Industry and Commerce for support and received a ready response. A new category of Corporate Subscribers was introduced which soon brought in more than the individual members' private subscriptions and allowed the activities both new and old to flourish.

Naturally these activities demanded a great effort from Bragg and the permanent staff of the Institution in developing the contacts with schools and with industry and, especially, in devising and mounting the large number of experimental demonstrations. Here Bragg was supported mainly by Ronald King, who had come to the Institution with Andrade and stayed on as Assistant Director of Research (and later Professor of Metal Physics), Kenneth Vernon the Librarian, and the Lecture Assistant. This important post was held first by Leonard Walden, who left in 1957 to rejoin Andrade, and then by W. A. (Bill) Coates, whom Bragg and King persuaded with some difficulty to leave the research laboratory for this more public role. Together they built up a wide repertoire of demonstrations, exploiting apparatus that had been accumulating in the Institution throughout its existence (208) and drawing on the advice and help of both staff and members. Bragg gave many of the lectures himself and the staff, research workers and office staff alike, would crowd into the gallery to watch him enthral, stimulate and provoke the packed audiences. One week there would be free-hand drawings of highland dances to illustrate the formation of ionic bonds and the next he would be seen lovingly caressing the Paget speech models and beaming with pleasure at every successful 'Ma-Ma'. With a wealth of everyday analogy and a complete avoidance of jargon he inspired a generation of schoolchildren in London and, through the television programmes that followed, the rest of the country.

The public activities of the Institution were also developed in other ways so that throughout his period of office at the R.I. Bragg was engaged continuously with the committees of members and the staff in considering and promoting ideas for new schemes. From 1955 there were 'Research Days' at which teams of workers from various laboratories described their work informally to parties of schoolteachers; television programmes were planned, rehearsed and recorded; films were made; and, towards the end of the period, a new series of lectures to Civil Servants was begun. Bragg's account of this venture (238) brings out well the continuous process of innovation and development in which he was involved with King and others as they sought to create a modern role for the Institution. The general support of industry during this time, the success of a subsequent general appeal, and the continuing vitality of the Institution under his successor are measures of his achievement.

The Friday Evening Discourses were, of course, continued in all their Victorian state accompanied by the traditional entertaining, skilfully managed by Lady Bragg. On each occasion the lecturer was entertained to dinner in the Resident Professor's flat together with a variety of other guests. Bragg attached great importance to this activity and in one of his reports to the Members he wrote: 'The invitations which we send to well known people to meet the lecturer as guests of the Royal Institution have a much greater importance than might perhaps be realized. My wife and I entertain some 120-150 guests in this way during the year. They provide an opportunity to make important people from all walks of life acquainted with the R.I. and its work.' Bragg no doubt enjoyed meeting these people and, for the most part, he also enjoyed listening to the Discourses which helped him keep abreast of the latest developments in science. He must have listened to some two hundred of these lectures and the experience helped to give final shape to his views on lecturing (239, 250). The lectures were often discussed during the following week in the laboratories and at the daily tea parties for all the staff that helped so much to create a family atmosphere in the Institution. Bragg's favourite criterion for judging the success of a lecture was whether a member of the audience could be expected to remember one idea from it the following morning. More than one failed this test. Bragg also deplored particularly lectures that were read, but he understood too well the difficulties of popularizing science to be over-critical and was always ready to commend a simple explanation or a good experiment.

Two of these Discourses in 1965 gave him particular pleasure. On 7 May, Lady Bragg, who had been a member of the Royal Commission on Marriage and Divorce (1951-55) and was Chairman of the National Marriage Guidance Council, lectured on 'Changing patterns in marriage and divorce'; and on 15 November, Bragg listened with evident pride to the Discourse on 'Oscillations and noise in jet engines' given by his engineer-son Stephen, who was then Chief Scientist at Rolls Royce Ltd and later became Vice-Chancellor of Brunel University.

During this period also Bragg kept up and extended his role as a world figure in science. In 1948 he had been invited to assume the Chairmanship of the Solvay Conferences on Physics and in that year he presided over a discussion of 'Elementary particles'. Conferences followed on 'The solid state' (1951); 'Electrons in metals' (1954); and 'The structure and evolution of the Universe' (1958) until in 1961 he presided over his last conference on the 50th anniversary of the inaugural meeting in 1911. The subject was 'Field theory'-which he confessed to finding 'completely unintelligible'. In 1958 he was the President of the International Science Hall at the Brussels Exhibition (207) and arranged the British contribution to it.

At this time he was also Chairman of the Soirée Committee of the Royal Society .This was one of the few committees that he enjoyed, and he greatly appreciated the contact that it gave him with the growing points of science. In this role he was responsible for organizing the exhibition at the Tercentenary Celebrations of the Society in 1960.

The Davy-Faraday research laboratories are an integral part of the Institution and in Bragg's time they were partly in the basement and partly in the upper floors of the house next door, connected to the main building at the level of the Resident Professor's flat. This intimate arrangement enabled Bragg to visit the laboratories whenever he had a moment to spare or needed relief from the discussion of some tedious difficulty, and he would announce his imminent arrival by a characteristic stamp on the ancient and creaking floor boards.

On his arrival in 1954 there was very little research still in progress. King and a small group were continuing the- research in metal physics started with Andrade, and U. W. Arndt was engaged in X-ray studies, partly technical and partly on proteins. Bragg had hoped to persuade Max Perutz or John Kendrew to move with him from Cambridge, but. at this promising moment in their work, they preferred to stay behind. They undertook instead to help Bragg build up protein research at the Royal Institution and they were each given the title of Reader in the Davy-Faraday Research Laboratory. With help and advice from them and from Dorothy Hodgkin and others a research team was quickly assembled. Helen Scouloudi, who had worked at Birkbeck with Bernal and Carlisle, came first and she was joined in the autumn of 1955 by D. W. Green, who had been a research student with Perutz and had contributed to the critically important development of isomorphous replacement in protein structure analysis, and by A. C. T. North, who had worked on collagen with Randall at King's College. J.D. Dunitz came back from the U.S.A. at the end of the year and D.C. Phillips returned from a post in Canada at the beginning of 1956. This rapid build up was made possible by the support of the Medical Research Council and the Rockefeller Foundation.

Dunitz continued with work on transition-metal compounds until his departure for Zürich in 1957 but the other members of Bragg's research group concentrated on proteins and worked closely for the first few years with their colleagues in Cambridge. Kendrew visited the laboratory nearly every week to keep everyone closely in touch with the rapidly developing work on myoglobin. Bragg was particularly interested in the development of diffractometer methods for measuring the diffraction data needed to produce the first three-dimensional image of a protein molecule (it reminded him of the early days with the X-ray spectrometer) and he sought ways of using such measurements to locate heavy atoms in the isomorphous derivatives that were the key to the analysis. The popular approach was based upon the use of Fourier methods which take into account all of the X-ray reflexions from the native and a derivative crystal at the same time. Bragg realized that determination of the small number of parameters defining a heavy-atom structure is potentially a simple problem, similar to the early analyses of mineral structure (55), and that careful consideration of a few well chosen and carefully measured reflexions might provide the required information. Using data from the studies of myoglobin and oxyhaemoglobin he set to work in characteristic style with pencil and graph paper and devised two new methods which were described in his last research paper (205). Although they have not been generally adopted, computer methods proved too powerful and appealing, they illustrate well Bragg's quick eye for the practical application of basic principles and one of them has proved invaluable in the study of tobacco mosaic virus.

Bragg (233) has given his own account of the period 1956-58 during which Kendrew produced the first image of a protein molecule in three dimensions: 'I remember well the thrill of that time. The collection of the vast body of data needed was shared between the laboratory at Cambridge and the Davy-Faraday Laboratory at the Royal Institution. I made a private test of my own. Kendrew supplied me with sets of data for the *hk*0 and 0*kl* projections, for which general phases had to be determined because they have no symmetry centres. I developed a method for getting the relative positions of the heavy atoms (205) and verified that the phases could be found by drawing vector diagrams, with a very convincing agreement between the results for the different ligands. This investigation played no part in the final analysis. Kendrew fixed the heavy atom positions by a more general and powerful analytical treatment aided by the electronic computer, and the phases for all hkl components were systematically determined. My investigation only had a meaning for myself because it showed that the problem had been solved, and that final success was now certain. Kendrew first determined the structure to a resolution of 6 A. It showed dense rods marking the stretches of α-helix and the flat disc of the haem group. It was a proud day when he brought the model to show it to me.'

In the following two years Phillips, joined by Violet Shore and a team of assistants, continued the collaboration with Kendrew to extend the resolution so that in 1960 a high-resolution image of the myoglobin molecule was obtained in which the detailed atomic arrangement could be seen. During the same period North helped Perutz to produce a low-resolution image of haemoglobin which showed that this large and complex molecule can be regarded roughly as four myoglobin molecules in a tetrahedral arrangement. Bragg was delighted and arranged to take a model of myoglobin with him When, in the autumn of 1960, he went to New Zealand to give the Rutherford Memorial Lecture at the University of Canterbury (219). But his glowing account includes one slightly regretful note: 'The new feature is that the element of guesswork has gone and been replaced by the handling of vast masses of measurements and calculations.'

This engagement in New Zealand gave Bragg a long-looked-for opportunity to take his wife on a visit to Australia where they saw again his favourite Aunt Lorna. It was a great success but there were disappointments: 'I promised my wife I would show her the shells and other fascinating marine life along the shores, where I knew the habitat of all the species, and we arranged to spend a week in a seaside place of my boyhood. Alas, it had all gone except for a few of the hardiest kinds. I suppose the pollution of extending Adelaide must have poisoned the sea for fifty miles along the coast.'

Retuming to London, Bragg had the twin pleasures of watching the growing recognition of the Cambridge work and encouraging the development of an independent research programme at the Royal Institution. The most dramatic advances were initiated in 1960 when Roberto Poljak, a visiting research worker, showed that he could prepare promising heavy-atom derivatives of hen egg-white Iysozyme. Bragg was immediately interested and encouraged Phillips to join in the work. With the strong support of the Medical Research Council the main resources of the laboratory were put into this study with the result that the complete structure of this enzyme, the first to be analysed, was ready to be presented to Bragg on his seventy-fifth birthday. He could not have been more delighted and immediately set to work, perched upon a stool in the dusty store room used for model building, making a drawing of the molecular structure for its first publication (x). The value of his constant advice, support and encouragement in this work can hardly be overestimated and his evident joy at the result gave the greatest possible pleasure to the people concerned. The work was described at a Royal Society Discussion meeting held at the Royal Institution on 3 February 1966 (245).

There can be little doubt that Bragg's years at the Royal Institution were his happiest despite the shadow of advancing age and illness. Although the Braggs lived most of the time at the Institution they had a family house at Waldringfield in Suffolk where they entertained their growing family-eventually there were ten grandchildren-and enjoyed a country life by the sea with plenty of opportunities for bird watching, gardening, sailing and painting. There too they entertained the members of the laboratory at memorable parties. The Braggs' elder son, Stephen, had married Maureen Roberts in 1951 and had three sons. Their two daughters were married from the Royal Institution, Margaret to Mark Heath a diplomat, and Patience to David Thomson, the son of G.P. and grandson of J.J.: David, their younger son who worked at the Seed Testing Station in Cambridge and was the artist of the family, was married later to Elizabeth Bruno. The Heaths had three children, two boys and a girl, and the Thomsons four, two boys and two girls; and Bragg, who was most at ease with children, spent happy hours entertaining his grandchildren (and any others he encountered) with animal drawings and fairy stories from an apparently inexhaustible store.

At the same time, through his lectures at the Royal Institution and the television series that followed from them, he became a popular lecturer and an admired and recognized public figure. As a University lecturer he had not been a very great success with undergraduates who expected to obtain detailed and complete expositions of important aspects of physics from his lectures whereas he was concerned to identify and explain general principles. In 1927 he had noted 'The air of detachment when one is explaining a general principle and the eager scribbling in notebooks when one comes out with a fact are well known to every lecturer' (49). Throughout his life he sought, in common with his father, to achieve as complete an understanding as possible of every physical phenomenon that he encountered and, although quantum effects presented some difficulty (85), this understanding provided the basis for his popular lectures. What made them unforgettable was his gift of illustration by analogy coupled with an infectious enthusiasm which engaged all but the most sophisticated.

During his time at the Royal Institution also his reputation as a scientist was finally assured. He was certainly aware that the changes he had promoted as Cavendish Professor had attracted criticism from physicists who hardly recognized crystallography, and certainly not molecular biology, as physics. In 1962, when he was critically ill after a serious operation, he heard the news of the Nobel Prizes awarded to Perutz and Kendrew, for their work on proteins, and to Crick, Watson and Wilkins, for their analysis of DNA. This recognition of molecular biology, coupled with exciting developments in the metal physics and radio astronomy which he had promoted, put his standing as Cavendish Professor beyond question.

The 50th anniversary of his Nobel Prize was marked on 15 October 1965 by a splendid party at the Royal Institution, which was attended by the Lord Chancellor and some twenty British Nobel-Prize winners. Later in the year, with Lady Bragg, he attended the Nobel Prize celebrations in Stockholm and was treated to all the acclaim he had missed 50 years earlier. In his lecture he reviewed his part in the early work, somewhat more bluntly than hitherto, and the subsequent growth of crystal structure analysis (241).

Bragg's retirement from his posts at the Royal Institution was announced in July 1965, to take place on 1 September 1966, but for some time previously he had been concerned to ensure that the members of his research group were found appropriate situations in which to continue their work. In the event, Green moved to Edinburgh where he continued his studies of β-lactoglobulin in the Department of Natural Philosophy, and Phillips, North, Blake, Scouloudi and others who had worked on lysozyme moved to Oxford where they set up a new Laboratory of Molecular Biophysics in the Department of Zoology .With Bragg's strong recommendation, this move to Oxford was supported by the Medical Research Council, but it almost failed at the last stage of the negotiations because of major differences between the salaries which had been paid to senior staff by the M.R.C. and those paid to university lecturers in Oxford. Lecturers in Oxford usually receive additional salaries as Fellows of Colleges but such Fellowships are awarded to meet teaching needs and, at this stage, no College believed that a need had been established for tutorial teaching in molecular biophysics. Nevertheless this suggested a mechanism for saving the situation. Bragg was advised that three Colleges would be prepared to create appropriate Fellowships if a suitable endowment could be found. At this he turned for help to his old friends Sir Kenneth Lee (243) and Harold Hernming and they generously provided the necessary money. In this way Bragg made sure that his last research workers could leave the Royal Institution for attractive new posts when he was succeeded as Resident Professor and Director by Sir George Porter.

In November 1966, when he had at last retired at the age of 76, Bragg was awarded the Copley Medal of the Royal Society, Bemal wrote to him (23 November 1966): 'This is only to congratulate the Society for giving you at last the Copley which you have deserved many times over. It cannot really at this stage mean much to you as you and the whole scientific world know what you have done. Crystal structure may seem now an old story, and it is, but you, its only begetter, are still with us. Three new subjects, mineralogy, metallurgy, and now molecular biology, all first sprang from your head, firmly based on applied optics. You can afford to look back on it all with justified feelings of pride and achievement.' Public recognition of his achievements was confirmed in the New Year Honours list of 1967 when he was made a Companion of Honour.

RETIREMENT, 1966- 71

After his formal retirement, Bragg continued to live in London most of the year and, as Emeritus Professor, he continued to lecture at the Royal Institution. He also lectured elsewhere, wrote a good deal and visited 'his' laboratories, the old one in Cambridge and the new one in Oxford.

During this time he saw his forecast about the study of proteins begin to come true as more and more structures were determined and patterns began to emerge in them. But he also saw and delighted in the application of physical methods and modes of thought to more complex biological problems. Perutz (y) has recorded an anecdote that vividly recaptures Bragg's style as a consultant: 'I took him to a young zoologist working on pattern formation in insect cuticles. The zoologist explained how disturbances introduced into these regular patterns pointed to their formation being governed by some kind of gradient. Bragg listened attentively and then exclaimed: "Your disturbed gradient behaves like a stream of sand running down hill and encountering an obstacle." "Good heavens," replied the zoologist, "I had been working on this problem for years before this simple analogy occurred to me and you think of it after twenty minutes."

This kind of insight into natural phenomena, especially those concerned with optics and three-dimensional relationships, underlay all of Bragg's scientific work and his lecturing and writing about it. Years of discipline had long overcome his early horror of writing and he was able to concentrate immediately on writing anything from a routine circular to a scientific paper. Perutz (u) again has described his approach to paper writing: 'He would illustrate his conclusions in a series of neatly drawn sketches, and then write the accompanying paper in a lucid and vivid prose. Some scientists produce such prose as a result of prolonged redrafting and polishing, but Bragg would do it in one evening, all ready to be typed the next day, rather like Mozart writing the overture to " The Marriage of Figaro" in a single night.' Bragg retained this hard-won facility into retirement, and his last book (M), which was barely complete when he died, displays all the old vigour and many examples of his ability to summarize a complex problem in a single polished paragraph. Thus, for example, after discussing the wave-particle problem which had so dominated the scientific discussions of his youth, he wrote: 'So the dividing line between the wave or particle nature of matter and radiation is the moment "Now". As this moment steadily advances through time, it coagulates a wavy future into a particle past.'

This book on *The Development of X-ray Analysis* is a history of X-ray crystallography and, since it concentrates particularly on the topics which had interested him most, it is almost Bragg's scientific autobiography. He was writing it in 1970 when crystallographers from all over the world met at the Royal Institution at a meeting to celebrate his eightieth birthday. Organized by W. H. Taylor, this 'Bragg Symposium 1970' was entitled 'X-ray analysis—past, present and future' and it gave many of his old friends and asSociates a chance to remember old triumphs together and to look to the future. Bragg was himself the liveliest participant. He listened attentively to every session and generally led the discussion—much to everyone's delight.

This Symposium illustrated very well Bragg's essential achievement. The sessions were devoted to most important advances in the forefront of mineralogy, metallurgy, chemistry and molecular biology, subjects which had been revolutionized or in some instances even created by his discoveries. But there were few papers on topics that would be universally recognized as physics by a modern audience of scientists. Centred in physics departments, his achievements had transformed understanding of the natural world and the descriptive sciences in terms of atomic arrangements but his revolution had coincided with others, especially in nuclear and quantum physics, which largely took over physics itself. Bragg remained essentially a classical physicist in the great tradition of those who thought in terms of tangible rather than mathematical models. But he had to struggle to be accepted as a physicist at all and his associates suffered from this, until at last many of them were recognized for their contributions to other fields.

Behind his conventional, even military, appearance as an establishment figure, Bragg had an artistic temperament with strong emotions normally kept in check by stem self-control. He reminded Colin Blake of Elgar, somewhat ironically since Bragg had no appreciation of music at all. But he delighted in painting, a skill he had learned from his mother, and many of his associates treasure examples of his country scenes and portraits. Literature was another of his loves — he instituted special courses in the humanities for physics students in Cambridge — and he delighted in referring to characters in the great novels: the 'philanthropoid' Mrs Norris in Mansfield Park was a great favourite. One of his last broadcasts was a personal choice of verse and prose which he presented 'With great pleasure' on 18 October 1970.

But most of all Bragg was a private family man. The draft autobiography that he was working on at the time of his death abounds with happy memories of family holidays, often sailing on the Broads, and adventures with his adored wife and children. Perutz (z) remembered that 'typical1y one would find him tending his garden, with Lady Bragg, children and grandchildren somewhere in the background, and before getting down to business he would proudly demonstrate his latest roses'. Even there his creativity was for ever bursting out in some new 'venture': only a short time before his death he was enthusiastically promoting a method for supporting tall plants, such as Michaelmas daisies, by letting them grow through sheets of wire netting.

Bragg was certainly one of the great creative scientists yet he often worried about his relative lack of more mundane gifts. Forgetful of names, uneasy on committees, reluctant to face personal problems or angry scenes, he depended a great deal on his wife who sustained him through all the triumphs and difficulties of a long public life. There is no doubt that he found peace at the last and the abiding affection of those that knew him best. He died in hospital near his home at Waldringfield on 1 July 1971.

My first thanks are due to Lady Bragg and to Stephen Bragg for their help in the preparation of this memoir and for their permission to consult and quote from the Bragg Archives at the Royal Institution. The Director of the Institution, Sir George Porter, F.R.S., and the Librarian, Mrs I. M. McCabe, have been most helpful. Mrs A. Caroe, Sir Lawrence Bragg's sister, whose biography of their father has been an invaluable help, has also given aid and counsel without stint and I am deeply indebted to her.

Very many of Bragg's colleagues and students have responded generously to my importunate appeals for help and I am grateful to them for their generous assistance and for allowing me to quote freely from their letters and articles. There are too many to mention all by name but I am particularly indebted to Mr Norman Tunstall for his account of the early days in Manchester, to Mr J. A. Ratcliffe, F.R.S., and to Dr M. F. Perutz, F.R.S. Most especially, however, am I grateful to Professor Henry Lipson, F.R.S. whose prompting, encouragement and help (which included writing his own account of the Manchester, N .P .L. and Cambridge periods) made a vital contribution to the completion of the memoirs.

Finally, I must record my indebtedness to Mrs C. C. F. Blake, whose researches into the details of Bragg's career were invaluable; and to my ex secretary, Miss Susan Partridge, who made an essential contribution to the literature survey and in many other ways.

GENERAL REFERENCES

(a) Caroe, G. M. 1978 William Henry Bragg, 1862-1942: Man and Scientist. Cambridge University Press.

(b) Royal Institution archives.

(c) Royal Society archives.

(d) Stuewer, R. H. 1971 William H. Bragg's corpuscular theory of X-rays and γ-rays. Br. J.Hist. Sci. 5, 258-281.

(e) Ewald, P. P. and numerous crystallographers 1962 Fifty years of X-ray diffraction. Published for The International Union of Crystallography by N. V. A. Oosthock's Uitgeversmaat-schappij, Utrecht, The Netherlands.

U) Forman, P. 1969 The discovery of the diffraction of X-rays by crystals; a critique of the myths. Arch. Hist. Exact Sci. 6,38-71.

(g) Bragg, W. H. 1912 Nature, Lond. 90,219.

(h) Schuster, A. 1909 An introduction to the theory of optics (2nd ed., revised). London: Edward Amold.

(i) Pope, W. J. 1908 A. Rep. Progr. Chem. pp. 258-279.

(j) Bragg, W. H. 1912 Nature, Lond. 90,360-361.

(k) Bragg, W. H. 1913 Nature, Lond. 90, 372.

(l) Moseley, H. & Darwin, C. G. 1913 Nature, Lond. 90,594.

(m) Heilbron, J. L. 1974 H.J. C. Moseley-the life and letters of an English physicist 1887-1915. Berkeley, Los Angeles and London: University of California Press.

(n) Bragg, W. H. 1913 Proc. R. Soc. Lond. A 89,246-248.

(o) Miers, H. A. 1918J. chem. Soc. 113, 363-386.

(p) James, R. W. 1952 Trans. R. Soc. S. Afr. 34,1-16.

(q) Armstrong, H. E. 1927 Nature, Lond. 120,478.

(r) Pauling, L. 1928 in Sommerfeld Festschrift. Leipzig: S. Hirgel. 1929 J. Am. chem. Soc. 51, 1010.

(s) Lipson, H. S. 1973 A. J. Bradley (1899-1972). Biogr. Mem. Fellows R. Soc. Lond. 19, 117-128.

(t) Patterson, A. L. 1935 Z.Kristallogr.Kristallgeom. 90,517-542.

(u) Perutz, M. F. 1970 Acta crystallogr. A26, 183-185.

(v) Crowther, J. G. 1974 The Cavendish Laboratory 1874-1974. London and Basingstoke: Macmillan.

(w) Watson, J. D. 1968 The double helix. London: Weidenfeld & Nicholson.

(x) Blake, C. C. F., Koenig, D. F., Mair, G. A., North, A..C. T., Phillips, D. C. & Sarma, V.R. 1965 Nature, Lond. 206,757-761.

(y) Perutz, M. F..1971 New Sci. & Sci.J. 8 July 1967.

(z) Perutz, M. F. 1971 Nature, Lond. 233,74-76.

BIBLIOGRAPHY

Books

(A) 1915 (With W. H. BRAGG) X-Rays and crystal structure. London: G. Bell & Sons Ltd. 2nd ed. 1916. 3rd 1918. 4th (revised) 1924. 5th 1925. Translated into Russian (1916 and 1929) and French (1921).

(B) 1930 The structure of silicates. 69 pages. Leipzig: kad. Verlag.

(C) 1933 The crystalline state. a general survey. London: G. Bell & Sons Ltd. Vol. I of The crystalline state. eds. W. H. & W. L. Bragg.

(D) 1936 Electricity. (The Royal Institution Christmas Lectures. 1934.) London: G. Bell & Sons Ltd. and U.S.A.: Macmillan Company. Translated: Swedish. 1937; Polish, 1939; Czech. 1940; Hungarian. 1948; Finnish, 1950; German. 1951; Japanese, 1951; Italian. 1953.

(E) 1937 Atomic structure of minerals. Ithaca. N. Y.: Cornell University Press and London: Oxford University Press.

(F) 1939 (With A. E. VAN ARKEL. U. R. EVANS & ~. P.~VANO) Chimie Minéral. Paris: Hermann & Cia.

(G) 1943 History of X-Tay analysis. London: British Council/Longmans. Revised edn 1946; German translation 1947.

(H) 1950 (With H. J. EMELEUS) Post-graduate Lectures (sponsored by the Oil and Colour Chemists Association). Cambridge: W. Heffer & Sons Ltd.

(l) 1965 (With G. F. CLARINGBULL) Crystal structures of minerals. London: G. Bell & Sons Ltd. Vol. IV of The crystalline state. ed. 'V. L. Bragg.

(J) 1967 The start of X-ray analysis. London: Longmans; Penguin. (Chemistry Background Books. The Nuffield Foundation.)

(K) 1970 Ideas and discoveries in physics. London: Longmans. (Longman Phys-ics Topics for 6th form pupils.) Translated: French, 19í4; Dutch 1969.

(L) 1971 (With A. H. DAWSON & H. H. HE.'IMING) Artillery survey in the First World War. London: Field Survey Association.

(M) 1975 The Development of X-Tay analysis. (ed. D. C. Phillips & H. Lipson). London: G. Bell & Sons Ltd.

Editor

The crystalline state. London: G. Bell & Sons Ltd.
Vol. 1(1933) (with W. H. Bragg)
Vol. II (1948)
Vol. III (1953)
Vol. IV (1965)

(with G. PORTER) The Royal Institution Library of Science: Physical Sciences (being the Friday Evening Discourses in Physical Sciences held at The Royal Institution. 1851-1939). Amsterdam: Elsevier Publishing Co. (1970). (10 volumes and Index.)

Consultant Editor

Contemporary Physics (1959-64).

Papers

(1) 1912 The diffraction of short electromagnetic waves by a crystal (lecture 11 November 1912). Proc. Camb. phil. Soc. 17.43-57. .Nature. Lond. 90,402.

(2) The specular reflection ofX-rays. Nature, Lond. 90.410.

(3) 1913 X-rays and crystals. Sci. Prog. 7.372-389.

(4) (With W. H. BRAGG) The reflection of X-rays by crystals. Proc. R. Soc. Lond. A 88. 428-438.

(5) The structure of some crystals as indicated by their diffraction of X-rays. Proc. R. Soc. Lond. A 89. 248-277.

(6) (With W. H. BRAGG) The structure of the diamond. Nature. Lond. 91,557.

(7) (With W. H. BRAGG) The structure of the diamond. Proc. R. Soc. Lond. A 89. 277-291. .

(8) 1914 Eine Bemerkung über die Interferenzfiguren hemiedrischer Kristalle. Phys. Z. 15. 77-79.

(9) X-rays and crystals. J. Röntgen Soc. 10, 70-78.

(10) 1914 The analysis of crystals by the X-ray spectrometer. Proc. R. Soc. Lond. A 89, 468-489.

(11) The crystalline structure of copper. Phil. Mag. (6) 27, 355-360.

(12) 1915 (With W. H. BRAGG) X-rays and crystal structure. (British AsSociation for the Advancement of Science, Manchester) Engineering, Lond. 100,305.

(13) (With W. H. BRAGG) Die Reflexion von Röntgenstrahlen aus Kristallen (translations of papers in Proc. Roy. Soc., etc., subsequently published by Leopold Voss, 1928). Z. anorg. Chem. 90, 153-296.

(14) 1919 Sound ranging. Manchester Memoirs 64, p. v and Nature, Lond. 104, 187.

(15) 1920 The crystalline structure of zinc oxide. Phil. Mag. (6) 39, 647-651.

(16) The arrangement of atoms in crystals. Phil. Mag. (6) 40, 169-189.

(17) Crystal structure (R. I. Discourse, 28 May 1920). Proc. R. Instn Gt Br. 23, 190-205 and Nature, Lond. 105, 646-648.

(18) 1921 (With R. W. JAMES & C. H. BOSANQUET) Uber die Streuung der Röntgenstrahlen durch die Atome eines Kristalles. Z. Phys. 8, 77-84.

(19) (With R. W. JAMES & C. H. BOSANQUET) The intensity of reflexion of X-rays by rock-salt. Phil. Mag. (6) 41, 309-337.

(20) (With R. W. JAMES & C. H. BOSANQUET) The intensity of reflexion of X-rays by rock-salt. II. Phil. Mag. (6) 42, 1-17.

(21) The arrangement of atoms in crystals. Nature, Lond. 106, 725.

(22) (With H. BELL) The dimensions of atoms and molecules. Nature, Lond. 107,107.

(23) The dimensions of atoms and molecules. Sci. Prog. 16, 45-55.

(24) 1922 (With R. W. JAMES & C. H. BOSANQUET) The distribution of electrons around the nucleus in the sodium and chlorine atoms. Phil. Mag. (6) 44, 433-449.

(25) The diffraction of X-rays by crystals. (Nobel Lecture, Stockholm, 6 September 1922.) Les Prix Nobel en 1921-1922. (Also published in: Nobel Lectures, Physics 1901-21. Amsterdam: Elsevier Publishing Co. (1967), pp. 370--382).

(26) (With R. W. JAMES) The intensity of X-ray reflection. Nature, Lond. 110,148.

(27) 1923 Sound. Manchester Memoirs 67, p. xii.

(28) 1924 The structure of aragonite. Proc. R. Soc. Lond. A 105, 16-39.

(29) The refractive indices of calcite and aragonite. Proc. R. Soc. Lond. A 105, 370--386.

(30) The influence of atomic arrangement on refractive index. Proc. R. Soc. Lond. A 106, 346-368.

(31) (With S. CHAPMAN) A theoretical calculation of the rhombohedral angle of crystals of the calcite type. Proc. R. Soc. Lond. A 106, 369-377

(32) Crystal structure. Manchester Memoirs 68, pp. xiii-xiv.

(33) 1925 The interpretation of intensity measurements in X-ray analysis of crystal structure. Phil. Mag. (6) 50, 306-310.

(34) Inorganic crystals (address delivered 17 September 1924 on the occasion of the Centenary of The Franklin Institute). J. Franklin Inst. 199, 761-772.

(35) Model gratings to illustrate the diffraction of X-rays by crystals. Manchester Memoirs 69,35-38.

(36) Model illustrating the formation of crystals. Manchester Memoirs 69, p. xvi.

(37) The sizes of atoms. Manchester Memoirs 70, p. viii.

(38) The crystalline structure of inorganic salts (R.I. Discourse, 1 May 1925). Proc. R. Instn. Gt. Br. 24, 614--620 and Nature, Lond. 116, 249-251.

(39) 1926 (With G. B. BROWN) Die Kristallstruktur von Chrysoberyll (BeAI2O.). Z. Kristallogr. Kristallgeom. 63, 122-143.

(40) (With G. B. BROWN) Die Struktur des Olivins. Z. Kristallogr. Kristallgeom. 63, 538-556.

(41) (With G. B. BROWN) The crystalline structure of Chrysoberyl. Proc. R. Soc. Lond. A 110, 34-63.

(42) (With J. WEST) The structure of beryl, Be3A12Si8O18. Proc. R. Soc. Lond. A 111, 691-714.

(43) (With C. G. DARWIN & R. W. JAMES) the intensity of reflexion of X-rays by crystals. Phil. Mag. (7) 1, 897-922.

(44) Interatomic distances in crystals. Phil. Mag. (7) 2, 258-266.

(45) X-my analysis of crystal structures and its relation with chemical constitution. 2ieme Cons. Chim. Inst. Intern. Chim. Solvay, 1926, 44--65.

(46) 1927 The structure of phenacite, Be2SiO. Proc. R. Soc. Lond. A 113,642-657.

(47) 1927 (With J. WEST) The structure of certain silicates. Proc. R. Soc. Lond. A 114,450-473.

(48) The structure of silicates (R.I. Discourse, 20 wlay 1927). Proc. R. Instn Gt Br.25, 302-310.

(49) Some views on the teaching of science (Presidential Address to the Manchester Library and Philosophical Society, 1927-28 Session). Manchester Memoirs 71,119-123.

(50) Some recent advances in the physics of the solid state. Manchester Memoirs 71,p. xü.

(51) Crystallography. Ann. Rep. Prog. Chem. 22, 257-279.

(52) (With W. H. BRAGG) Stereoscopic photographs of crystal models, to illustrate the results of X-ray crystallography. 2 series (1927, 1930). London: Adarn Hilger Ltd.

(53) 1928 The diffraction of short electromagnetic waves by a crystal. Atti Congr. Internazionale dei Fisici (Corno, Septembre 1927 (V», 171-180.

(54) L'Intensité de Réflexion des Rayons X. Inst. Intern. Phys. Solvay. 5ieme Cons. De Physique, 1927, 1-43.

(55) (With J. WEST) A technique for the X-ray examination of crystal structures with many parameters. Z. Kristallogr. Kristallgeom. 69, 118-148.

(56) (With B. WARREN) The structure of diopside, CaMg(SiO3)2. Z. Kristallogr. Kristallgeom. 69, 168-193.

(57) 1929 (With R. W. JAMES, J. D. BERNAL & A. J. BRADLEY) Crystallography. Ann. Rep. Prog. Chem. 25, 275-302.

(58) X-ray optics (the 9th Mackenzie Davidson memorial Lecture). Br. J. Radiol. 2, 65-71.

(59) Atomic arrangement in the silicates. Trans. Faraday Soc. 25, 291-314.

(60) The determination of parameters in crystal structures by means of Fourier series. Proc. R. Soc. Lond. A 123, 537-559.

(61) An optical method of representing the results of X-ray analysis. Z. Kristallogr. Kristallgeom. 70, 475-492.

(62) An optical method of displaying the results of X-ray examination of crystals. Manchester Memoirs 72, pp. xii-xiii.

(63) The diffraction of short electromagnetic waves by a crystal. Scientia March 1929, pp. 153-162.

(64) Diffraction of X-rays by two-dimensional crystal lattice. Nature, Lond. 124, 125.

(65) 1930 (With J. WEST) A note on the representation of crystal structure by Fourier series. Phil. Mag. (7) 10, 823-841.

(66) (With W. H. ZACHARIASEN) The crystalline structure of phenacite, Be2SiO, and willemite, Zn2SiO,. Z. Kristallogr. Kristallgeom. 72, 518-528.

(67) The structure of silicates. Z. Kristallogr. Kristallgeom. 74,237-305.

(68) (With B. E. WARREN) The structure of chrysotile H,Mg_3Si_2O. Z. Kristallogr. Kristallgeom. 76, 201-210.

(69) Die Untersuchung der Atornanordnung mittels Röntgenstrahlen. Metallwirt schaft 9, 461-465.

(70) Bau der silikate. Glastech. (Frankfurt) 8, 449-453.

(71) The structure of silicates (lecture to the Mineralogical Society, 18-March 1930). Nature, Lond. 125,510-511.

(72) Structure of silicates. J. Soc. Glass Technol. 14, 295-305.

(73) X-ray optics. Photogr.J. 70,179-186.

(74) 1931 The architecture of the solid state (the 22nd Kelvin Lecture, 30 April 1931). J. Instn elect. Engrs 69,1239-1244 and Nature, Lond. 128,210-212 and 248-250.

(75) (With C. GOTTFRIED & J. WEST) The structure of 3 alumina. Z. Kristallogr. Kristallgeom. 77,255-274.

(76) (With F. KIRCHNER) The action of a crystal as a two-dirnensional lattice in diffracting electrons. Nature, Lond. 127,738-739.

(77) 1932 The structure of molecules: the solid state. In: Chemistry at the Centenary (1931) Meeting of the British Association for the .4dt,.ancement of Science, pp. 255-256. Cambridge: W. Heffer & Sons Ltd.

(78) 1932 The application of X-ray methods to industrial problems. (lecture at the University of Manchester, 11 July 1932), pamphlet, 19 pages.

(79) 1932 (With J. A. DARBYSHIRE) The structure of thin films of certain metallic oxides Trans. Faraday Soc. 28, 522-529.

(80) Structure of complex ionic compounds (lecture 20 November 1931). Trans. Oxf. Univ. jr scient. Club. Fifth series, no.5, 151-153.

(81) 1933 The structure of alloys (R.I. Discourse, 17 March 1933). Proc. R. Instn Gt Br 27,756-784 and Nature, Lond. 131,749-753.

(S2) Development of Röntgen-ray analysis of crystals (a review). Usp. fiz. Nauk. 13, 195-208.

(83) 1934 (With E. J. WILLIAMS) The effect of thermal agitation on atomic arrangements in alloys. Proc. R. Soc. Lond. A 145, 699-730.

(84) The exploration of the mineral world by X-rays. (Evening Discourse to B.A., 10 September, 1934.) Rep. a. Meet Br. Ass. Advmt Sci. 104, 437-444 and Nature, Lond. 134, 401-404.

(85) The Physical Sciences (introductory lecture as non-resident Lecturer in Chemistry at Cornell University). Science, N. Y. 79, 237-240.

(86) 1935 (With E. J. WILLIAMS) The effect of thermal agitation on atomic arrangement in alloys-II. Proc. R. Soc. Lond. A 151, 540-566.

(87) The new crystallography. Proc. R. Soc. Edinb. 55, 62-71.

(88) Atomic arrangement in metals and alloys (25th May Lecture to the Institute of Metals, 8 May 1935).J. Inst. Metals 56,275-299.

(89) 1936 L 'exploration du monde minéral a l'aide des rayons X. (Conference faite devant la Societé Francaise de Physique, 21 April 1936.) J. Phys. Radium, Paris 7, (série VII) 321-325.

(90) (With H. Lipson) The employment of contoured graphs of structure-factor in crystal analysis. Z. Kristallogr. Kristallgeom. 95, 323-337.

(91) Structure-factor graphs for crystal analysis. Nature, Lond. 138, 362-363.

(92) Anordnung der Atome in den Metallen und Legierungen. Usp. fiz. Nauk. 16, 977-1000.

(93) 1937 Alloys. Jl R. Soc. Arts 85,431-447.

(94) (With W. H. BRAGG) The discovery of X-ray diffraction. Curr. Sci. 7 (suppl., special number, 'Laue Diagrams'), 9-13.

(95) Alloys. (Report of lecture to meeting, 2 February 1937.) Manchester Me1tIOirs 81, p. xii.

(96) (With C. SYKES & A. J. BRADLEY) A study of the order-disorder transformation. Proc. phys. Soc. 49, 96-102 and 108-109.

(97) 1938 The atomic structure of alloys (Watt Anniversary Lecture for 1938). Pap. Greenock phil. Soc. 1938.

(98) A discussion on plastic flow in metals (opening address). Proc. R. Soc. Lond. A 168, 302-303.

(99) Forty years of crystal physics. In: Background to modern science (eds. W. Pagel & J. Needham), pp. 77-92. Cambridge University Press.

(100) The Physics Department, Manchester University. J. Univ. Manchester 1, 35-41.

(101) (With H. LIPSON) Structure of metals. Nature, Lond. 141, 367-368.

(102) General features of atomic structure of silicates : inferences to be drawn from them as to the structure of clay minerals (summary). Rep. a. Meet. Br. Ass. Advmt Sci. 108, 403.

(103) The structure of alloys (being the 39th Robert Boyle Lecture delivered before the Oxford University Junior Scientific Club, 11 June 1937). O.U.P. 1938 (5 pages).

(104) The structure of alloys. Der Feste Korper (1938), pp. 24-41. Leipzig: Hirzel.

(105) Röntgenstrahlen in der Industrie. Indian east. Engr. 82, 219.

(106) 1939 Magnets (R.I. Discourse, 5 May 1939). Proc. R Instn Gt Br. 30,783-787 and Engineering 147, 595-596.

(107) Patterson diagrams in crystal analysis. Nature, Lond. 143, 73-74.

(108) A new type of .X-ray microscope'. Nature, Lond. 143, 678.

(109) Atomic patterns of metals (Fourth Edward Williams Lecture, Institute of British Foundrymen, 13 June 1939). Foundry J. 12-17 June 1939, pp. 25-31; Fndry Trade J. 60, 506-508; Proc. Instn Br. Foundrym. 32, 25-31 and Engineering 147,788.

(110) 1940 The structure of a cold-worked metal. Proc. phys. Soc. 52, 105-109.

(111) 1940 (With A. J. BRADLEY & C. SYKES) Researches into the structure of alloys. J. Iron Steel Inst. 141, 63P-156P.

(112) The symmetry of patterns (title only) (R.I. Discourse, 3 May 1940). Proc. R. Instn Gt Br. 31, 149.

(113) (With A. J. BRADLEY & C. SYKES) The structure of alloys: X-ray and thermal analysis. Iron Steel, Lond. 13, (no.9), 305-307.

(114) (With A. J. BRADLEY) Part I-Investigation of equilibrium diagrams and theory of order-disorder transformation. Iron Steel, Lond. 13, (no.9), 308-310.

(115) 1941 Diffraction of monochromatic X-rays by crystals at high temperatures. Proc. R. Soc. Lond. A 179, 61-64.

(116) X-ray analysis in industry. J. scient. Instruln. 18, 69.

(117) 1942 Physicists after the War (Afternoon Lecture at the R.I., 26 March 1942). Proc. R. Instn Gt Br. 32, 253-271 and Nature, Lond. 150, 75-80 and 374.

(118) A model illustrating intercrystalline boundaries and plastic flow in metals. J. scient. Instrum. 19, 148-150.

(119) The X-ray microscope. Nature, Lond. 149, 4i0-471.

(120) A theory of the strength of metals (R.I. Discourse, 31 March 1942). Nature, Lond. 149,511-513.

(121) Index of X-ray diffraction data. Nature, Lond. 150, 738.

(122) 1943 Seeing ever-smaller worlds (R.I. Discourse, 12 March 1943). Proc. R. Instn Gt BT. 32,475-481 and Nature, Lond. 151, 545-547.

(123) (With H. LIPSON) A simple method of demonstrating diffraction grating effects. J. scient. Instrum. 20,110-113.

(124) Tensile strength of metals. Tek. Tid. i3, 403-407.

(125) (With R. H. PICKARD & A. FINDLAY) The place of scientists in the community. Chemy Ind. 62, 263.

(126) Metals. Endeavour II, 43-52; Trans. Can. Inst. Min. Metall. 46, 291-304; and Can. Mach. 54, 95-100.

(127) 1944 The mechanical strength of metals. General discussion on radiological testing. (Introductory Contribution.) Trans. VE. Cst Instn EngTs Shipbldrs 60, 299-306.

(128) Lightning calculations with light (R.I. Discourse, 24 March 1944). PROC. R. Instn Gt BT. 33, 107-113 and Nature Lond. 154, 69-72.

(129) The spirit of science. The Listener 10 February 1944, p. 147.

(130) Organization and finance of science in universities. The Political Quarterly 15, 330-341.

(131) Mr. F. Lincoln and the Cavendish Laboratory. Nature, Lond. 154,643.

(132) Metalle. Metallurgia Elect. 8, 20-26.

(133) 1945 Some problems of the metallic state (14th Andrew Laing Lecture). TTans. NE. Cst Instn Engrs Ship. 62,25-34 and Iron Steel, Lond. 18, 531-535.

(134) Magnetic materials (Lecture to Measurement Section, 18 May 1945). J. Instn. elect. Engrs 92, (Part I, General), 444-451.

(135) X-ray analysis: past, present and future (R.I. Discourse, 11 May 1945). Proc. R. Instn Gt BT. 33, 393-400.

(136) (With A. R. STOKES) X-ray analysis with the aid of the .fly's eye'. Nature, Lond. 156, 332-333.

(137) La Cohésion des Métaux Revue Métall, PaTis 42, 187-193.

(138) 1946 The Austin Wing of the Cavendish Laboratory. Nature, Lond. 158, 326-327.

(139) (With E. B. BOND) The Rutherford Papers in the Library of the Cavendish Laboratory. Nature, Lond. 158, 714.

(140) X-ray analysis in research and practice today (R.I. Discourse, 24 May 1947) Proc. R. Instn. BT. 33, 649-661.

(141) X-rays' part in metallurgical research. In: Science lifts the veil, pp. 33-37. London: British Council/Longmans, Green & Co.

(142) 1947 The relationship of the university and the technical college (lecture to the Association of Technical Institutions, 28 February 1947). Association of Technical Institutions: miscellaneous pamphlets: London (1947).6 pages.

(143) The conversion factor for kX units to Angstrom units. J. scient. Instrum. 24, 27; Phys. Rev. 72, 437; and (with E. A. WOOD) J. Am. chem. Soc. 69, 2919.

(144) 1947 Working models of crystals (R.I. Discourse, 9 May 1947). Proc. R. Instn Gt Br. 34, 103-108.

(145) 1947 (With J. F. NYE) A dynamical model of a crystal structure. Proc. R. Soc. Lond. A 190, 474-481 and Naturwissenschaften 34, 328-336.

(146) Effects associated with stresses on a microscopic scale. In: Proceedings of Symposium on Internal Stresses in Metals and Alloys, 15-16 October 1947. Inst. of Metals Monograph and Report Series No.5, 221-226.

(147) Recent advances in X-ray analysis. Paint Technol. 12, 421-425.

(148) 1948 Current researches in the Cavendish Laboratory (lecture to the Manchester Association of Engineers, 9 January 1948). Pamphlet, 10 pages.

(149) Organisation and work of the Cavendish Laboratory (course of 3 lectures given at the R.I., 4, 11,18 March 1948). Nature, Lond. 161, 627-628.

(150) The standards of advanced studies and research in science and technology. (Address to the 19th Annual Convention of the Yorkshire Council for Further Education, May 1948.) Pamphlet no. 36, 10 pages.

(151) Atomic rearrangement in the metallic state. Festkr. J. Arvid Hedvall 1948, 75-81.

(152) Recent advances in the study of the crystalline state. Science N. Y. 108, 455-463 and Advmt Sci., Lond. 5. 165-174.

(153) The Cavendish Laboratory (19th Autumn Lecture, Institute of Metals. 16 September 1948). J. Inst. Metals 75, 107-114.

(154) The yield point of a metal. Rep. Bristol Conf. .Strength of Solids' (1947). Phys. Soc. Lond. (1948) 26-29.

(155) 1949 Acceptance of the Roebling Medal of the Mineralogical Society of America. Am. Miner. 34, 238-241.

(156) (With W. M. LOMER) A dynamical model of a crystal structure. II. PROC. R. Soc. Lond. A 196, 171-181.

(157) Giant molecules (R.I. Discourse, 27 April 1949). PROC. R. Instn Gt Br. 34, 395-405 and Nature, Lond. 164. 7-10.

(158) Slip in metals. Physica 15, 83-91.

(159) The place of technological education in university studies. Conference on the Home Universities (1949). London: Association of Universities of the British Commonwealth, pp. 72-77 (and 24-25, 56-58).

(160) The strength of metals. PROC. Camb. phil. Soc. 45, 125-130.

(161) 1950 (With J. C. KENDREW & M. F. PERUTZ) Polypeptide chain configurations in crystalline proteins. PROC. R. Soc. Lond. A 203, 321-357.

(162) Famous experiments in the Cavendish Laboratory (R.I. Discourse, 12 May 1950). Proc. R. Instn Gt BT. 34, 626-633 and Nature, Lond. 166, 7-9.

(163) Microscopy by reconstructed wave-fronts. Nature, Lond. 166, 399-400.

(164) Science and the adventure of living (the Radford Mather Lecture, 25 October 1950 at the R.I.). The British Association of the Advancement of Science 7, 279-284. (Also Norwegian translation, 1952.)

(165) Riesenmoleküle. Usp. fiz. Nauk. 40, 108.

(166) Address at Electrical Research Association Annual Lunch, 1950. E.R.A./R720. pamphlet, 18 pages.

(167) 1951 (With G. L. ROGERS) Elimination of the unwanted image in diffraction microscopy. Nature, Lond. 167, 190-191.

(168) Crystallographic research in the Cavendish Laboratory (R.I. Discourse, 16 March 1951). Proc. R. Instn Gt BT. 35, 103-113 and Mitt. natuTf. Ges. Bern. 8, xi-xiii.

(169) 1952 The atomic patterns of everyday materials (the Keith Lecture, 2 April 1952. Royal Scottish Society of Arts). Edinb. J. Sci. Technol. photogr. Art, pp. 47-56.

(170) The Cavendish Laboratory archives (R.I. Discourse, 28 March 1952). Proc. R. Instn Gt BT. 35, 299-304 and Nature, Lond. 169, 684-686.

(171) (With E. R. HOWELLS & M. F. PERUTZ) Arrangement of polypeptide chains in horse methaemoglobin. Acta Crystallogr. 5, 136-141.

(172) (With M. F. PERUTZ) The external form of the haemoglobin molecule. I. Acta crystallogr. 5, 277-283.

(173) (With M. F. PERUTZ) The external form of the haemoglobin molecule. II. Acta crystallogr. 5, 323-328. .

(174) A device for calculating structure factors. Acta crystallogr. 5,474-475.

(175) 1952 (With. M. F. PERUTZ) The structure of haemoglobin Proc. R. Soc. Lond. A 213. 425-435.

(176) 1952 X-ray analysis of proteins (36th Guthrie Lecture, 12 March 1952). PROC. phys. Soc. 65B, 833-846.

(177) The Cavendish Laboratories of Cambridge University. Nucleo (Barcelona) 7,447-449.

(178) 1953 The discovery of X-ray diffraction by crystals (R.I. Discourse, 22 May 1953). PROC. R. lnstn Gt BT. 35, 552-559.

(179) X-ray analysis of the haemoglobin molecule. (In: A Discussion on the Structure of Proteins.) Proc. R. Soc. Lond. B 141, 6i-69.

(180) Budgets of the scientific departments of the University of Cambridge. Nature. Lond. 171.642-643.

(181) The X-ray analysis of protein molecules (Fison Memorial Lecture, 15 May 1953) Guy.s Hosp. Gaz. (new ser) 57, 242-2J,6.

(182) (With A. B. PIPPARD) The form birefringence of macromolecules. Acta crystallog 6.865-867.

(183) The bubble model of a metal structure (lecture delivered in Johannesburg, 18 July 1952). A. Proc. ass. tech. Socs S. .4fT. (July 1953), pp. 33-44.

(184) A centre of fundamental research. Physics to-day 6, 18-19.

(185) X-ray analysis of protein structure. lnst. intern. Chim. Solvay. Conseil. Chim., 9th Conseil. Brussels. 1953, pp. 100-109.

(186) 1954 X-ray studies of biological molecules (R.I. Discourse. 29 January 1954). Proc. R. lnstn Gt BT. 35, 685-696 and Nature. Lond. 174. 55-59.

(187) Models of metal structure (R.I. Discourse, 19 November 1954). PROC. R. Lnstn Gt BT. 35, 844-852.

(188) (With E. R. HOWELLS) X-ray diffraction by imidazole methaemoglobin. Acta crystallogr. 7, 409-411.

(189) (With E. R. HOWELLS & M. F. PERUTZ) The structure of haemoglobin, II. Proc. R. Soc. Lond. A 222. 33-44.

(190) (With M. F. PERUTZ) The structure of haemoglobin. VI. Fourier projections on the 010 plane. Proc. R. Soc. Lond. A 225. 315-329.

(191) (With M. F. PERUTZ) I. A Fourier projection of haemoglobin on the 010 plane. II. Sign determination by the isomorphous replacement method (abstracts of the meeting of the International Union of Crystallography, 21-28 July 1954). Acta crystallogr. 7, 653.

(192) 1955 X-rays and the molecule (Dunn Memorial Lecture to the Newcastle Section of Soc. of Chemical Industry, 11 May 1955). Chemy lnd. 1955, pp. 1164-1169.

(193) Obituary Notice: George Frederick Charles Searle. Physical Society Year Book, p.72.

(194) The Royal Institution-maintaining standards of popular exposition. The Times Educational Supplement 3 June 1955. p. 597.

(195) 1966 Masters of modern science. No. V: Michael Faraday-Our greatest experimentalist The Times Educational Supplement 9 March 1956. p. 302.

(196) The discovery of useful electricity (R.I. Discourse, 9 December 1955). PROC. R. lnstn Gt BT. 36, 278-289.

(197) The diffraction of X-rays (3J,th Silvanus Thompson Memorial Lecture). BT. J. Radiol. 29. 121-126.

(198) Information centre for science teachers. The Schoolmaster 6 July 1956.

(199) 1957 Schools Lectures at The Royal Institution: a new venture. Discovery 18. 66-67 (February 1957).

(200) The interference of waves (Trotter-Patterson Memorial Lecture). Trans. ilum. Engng Soc. Lond. 22, 175-181.

(201) Experimental demonstrations (R.I. Discourse. 29 March 1957). Proc. R. lnstn. Gt BT. 36.657-664 and Nature, Lond. 179, 1211-1212.

(202) X-ray analysis. New Scient. 3. 19-21 (21 November 1957).

(203) 1822-1957-135 Years of British Achievement: three great men of physics. The Sunday Times 14 July 1951. p. 15.

(204) 1958 Gemstones (R.I. Discourse, 31 January 1958) Proc. R.lnstn Gt BT. 37.1-15.

(205) The determination of the coordinates of heavy atoms in protein crystals. Acta crystallog.11,70-75.

(206) 1958 Interpretation of science to the public. Nature, Lond. 181, 807-808.

(207) 1958 An international survey of recent scientific research. New Scient. 3, 16-17 (27 March 1958).

(208) 1959 Treasures in the collections of apparatus at The Royal Institution (R.I. Discourse, 24 October 1958). PROC. R. Instn Gt BT. 37, 259-275 and Nature, Lond. 182,1541-1543 (1958).

(209) The diffraction of Röntgen rays by crystals. In: BeitTiige ZUT Physik und Chemie des 20 JahThunderts (ed. O. R. Frisch, F. A. Paneth, F. Laves & P. Rosbaud), pp. 147-151. Braunschweig: Friedr. Vieweg & Sohn.

(210) The contribution of the Royal Institution to the teaching of science (Presidential Address to the Science Masters' Association, 30 December 1958). Sch. Sci. Rev. 40, (no.141), 240-245.

(211) Talking and writing about science. (Based on an address given on the occasion of the award of the first Waverley Gold Medalby Butterworths Scientific Publications, 15 October 1956.) I.R.E. Trans. EWS-2, 69-72.

(212) 1960 Atoms and molecules (R.I. Discourse, 20 November 1959). Proc. R. Instn Gt BT. 38, 87-92 and Contemp. Phys I, 390-393.

(213) William Henry Bragg. New Scient. 7,718-720.

(214) British achievements in X-ray crystallography. Science N. Y. 131, 1870-1874.

(215) The Schools Lectures at the Royal Institution. Public Schools Appointments Bureau, bulletin no.89, July 1960, pp. 25-27.

(216) What constitutes life? The Times 19 July 1960, p. xiv (special number on The Royal Society Tercentenary).

(217) The nature of light. Trans. Illum. Engng. Soc. Lond. 25,6-10.

(218) Achievements in X-ray crystallography. Proc. K. ned. Akad. Wet. B. 63, 210-220.

(219) 1961 The development of X-ray analysis (Rutherford Memorial Lecture, 1960). PROC. R. Soc. Lond. A 262, 145-158.

(220) The development of X-ray analysis (R.I. Discourse, 3 March 1961). Proc. R. Instn Gt BT. 38, 526-543.

(221) Memoir of Maurice, Duc de Broglie. Proc. phys. Soc. 77, 1232.

(222) Adventures of the mind: what is life made of? The Saturday Evening Post, 7 October 1961, Vol. 234 (no.40), pp. 34-35, 54,62, 64.

(223) 1962 (With Mrs G. M. CAROE) Sir William Bragg, F.R.S. (1862-1942). Notes Rec. R. Soc. Lond. 17, 169-182.

(224) The analysis of protein molecules by X-rays. (Discourse at the R.I., 26 September 1961, to delegates attending the 9th General Assembly of I.C.S.U.) ICSU Rev. 4, 33-41.

(225) The growing power of X-ray analysis. In: Fifty years of X-Tay diffraction (ed. P. P. Ewald), pp. 120-135. N. V. A. Oosthoek's Uitgeversmaatschappij, Utrecht.

(226) Personal reminiscences. In Fifty years of X-Tay diffraction, pp. 531-539.

(227) 1963 X-ray analysis of biological moleculs (Presidential Address to the Madras Symposium, 1963). In: Aspects of protein structure (ed. G. N. Ramachandran), pp. 1-9. London: Academic Press.

(228) 1964 The start of X-ray analysis. Chemistry 40, 8-13.

(229) The difference between living and non-living matter (Saha Memorial Lecture).Sci. Cult. 30, 161-167.

(230) Minerals (R.I. Discourse, 28 February 1964). Proc. R. Instn Gt BT. 40,64-81.

(231) Reginald William James (1891-1964) (Obituary). Acta crystallogr. 17, 1615-1616.

(232) 1965 Reginald William James, 1891-1964. Biogr. Mem. Fellows R. Soc. Lond. 11, 115-125.

(233) First stages in the X-ray analysis of proteins. Rep. Prog. Phys. 28, 1-14.

(234) The history of X-ray analysis. Contemp. Phys. 6, 161-171 and Phys. Teach. 3, 295-300.

(235) The Schools Lectures at The Royal Institution. Science, N. Y. 150, 1420-1423.

(236) Birthday Greeting to F. Machatschki. (Foreword to 70th birthday tribute). Tschermaks miner. petrog. Mitt. 10, (3rd ser.) p. 3.

(237) The two cultures. Overseas vol 50, no.504, pp. 3, 5, 9. October 1965 (pub. Royal Overseas League). .

(238) 1966 The Royal Institution Lectures in Science for Members of the Administrative Class of the Civil Service. Contemp. Phys. 7,358-361.

(239) 1966 The art of talking about science. Science, N. Y. 154, 1613-1616.

(240) 1966 Reminiscences of fifty years' research (R.I. Discourse, 11 March 1966). Proc. R. Instn Gt Br. 41, 92-100.

(241) Half a century of X-ray analysis. Nobel Guest Lecture, I (read 1 June 1966). Ark. Fys. 40, 585-603 (published in 1974).

(242) 1967 The art of ta1king about science. Warine Technol. 4, 258-261.

(243) Sir Kenneth Lee. (Obituary) Nature, Lond. 216, 945.

(244) The spirit of science (James Scott Lecture). Proc. R. Soc. Edinb. A 67, 303-308.

(245) Introduction to c A Discussion on the Structure and Function of Lysozyme'. Proc. R. Soc. Lond. B 167, 349.

(246) Reminiscences of fifty years of research (Redding Lecture, Annual General Meeting of the Franklin Institute, 18 January 1967). J. Franklin Inst. 284, 211-228.

(247) William Henry Bragg. The Encyclopaedia Americana.

(248) 1968 Professor P. P. Ewald (appreciation on 80th birthday). Acta crystallogr. A24, 4.

(249) X-ray crystallography. Scient. Am. 219 (1), 58-70.

(250) More on the art of talking about science (editorial) Nucl. Appl. Technol. 4, (no.5), 282-283.

(251) Foreword to The double helix by J. D. Watson. New York: Atheneum.

(252) How a secret of life was discovered (article about the double helix).The Times May 1968, p. 13.

(253) The white-coated worker. Punch. 01. 255 (no.6679), 11 September 1968, pp.352-354.

(254) 1969 The early history of intensity measurements. Acta crystallogr. A 25, 1-3.

(255) The history of X-ray analysis. In: Sources of physics teaching pt 3, pp. 74-84. London: Taylor & Francis.

(256) What makes a scientist. (Lecture for Civil Servants, 11 December 1968.) Proc. R. Instn Gt Br. 42, 397-410.

(257) 1972 Dame Kathleen Lonsdale. (Obituary.) Acta crystallogr. A 28,226.

(Some additional articles and reports of lectures are listed in William Henry Bragg and William Lawrence Bragg: A Bibliography of their Non-Technical writings (1978), Office for History of Science and Technology, University of California, Berkeley.)

PART II PETER DEBYE

Peter Debye — A Life for Science

Eric Courtens

Laboratorie des Verres, UMR 5587 CNRS, Université Montpellier 2
F-34095 Montpellier Cedex 5, France

I. INTRODUCTION

Any graduate student in physics or chemistry should have encountered the name Debye frequently. Some might be surprized to learn that it always refers to the same Peter Debye. In solid state physics there is the well known Debye theory for the specific heat of insulators, with the related Debye temperature and Debye frequency. In diffraction and scattering, there are the Debye-Scherrer rings of powder diagrams and the Debye-Waller factor, of daily use in spectroscopy. The practical unit for the dipole moment of molecules is the debye, 1 D = 10^{-18} e.s.u. = 3.34×10^{-30} C·m. Another common concept is that of Debye relaxation. Chemists will have studied the standard Debye-Hückel theory of electrolytes and the associated Debye length. The diffraction of light produced by an ultrasonic grating is named after Debye and Sears. In cryogenics, Debye was also first to propose adiabatic demagnetization as a practical method to achieve cooling well below 1 K. Finally, at Cornell University, Debye formed a group which produced much of the seminal work on light scattering before the advent of the laser.

These are but a few of the highlights. In fact, Debye substantially influenced the early developments of quantum mechanics, phonon physics, X-ray scattering, and molecular physics. One of his recurring themes was the interaction of waves with matter. A large part of his work was performed in direct relation with experiments. Hence, Debye also invented or stimulated important new experimental techniques. It is very proper to have selected him for the present occasion. Among the many outstanding physicists of his time, he certainly was a leader in the birth of material science. Furthermore, in 1912 he discovered the permanent electric dipole [1]. In the same paper, which he himself carefully called 'preliminary', he evoked the possibility of a spontaneous polarization. His formula did not include the reaction field, later discovered by Lars Onsager [2]. This however was ten years before the observation in Rochelle salt [3], usually considered as the birth of Ferroelectricity.

Besides his exceptionally fruitful research, Debye was also an outstanding teacher, as direct witnesses have reported. The remarkable clarity of Debye's mind is evident upon reading any of his long papers. As this fact was widely known, Debye received many invitations to deliver lectures series at prestigious institutions, this already much before he was awarded the 1936 Nobel Prize for Chemistry.

Finally, Debye was also a very efficient leader and organizer. This was recognized early in his career as he was asked to direct the entire Physics Institute in Göttingen at age 32. He attracted excellent coworkers and knew how to stimulate them. He also stimulated many people of his less immediate surroundings. He later was always called to direct large laboratories, in particular at the Federal Polytechnic in Zurich (Fig. 1) and at the University in Leipzig. The most notable instance was certainly the foundation and building of what Debye himself named the '*Max Planck Institute*' in the Berlin suburb of Dahlem. This construction was only completed in 1938, while Debye, compelled by the political situation in Germany, and presumably with considerable regret, finally emigrated to the United States early in 1940.

There exist only a few biographies of Peter Debye, notably that by Mansel Davies [4], and in German that by Georg Busch [5] which may be of special interest to ferroelectricians. A recent collection of articles in three languages (Dutch, German, and English) offers additional information, particularly on the early years [6]. A publication of selected *Collected Papers of Peter J. W. Debye*, translated into English, appeared on the occasion of his 70th birthday [7], while a fairly complete list of scientific publications is found in [4]. The publications of course provide first hand information on the scientific work. References to other aspects are found in

[4-6]. Additional direct sources can be consulted in numerous archives. I have studied those of the Max-Planck-Gesellschaft (MPG) in Berlin, of the ETH in Zurich, and I read some correspondence with Arnold Sommerfeld kept at the Deutsches Museum in Munich. Unless otherwise noted, the information below can mostly be found in [4-7] or in good dictionaries. In the few instances were these sources disagree, the hopefully correct version was checked more directly.

II. A BIOGRAPHICAL SUMMARY

Pierre Joseph Wilhelm Debije was born in Maastricht, the southernmost tip of the Netherlands, on the 24th of March 1884. He was the first son of a young worker family. His father, Joannes Wilhelm Debije (1859-1937), was ironsmith at a small factory. His mother, Maria Ruemkens (1859-1940), had a dominant character, eager of social ascent. She taught Pierre to read when he was only four. He went to a catholic primary school where he was quickly recognized as a talented child. The brothers prepared him for admission to the local high school, which apparently was an excellent one. Pierre finished there at age 17 and he was ranked first for the entire Province of Limburg. Rather than taking a clerk job, as first intended, he started engineering studies at the nearby Technische Hochschule in Aachen, just across the border to Germany.

He registered in Machine Engineering with specialization in Electrotechnics, under the name Peter Debye. He discovered that he had more interest in sciences than in engineering, and spent extra time in the physics laboratory of Max Wien. He was also remarked by Arnold Sommerfeld (1868-1951) who taught Technical Mechanics. Sommerfeld invited Debye and another student to his house for discussions on physics, and he became very impressed by Debye's capabilities. When he first could hire an assistant in 1904, he selected the 20 year young Debye who was still a student. Debye received his engineering degree with excellence in 1905. He continued as Sommerfeld's assistant. In 1906 Sommerfeld was called to the chair of Theoretical Physics at the University of Munich. He asked Debye to join, and he suggested to him to start a doctorate.

Debye obtained the Doctor's degree *summa cum laude* in 1908 and the Habilitation in 1910, both from the Ludwig-Maximilian University in Munich under the supervision of Sommerfeld. He became *Privat-Dozent*. Munich was an exceptionally stimulating place for physics. Debye was participating in full to the scientific life, with interests both in theory and experiments. Indeed, Sommerfeld had his own experimental laboratory. However, in spring 1911, the Associate Professorship for theoretical physics that Einstein had occupied at Zurich University became vacant. Debye was supported by Sommerfeld and Röntgen. The position was offered to the young Debye who moved to Zurich in April 1911. This is where he wrote the theory of specific heats and discovered the permanent dipole moment of molecules. He relates the latter in a touching postcard addressed to his master Sommerfeld on the very same day that the paper (1) was mailed to the journal (Fig. 2).

A year later he obtained a Full Professorship in Theoretical Physics at the University of Utrecht where he moved in April 1912. During that period, Debye married in Munich a Bavarian lady, Matilde Alberer (1887-1977), on April 10, 1913. In Utrecht, he worked out the theory of dielectric dispersion and he wrote a very important paper on the effect of lattice vibrations on diffraction intensities. However, Debye was not fully happy in Utrecht. He had stated in his inaugural lecture that '*mathematical physics is in the first place physics and it could not exist without experimental investigations*'. However, contrary to earlier hopes, he could not obtain in Utrecht his own laboratory facilities any better than in Zurich.

An occasion came in the course of 1913 from the prestigious University in Göttingen. Voigt wanted to discontinue his duties as head of the theoretical section. A personal professorship was created for Debye starting in Sep. 1914. As Riecke who directed the entire Institute passed away in June 1915, Debye obtained then a chair for Theoretical *and* Experimental Physics in Feb. 1916, together with the direction of the Institute. His son Peter Paul Rupprecht was born on the 7th of March 1916. Debye could start experimental work on X-ray diffraction. He investigated powders together with his Swiss assistant Paul Scherrer (1890-1969). On the theoretical side, he calculated hydrogen-like spectra, he produced a quantum theory of the Zeeman effect predicting the space quantization of electron orbits six years before the Stern-Gerlach experiment, and he worked out a theory for the van der Waals cohesion between molecules. Unfortunately, by the end of the First World War, the conditions in Göttingen had deteriorated on several counts, and in particular they became very precarious materially.

At the ETH in Zurich, Pierre Weiss was directing the Physics Institute. He had accepted a position in Strasbourg in 1919. After an unsuccessful public call for candidates, the authorities personally invited Debye. He first set conditions to improve material facilities and the number of staff members before he would accept the appointment. Debye became Director of the Institute in 1920. On the 8th of March 1921 his daughter Mathilde Maria was born. Debye considerably upgraded physics teaching and research at the ETH. He developed laboratory facilities, built up lecture demonstrations, and established a strong colloquium tradition. A memorable event was the 1926 lecture for which Erwin Schrödinger from the University, on the invitation of Debye, had worked out his wave equation. Debye main subjects in Zurich were the theory of electrolytes together with Erich Hückel (1896-1980), and the structure of molecules and their dipole moments. He also wrote a paper on demagnetization at low temperatures in which he proposed for the first time to use adiabatic cooling to achieve sub-Kelvin temperatures.

The next call came from Leipzig in 1927. Both the theoretical and experimental physics positions had become free there. The former was offered to Heisenberg (1901-1976), the latter to Debye together with the Direction of the Physics Institute. This was an attractive offer which Debye first used to press for further improvements at the ETH, among which a lighter teaching load for himself. As this did not succeed, Debye joined Leipzig by the end of September 1927. His fame attracted a large number of students to his lectures. His research turned almost completely to physical chemistry and molecular physics, in particular to diffraction and dielectric studies of molecular structures and of liquids. It is there that Debye first obtained clear X-ray diffraction pictures of molecules. He also wrote his first book on '*Polare Molekeln*' (Leipzig, S. Hirzel 1929). He initiated yearly 'small and intimate' discussion meetings that became known as the '*Leipziger Vorträge*'. These were published both in German and in English and ran from 1928 (on *Quantum Theory and Chemistry*) to 1933 (on *Magnetism*). Also during these years the Debye-Sears effect was first observed, together with Francis W. Sears, then Assistant Professor at M.I.T. (1932).

The last Leipzig years of Debye coincided with a political cataclysm that eventually produced a worldwide disaster. Hitler became chancellor on Jan. 30, 1933. He rapidly took full power and ruthlessly enforced his racial biases. Within two months there was a major exodus of Jewish scientists. Altogether, physics lost from 20 to 30 percent of its staff, among which many Nobel Prize winners. Debye became dissatisfied in Leipzig. Being neither German nor Jew, and optimist by nature, he did not feel personally threatened. However, as Institute Director he certainly noticed the dismissals of Jewish civil servants that were taking place. On Nov. 28, 1933, Max Planck from Berlin sent an attractive offer to Debye. Planck presided the Kaiser-Wilhelm-Gesellschaft (KWG), the research organization that transformed into the MPG after the Second World War. He asked Debye whether he would consider leading the Kaiser-Wilhelm-Institute (KWI) for Physics. The former director was Albert Einstein, who ran the Institute essentially as a grant agency from his home study. Einstein was abroad when Hitler took power, and he never returned to Berlin. Max von Laue, who seconded Einstein, effectively became acting director. The offer was attractive as the Rockefeller Foundation in New York had promised in 1930 a large sum of money to build an Institute for Physics in the Berlin suburb of Dahlem. Debye took a week to consider Planck's letter, for him an unusually long thinking time. He responded favorably on Dec. 8, 1933, explaining that he had good feelings about the importance and outcome of the research that could be achieved at such a facility, and that he would be pleased to direct it, especially if he could find more time for research [8]. The directorship was to be connected with a professorship at Berlin University, the one of Nernst who had just retired, and Debye insisted for a minimal teaching load. By his promise to Planck, whom he highly respected, Debye had tied himself for what became a very long time, as he definitely wanted to keep his word.

The acceptance of Debye was for Planck just a first step. He had to convince the Rockefeller Foundation to make the donation in spite of Hitler. He also had to obtain the commitment from the Ministry for the running costs. Finally, he had to see that all went fine with Debye's professorship. In a letter to Debye on Feb. 11, 1935, Planck first announces the success [9]. Debye was then on leave to the University of Liège in Belgium where he held a Francqui Chair from Oct. 1934 to the end of March 1935. On Oct. 1, 1935, Debye became officially the Director of the KWI for Physics, and only in December did he obtain a satisfactory contract from the University. During the winter semester, Debye was asked to remain director of the Institute in Leipzig, without teaching. By then the construction in Dahlem had started. Debye moved into the director's villa, adjacent to the future Institute, on July 15, 1936.

In November 1936, Debye was awarded the Nobel Prize for Chemistry '*for his contributions to our knowledge of molecular structure through his investigations on dipole moments and on the diffraction of X-rays and electrons in gases*' [10]. Among the heap of correspondence there was a telegram from Sommerfeld which Debye answered immediately. About the

Prize, he wrote '*...I am very pleased, as chemistry has always been my secret love*', and further '*... my parents seem to enjoy it deeply. That they are so happy ... rejoices me the most of all.*'[11].

These years were not so good for research. Only one publication from 1934 was selected for the *Collected Papers* [7], and none from 1935 to 1941. Not only the political situation must have interfered, but Debye was absorbed with his Dahlem plans. He pursued work on molecular and dielectric properties. In Liège he wrote on molecular rotations in liquids. Another subject was adiabatic cooling, of direct application as he wanted a full cryogenic installation in the new Institute. Interestingly, he wrote on nuclear physics, also of interest for Dahlem where a high voltage tower capable of 2.8 MV at 3 mA was being planned. In a paper of 1937, Debye describes the superb technical installations of the new KWI [12]. The Institute was officially inaugurated on May 30, 1938. Above the main entrance Debye had a large inscription placed: '*Max Planck Institut*'. It must have pleased Max Planck who was present and had by then transmitted to Carl Bosch the Presidency of the KWG. It however highly displeased the Nazi officials who knew of Planck's political opinions and of his disagreement with Hitler.

The research activities of the Institute were divided into three groups. Debye ran very high voltage experiments, the cryogenic laboratory, dipole measurements, electron diffraction from gases, and thermo-diffusion in liquids. Laue, who remained adjunct director, investigated X-ray diffraction. Schüler performed spectroscopy on magnetic and quadrupolar moments of nuclei. During that period Debye wrote papers on electron-spin resonance and relaxation, on the quasi-crystalline structure of liquids, and on techniques to measure electron interferences. In 1939 he wrote a theory for isotope separation by thermo-diffusion, a method due to Clusius, an experimentator then in Munich. All this reflects the large variety of Debye's research and the fact that his subjects were always of great current interest and at the forefront of the scientific knowledge. For a brief lapse of time Debye gave his usual magnificent thrust to his colleagues and he must have felt very satisfied with the realisation of his grand project. However, this was without counting with the evil regime and the folly of men.

On Sep. 1, 1939, Germany invaded Poland. On Sep. 3, France and Great Britain declared war to the Reich. On Sep. 16, Dr. Telschow, the Secretary of the KWG visited Debye at the Institute to ask him whether he would be willing to abandon the Dutch citizenship. Would he not, he would have to resign as Director. The armed forces wanted to use the Institute for their research and could not tolerate a foreign director. They obviously had in mind the development of an atomic bomb [13]. Debye answered that he would definitely not abandon his citizenship, and that he also refused to resign. A proposed alternative was then that Debye would '*write a book*' without access to the Institute. Such an idleness he could not accept. A compromise was then found. Debye had kept since at least Nov. 1938 a standing invitation from John Kirkwood to deliver the Baker Lectures at the Chemistry Department of Cornell University in Ithaca, New York [14]. He had already answered in Apr. 1939 that he would accept the invitation for the second term 1939-1940 [14]. He immediately reactivated this contact. By October 31st, he obtained for this purpose a paid leave from the Ministry. This was further confirmed by President Bosch who granted the leave for 9 months, from December to August. In a long letter to Sommerfeld, on Dec. 30, 1939, Debye describes the situation [15]. About his refusal to resign he says '*I will not let people say that I ran away from it*'. According to that letter, he was not yet certain to leave, as another proposal was for him to use the separate cryogenic laboratory, but the financing was unclear. He finally left Berlin during the first half of Jan. 1940. He travelled over Switzerland to Italy, and sailed on an Italian liner, the '*Conte di Savoia*', from Genoa to New York on Jan. 22, 1940 [16]. Wife and daughter stayed behind, while his 23 year old son, also a scientist, was already in the USA where he had spent summer vacations in Connecticut and had moved to Oberlin College near Cleveland following the outbreak of war [14]. Debye's wife became ill and spent an extended period in Switzerland before joining him in Cornell. His daughter spent the war time in Germany, for a few years in their Berlin house.

The Baker Lectures at Cornell were for a full term, *i.e.* sixteen weeks, at two hours per week. Debye spoke about the determination of molecular structures by means of X-ray and electron scattering. In the meantime, Germany had invaded neutral Holland in May 1940. Soon Debye received offers of professorships from various American universities. In July 1940, he decided to stay at Cornell and to accept for three years the Chairmanship of the Chemistry Department. In a letter to Dr. Telschow he explains this situation [16]. His paid leave was then extended, but only until the end of March 1941. The correspondence shows that Debye deeply regretted his Institute in Berlin. The German authorities were also cautious to remain correct with Debye, and they never demoted him. During 1942, the military lost interest in the KWI for Physics. It became clear that no bomb could be made with the then feasible efforts, but only an 'uranium burner' as a nuclear reactor was called [13]. In consequence, Heisenberg was named Director *at* the Institute, while Debye remained formally Director *of* the Institute, this until it was transferred in 1946 to the newly founded MPG.

In Cornell, Debye was Professor of Chemistry until his retirement in 1952. He also chaired the Department until he resigned in 1950. As a foreigner, and owing to his remaining contacts with Germany, it seems he was carefully kept out of classified projects, such as atomic research. However, he was soon called as consultant on important issues, most notably for work on rubber substitutes. In Dec. 1941, Japan bombed Pearl Harbor forcing the United States into the war. Malayan rubber was under Japanese control, and artificial substitutes became strategic. Debye started using light scattering to determine molecular weights and shapes of polymers in solution. This was twenty years before the laser, so early in fact that much of this excellent pioneering work seems to have been forgotten by many later researchers. The early papers mention the 'Office of Rubber Reserve' and contracts with the 'Synthetic Rubber Program' of the US Government. The experimental setups were constructed to a large part by Debye's son, Peter Paul. In 1946 Debye took American citizenship, and he became Todd Professor of Chemistry in 1948. His work extended to colloids, in particular to micelles. It was entirely supported by contracts, on which Cornell imposed the usual hefty overhead. When Debye became emeritus, he had not accumulated enough service at Cornell to obtain a pension. By necessity, but also by incline, he continued his contract work unabated.

This last period of Debye's life is marked by unusual vigor and creativity. At age 80, Debye still employed four researchers out of his grants. His interests turned in part to critical phenomena, in particular to critical opalescence. He wrote well over fifty papers after retirement, among which some major ones on binary mixtures or on the spectral widths produced by density fluctuations. This activity was recognized by not fewer than five distinctions which he received when he was over seventy years old. He did all this together with a remarked presence at many international meetings, and an average of two European tours per year. In April 1966 he suffered a heart failure on departing from Kennedy Airport. He recovered from that strong warning, but was hit again in October. It is said that at one point he insisted to have a telephone installed under his oxygen tent in order to keep contact with his laboratory. Debye finally passed away on November 2, 1966.

This final chapter amply justifies the title 'A life for science'. It would be wrong to conclude that science was Debye's only interest. He was very much a family man. In 1951, the Debye's even took on the education of their two young grandsons, Norwig and Nordulf, who had arrived from Germany. At that time, Debye insited on a clear separation between work and home [17]. He also kept several hobbies, like fishing and gardening. Debye certainly had an exceptionally full life. In closing, it is probably most appropriate to cite Prof. Henri Sack, who was close to him for several decades [4] '... *I believe I can speak in the name of all those who had the privilege of having been closely associated with Professor Debye, in saying that we are grateful for all he has given us and that he will live in our memory as a brillianr scientist, a great teacher, a fatherly and helpful adviser, and, above all, as a happy man.*'

III. THE MAIN SCIENTIFIC CONTRIBUTIONS OF PETER DEBYE

The contributions of Debye have been numerous and rather diverse. A few very important ones are even not widely known as being originally due to Debye. It is thus quite useful to present a recapitulative list.

Quantum Theory

In justifying the blackbody radiation law, which he had first obtained empirically, Planck invoked quantized energy exchanges with the 'walls' of the blackbody. It is Debye who first wrote a treatment [*Ann. Phys.* **33**, 1427 (1910)] where it is the modes of the vacuum field that are quantized. He used in 1912 a related concept for the phonons in his theory of the specific heat [*Ann. Phys.* **39**, 789 (1912)].

Debye was also first in writing the quantum condition usually attributed to Bohr that the integral of action on a closed loop must be an integer multiple of Planck's constant h (on Feb. 10, 1913, cited in [4]).

In his remarkable paper on X-ray intensities [*Ann. Phys.* **43**, 49 (1914)], Debye showed that the measurement of the temperature dependence of the scattering would settle the question of the existence of the zero point energy, which was then much debated.

In a quantum treatment of the Zeeman effect, Debye was first to show, simultaneously with Sommerfeld, the space quantization of the electron orbits [*Phys. Z.* **17**, 507 (1916)]. At the same time [*Phys. Z.* **17**, 512 (1916)] he produced a very clear theory of hydrogen-like spectra.

Finally, as soon as Compton reported his observation of inelastic X-ray scattering, Debye published a crisp explanation for it [*Phys. Z.* **24**, 161 (1923)] including the wave-particle duality and the relativistic calculation of the conservation of energy and momentum. Debye says he had this ready in his drawer. One might speculate that he delayed publishing possibly because he was waiting for an experiment at the ETH. In any case, Compton produced the explanation simultaneously, for which he shared the Physics Nobel Prize with Wilson in 1927.

Phonon Physics

Debye wrote the theory of specific heats soon after taking Einstein's position in Zurich [*Ann. Phys.* **39**, 789 (1912)]. He was not aware of the work of M. Born and T. von Kármán in Göttingen, which took place simultaneously. A year latter Debye showed the importance of anharmonic potentials in producing thermal expansion and in causing the phonon mean free path that controls thermal conductivity. This he presented at the Apr. 1913 Wolfskehl Conference in Göttingen [*Math. Vorlesgn. Göttingen* **VI**, B.G. Teubner, Leipzig]. It so impressed David Hilbert and other faculty members that it resulted in Debye's call to Göttingen. A fine historical account was given by Debye himself at the 1963 Lattice Dynamics Conference in Copenhagen. It is found in the Proceedings [R.F. Wallis, Ed. (Pergamon Press, Oxford, 1965) p. 9].

X-Ray Diffraction

This is one of the main themes of Debye. It started with a very detailled description of the effects of thermal motions on scattering intensities [*Ann. Phys.* **43**, 49 (1914)]. Debye's treatment needed correction by a factor 2 [18], as later shown by I. Waller in his dissertation (Uppsala, 1925). Hence the name Debye-Waller factor which is now in use. Debye was also first to start working on atomic form factors, first alone [*Ann. Phys.* **46**, 809 (1915)] and later together with Scherrer [*Phys. Z.* **19**, 474 (1918)], a paper in which he asks the question of the charge distribution between the various atoms of a solid. On the experimental front, there is of course the invention of the powder method with Scherrer [*Phys. Z.* **17**, 277 (1916); *id.* **18**, 291 (1917)], but also the work on gases which demanded considerable apparative developments, together with Bewilogua and Ehrhardt [*Phys. Z.* **30**, 84

(1929)]. In that last paper the first clear observation of x-ray interferences from isolated molecules is reported. Finally there is the important work together with Menke on the inner structure of liquids [*Phys. Z.* **31**, 797 (1930)].

Dipole Moments, Relaxation, and Polarizability

This is another main theme of Debye. It started in Zurich (1) with the proposal that dielectrics can contain permanent electric dipoles. In the following year, Debye used this idea to describe dielectric relaxation [*Verh. dt. phys. Ges.* **15**, 777 (1913)]. The expressions he obtained have become classic. Debye also treated the origin of the van der Waals attraction for quadrupolar molecules [*Phys. Z.* **21**, 178 (1920); *id.* **22**, 302 (1921)]. It does not apply to noble gases, which have no quadrupole, and for which the attraction is due to the dispersion force as later shown by London in 1926. In a long article for the *Handbuch der Radiologie* [Vol.**VI/2**, E. Marx, Ed., Leipzig 1925], Debye gives a full exposition of the dipole moment theory and he extends Langevin's treatment of the Kerr effect, showing how it gives information on molecular anisotropies. The study of dipole moments became a classic for the determination of molecular structures, and much later Debye himself used it in his work on polymers [*J. Chem. Phys.* **19**, 589 (1951), with F. Bueche].

Electrolytes

This is the fourth subject to remain forever associated with the name of Debye. It started in the second Zurich period, with two long papers written together with his assistant Erich Hückel on the theory of strong electrolytes [*Phys. Z.* **24**, 185 (1923); *id.* **24**, 305 (1923)]. The problem was that the van't Hoff laws for the osmotic pressure and related properties, and the Arrhenius law for the conductivity, which applied to weak electrolytes, were strongly in error for fully dissociated ones like the usual salts or the strong acids. Debye and Hückel combined Poisson's equation and Boltzmann's equilibrium condition to obtain a characteristic length which is the basic parameter of the problem. It is now called the Debye screening length and it finds application in other problems as well, such as in semiconductors or for unbound vortices. The first paper calculates the osmotic pressure, the vapor pressure lowering, the freezing point depression, and the boiling point elevation for the solutions. The second paper evaluates the concentration dependence of the ion mobilities, a much harder problem, as it involves deformation of ionic clouds and electrophoretic forces due to the counter movement of the ions. The paper includes an impressive comparison with the available experimental results of the time.

There followed many other papers on electrolytes. There is one with Linus Pauling [*J. Am. Chem. Soc.* **47**, 2129 (1925)] and another with McAulay [*Phys. Z.* **26**, 22 (1925)], the latter explaining the 'salting-out' of organic solutes upon addition of salt to water. That problem had already been addressed by Debye in an earlier publication in Dutch [*Hand. v. h. XIX Nederlansch. Natuur en Geneeskung, Congres*, Maastricht (1923)]. Together with Falkenhagen, Debye treated the frequency dependence of the conduction, an aspect that later gave information on the relaxation times of the ionic clouds [*Phys. Z.* **29**, 121 (1928); *id.* 401 (1928)]. With all this work, Debye had definitely entered the realm of physical chemistry where he made an everlasting impact.

Magnetism and Cryogeny

The name of Debye is not often associated with studies on magnetism. As already mentionned above, his first paper on the subject was however on the quantization of electron orbits in the Zeeman effect, a far from trivial frontier subject in 1916. There followed a series of papers on magnetism in the final years at the ETH, in particular on paramagnetic molecules [*Physica, Eindhoven* **5**, 377 (1925)] and on a magnetoelectric effect [*Z. Phys.* **36**, 300 (1926)]. The long article for the *Handbuch der Radiologie*, already cited above, also describes magnetic molecular properties. However, the most remarkable was a short paper containing 'Some remarks on magnetization at low temperatures' [*Ann. Phys.* **81**, 1154 (1926)]. It is in this work that Debye proposes for the first time to use adiabatic demagnetization to reach very low temperatures, this with gadolinium sulfate whose magnetization had been measured by Kammerlingh Onnes at 1.3 K. He returned to this theme repeatedly, in *Phys. Z.* **35**, 923 (1934), in *Res. Prog.* **2**, 89 (1936), and finally reporting results in *Ann. Phys.* **32**, 85 (1938). At the same time he developped interest in paramagnetic relaxation [*Phys. Z.* **39**, 616 (1938)] and he even calculated spin resonance spectra in a cubic crystal field [*Ann. Phys.* **32**, 85 (1938)].

Light Scattering

Debye's first work on light scattering was performed together with Sears 'On the scattering of light by supersonic waves' [*Proc. Nat. Acad. Sci. U.S.A.* **18**,

409 (1932)]. He suggested this experiment to Sears during a visit at M.I.T. in April 1932. One must remember that Brillouin's seminal paper had been published in 1922. The first observation of spontaneous Brillouin light scattering by sound only occured in 1930, by E. Gross. Indeed, this was a difficult experiment at the time, making severe demands on optical sources and instruments. Debye's experiment with Sears, the diffraction of light by ultrasound launched from a quartz crystal into a liquid, turned out to be a relatively easy one in comparison. It gave an immediate measurement of the speed of sound. As Debye understood perfectly all the interference phenomena involved, he provided a crisp and complete explanation for the observations.

In Cornell, Debye started by developing light scattering for the study of polymers in solution. The first paper was on reaction rates [*Trans. Electrochem. Soc.* **82**, 265 (1942)]. It was followed by studies of molecular weights [*J. phys. colloid Chem.* **51**, 18 (1947)] and shapes {report in [7]}. Debye then turned to soap solutions and the study of micelles [*e.g.*, *Ann. N.Y. Acad. Sci.* **51**, 575 (1949)], but he intensely pursued the work on polymers, with at least 20 original papers on that subject until 1960. In 1959 he turned his attention to critical opalescence [*J. Chem. Phys.* **31**, 680 (1959)]. Citing Prof. B. Widom [7], Debye '*essentially rederived (without fully realizing it) the Ornstein-Zernike theory*'. He then initiated systematic light scattering measurements of this phenomenon, in particular in binary mixtures together with B. Chu and H. Kaufmann [*J. Chem. Phys.* **36**, 3378 (1962)]. There followed many papers, inparticular one on the spectral widths of critical opalescence [*Phys. Rev. Lett.* **14**, 783 (1965)]. Debye clearly perceived light scattering as a tool, and on the year he passed away, he published a paper with precisely that title [*Pure Appl. Chem.* **12**, 23 (1966)].

IV. LECTURESHIPS, AWARDS AND OTHER RECOGNITIONS

This section is principally based on a document that Debye himself wrote in Feb. 1965 [19], only one-and-a-half year before he died. It has just been slightly complemented with other reliable sources.

Extended Lecture Engagements

Debye held at least a dozen lecture series besides his permanent teaching duties. He lists lectures at the following institutions:

- Massachusetts Institute of Technology, Cambridge, MA
- California Institute of Technology, Pasadena, CA

- Columbia University, New York, NY
- University of California
- University of Paris, Paris, France
- University of Liège, Belgium
- University of Oxford, Oxford, England
- University of Notre Dame, Notre Dame, IN
- Harvard University, Cambridge, MA
- University of Southern California, Los Angeles, CA
- University of Michigan, Inst. of Science & Tech., Ann Arbor, MI

Medals, Prizes, and Orders

- Rumford Medal, Royal Society, London (1930)
- Faraday Medal, Royal Society of Chemistry, London (1933)
- Lorentz Medal, Royal Dutch Academy, Amsterdam (1935)
- Nobel Prize in Chemistry, Sweden (1936)
- Franklin Medal, Franklin Institute, Philadelphia (1937)
- Knight of the Order of the Dutch Lion (1937)
- Willard Gibbs Medal, American Chemical Society, Chicago (1949)
- Max Planck Medal, German Physical Society (1950)
- Commander of the Order of Leopold II (1956)
- Kendal Award, American Chemical Society, Miami (1957)
- Nichols Award (1961)
- Priestley Medal, American Chemical Society (1963)
- High-Polymer Physics Prize (1965)
- National Medal of Science, USA (1965)

Membership in Academies

- National Academy of Sciences, Washington, DC
- New York Academy of Sciences, New York, NY
- American Academy of Arts and Sciences, Boston, MA
- Franklin Institute, Philadelphia, PA
- Royal Dutch Academy, Amsterdam, Holland
- Royal Institution of Great Britain, London, England
- Royal Society, London, England
- Royal Danish Academy, Copenhagen, Denmark
- Academies of Berlin, Göttingen, and Munich, Germany
- Academy of Brussels and Liège, Belgium
- Royal Irish Academy, Dublin, Ireland

- Papal Academy, Rome
- Indian Academy, Bangalore, India
- National Institute of Science, India
- Real Sociedad Española de Física y Química, Madrid, Spain
- Academia de Ciencias, Madrid, Spain
- American Philosophical Society, Philadelphia, PA
- Academy of Sciences of the USSR, Moscow, USSR
- Hungarian Academy of Sciences, Budapest, Hungary
- Argentine Academy of Sciences, Buenos Aires, Argentina.

Honorary Doctor Degrees

- University of Brussels, Belgium
- University of Liège, Belgium
- Oxford University, England (1935)
- University of Sofia, Bulgaria
- Harvard University, Cambridge, MA (1936)
- Polytechnic Institute of Brooklyn, Brooklyn, NY
- St. Lawrence University, Canton, NY
- Colgate University, Hamilton, NY
- Eidgenössische Technische Hochschule, Zurich, Switzerland
- Boston College, Boston, MA
- University of Notre Dame, Notre Dame, IN
- Providence College, Providence, RI
- Clarkson College of Technology, Potsdam, NY
- Technische Hochschule Aachen, Germany
- Gustavus Adolphus College, St. Peter, MN
- Seton Hall College, NJ
- College of the Holy Cross, Worcester, MA
- University of Mainz, Germany.

Acknowledgements

References and additional information were kindly provided by Dr. Peter Paul Debye, and Profs. Nordulf Debye (Towson State U.), Hans-Friedrick Eicke (U. of Basel), Robert Pohl (Cornell U.) and Gijs van Ginkel (Debye Institute, U. of Utrecht). Original documents were consulted at the Archives of the Max-Planck-Gesellschaft in Berlin under the guidance of Dr. Marion Kazemi. Documents were obtained from the Deutsches Museum München with the help of Dr. Michael Eckert. A few documents were also consulted at the Science Historical Collection of the ETH Library in Zurich with the assistance of Dr. Yvonne Voegeli. Photographs were obtained from these various sources. All this has been extremely helpful, and the persons involved are most sincerely thanked for their invaluable help.

BIBLIOGRAPHY

[1] P. Debye, *Phys. Z.,* **13,** 97 (1912) [dated: Zurich, 10 Dec. 1911].
[2] L. Onsager, *J. Am. Chem. Soc.,* **58**, 1486 (1936); for the local 'mean field', see H. Thomas and R. Brout, *J. Appl. Phys.,* **39**, 624 (1968).
[3] J. Valasek, *Phys. Rev.,* **15**, 537 (1920).
[4] M. Davies, *Memoirs of Fellows of the Roy. Soc.*, **16,** 175 (1970).
[5] G. Busch, *Vierteljahrsschr. Naturforsch. Ges. Zür.*, **130**, 19 (1985).
[6] *Pie Debije – Peter Debye 1884 – 1966*, Stichting E. Hustinx und C. Bremen Eds. (Gardez! Verlag, D-53757 St. Augustin, Germany) 2000.
[7] *The Collected Papers of Peter J. W. Debye* (Interscience Pub., New York, N.Y.) 1954.
[8] P. Debye to M. Planck, 8.XII.1933, *MPG-Archiv*, I Abt., Rep. 0001A, Nr. 1651.
[9] M. Planck to P. Debye, 11.II.1935, *MPG-Archiv*, I Abt., Rep. 34, Nr. 9.
[10] *Nobel Lectures, Chemistry* 1922-1941.
[11] P. Debye to A. Sommerfeld, Dahlem 18.XI.1936, *Deutsches Museum München, Archiv HS* 1977-28/A061
[12] P. Debye, *Naturwissenschaften*, **25**, 257 (1937).
[13] W. Heisenberg, *Naturwissenschaften*, **34**, 325 (1946).
[14] P. Debye to J. Kirkwood, Dahlem 21.XI.1938 and 6.X.1939; P. Debye to J. Papish, Dahlem 5.IV.1939; *MPG-Archiv*, III Abt., Rep. 19, Nr. 432.
[15] P. Debye to A. Sommerfeld, Dahlem 30.XII.1939, *Deutsches Museum München, Archiv HS* 1977-28/A061(18.
[16] P. Debye to E. Telschow, Genoa 22.I.1940, and Ithaca 28.VII.1940; *MPG-Archiv*, II Abt., Rep. 1A, Nr. 7.
[17] Nordulf Debye, private communication.
[18] A.H. Compton and S.K. Allison, *X-Rays in Theory and Experiments* (Van Nostrand, Princeton) 1935, p.192.
[19] *Lebenslauf* dated 8.II.1965, *MPG-Archiv*, II Abt., Rep. 1A, Nr. 1.

Peter Joseph Wilhelm Debye (1884 – 1966)

Mansel Davies

Memoirs of Fellows of the Royal Society

PETRUS JOSEPHUS WILHELMUS DEBIJE (PETER DEBYE) was born on 24 March 1884 at Maastricht in the Limburg province of the Netherlands, a town bearing witness to much in the history and culture of at least three of Europe's present national groups. Some other notable physical scientists who share the same birthday are Georg Bauer Agricola (O.S. 1494: De re metallica), Joseph Priestley (1733), Josef Stefan (1835), and Adolf Butenandt (1903), another Nobel prizewinner in Chemistry. Debye died at lthaca, the city where Cornell University is located in upper New York State, on 2 November 1966. In the three years that have since elapsed, a number of brief obituary notices have appeared: these and an earlier article by the writer have been used in preparing the following account. The sequence is essentially chronological with the subdivisions indicating Debye's academic location.

EARLY YEARS

The burgomaster of Maastricht has made available a genealogy which is essentially complete to the fourth generation. One of the eight great-great-grandfathers is given as Joannes Debeij, born in 1752, the great-grandfather (1783-1854) being Pieter Eoduryk Debije. The whole family tree is firmly planted in Maastricht: three of the scientist's grandparents were born in and died at Maastricht, as did his parents. They were all also Roman Catholics. His parents, Joannes Wilhelmus Debije (1859-1937) and Maria Anna Bar-

bara Ruemkens (1859-1940) were married on 23 May 1883, his father's twenty-fourth birthday. Records describe his father as a smith or foreman (smid: Werkmeister) in a metal-ware manufacturer's (J. G. Lambriex) at Maastricht which made items (including tinned-ware) for general use, from gates to kettles: he was well respected by his fellow-workmen and by the other citizens of Maastricht. The Sommerfelds met the parents at Aachen and Frau Sommerfeld carried a memory of the father's bright eyes. There was only one other child, a sister four years younger than the scientist. The mother was a dominant character with a keenness in money matters. She was for many years the cashier of the theatre which was an important centre in the life of Maastricht. Amongst Debye's most cherished memories (and he had a very retentive memory) were those of the opera music he heard as a youngster in Maastricht. He could recall and whistle many of the tunes even from works which have long since ceased to be performed.

Debye himself made it known that until he was five or six years old, and therefore regularly attending school, he very little knew or used the Dutch language: his medium was the Maastricht patois. This he fully retained throughout his life and not only spoke, but corresponded in it with his Maastricht friends-the patois being appreciably different from Dutch. This serves to illustrate a significant aspect of Debye's character. There is no doubt but that he thought of himself as a Maastricht man, and Maastricht has, as the centre of the province of Limburg, some considerable independence of character in the Netherlands. Before the liberation from Spanish hegemony it was ruled jointly by the Duke of Brabant and the Archbishop of Liege. It has remained firmly Roman Catholic and its citizens still retain a feeling of some independence with respect to Holland in particular and the Netherlands in general. This lack of a firm national attachment added to the fact that his parents were fond of making family excursions at short notice to other, often distant, European cities, undoubtedly contributed to Debye's independence from national concerns and national politics.

Debye attended the Hoogere Burger School at Maastricht for five years from the age of twelve. His record in the school-leaving examination, which involved both written and oral tests in most subjects, can be summarized: algebra (8), geometry (10), trigonometry (8), projective drawing and geometry (8), mechanics (10), physics (10), chemistry (9), natural history (10), cosmography (10), history (6), geography (8), government and public administration (7), economics (6), Dutch (9), French (8), German (8), English (5). This record, achieved without any particular effort on his part, put him at the head of the Limburg list. It is impressive also for the range of subjects studied which, however, did not include Greek and Latin. This meant Debye could not then have entered a Dutch university. But it had been decided that he would take the job he had secured with Jurgens, the fats and margarine firm which later formed part of Unilever. Between leaving school and starting work the possibility of further education arose. Seventy years ago a fully-developed secondary education was itself by no means common for boys of such modest homes as Debye's: it implies unusual enlightenment on the part of his parents. The father, it seems, now averred that he 'would work night and day' were it necessary in paying for the further education out of his weekly earnings. The choice was between the Technical University at Delft and the Technische Hochschule at Aachen. Cost and the possibility of living at home made Aachen the choice and so Debye for some years left his bed soon after five to catch the early morning train to Aachen. In so doing he had already taken the most critical step of his whole career.

Aachen was thirty kilometres away and across the border in Germany. This transfer may have led to the restyling of his surname. The original form is Debije, variously pronounced in the Netherlands but often as Deb-ay: outside Dutch sources he almost invariably used Debye which, in English, had always been pronounced Deb-eye: on his birthplace (rebuilt as a shopping centre) he is Pie Debye, with the Maastricht familiar form of Peter by which he was known there. This diversity of name-forms in part reflects the diversity of languages which is almost the birth-right of a Netherlander and which he early acquired and which (with Italian) he used to excellent effect in promoting (and translating) exchanges at conferences. But it was at Aachen that Debye first acquired full facility in the most powerfullanguage at his command.

He was fortunate to have amongst his teachers at Aachen, Max Wien (experimental physics) and Arnold Sommerfeld (theoretical physics). His initial course was for the diploma in electrotechnology. In view of his lifelong mastery in theoretical developments, it is relevant to note he can have devoted no great time to learning mathematics as such, but he did have an immediate immersion in its applications. Later he was to comment (in comparing matrix- and wave-mechanics) that he was far more familiar with differential equations than with matrices. His diploma work included a theoretical treatment of Foucault currents in a rectangular conductor and presented a mathematically elegant solution based on Green's theorem. This became his first published work. He was not, however, interested in electrical technology but rather in physical theory. Sommerfeld, who was later to describe the young student as 'a charming boy who looked out on the world and on life with intelligence and curiosity', had immediately recognized his ability. 'When I had the first possibility to appoint an assistant, there was no doubt in the choice. He went with me as an assistant to Munich in 1906. From here he set out on his victorious progress through Physics and Chemistry.' (1)*

MUNICH, 1906-1911

Debye's doctoral thesis was completed at Munich (July 1908). He had at Aachen already studied the diffraction of light by cylindrical and spherical particles and the original title of the thesis was 'Über den Regenbogen (Concerning the rainbow)'. It dealt with the radiation pressure experienced by spherical particles of varied refractive properties. Not only did it include further evidence of mathematical ability of a high order in its original and extended treatment of Hankel functions and the asymptotic representation of Bessel functions, but it also gave Debye a grasp of what became a continually recurring theme in his career, the refinements of the interaction of radiation with matter. Another publication of this early period with Hondros (1910) was far ahead of its time in its practical value: it dealt with propagation problems of significance to radar and waveguide systems. In much of his later work Debye was often concerned with physically significant treatments and especially with deriving adequately sound and immediately usable relations which allowed of the quantitative evaluation of molecular systems. In these treatments mathematical elegance and rigour could take second place, but few will doubt Debye's ability as a mathematician. David Hilbert was certain of it.

The five years at Munich (1906-1911) probably had a determinative role in orientating Debye towards his life-long interest in diffraction and the interaction of radiation with atomic and molecular systems. Röntgen was professor of experimental physics and, with selected students (e.g. Joffé, Pringsheim) continued careful studies of X-rays. In this environment and almost as a sequel to Debye's thesis, Sommerfeld suggested the problem of finding

* A reference given as a running number relates to the list of References: given as a year it relates to the Debye paper of that date, the tides of which appear in the Bibliography: if the year carries an asterisk, it means that the quotation given from the paper is taken, with the publishers' permission, from its translation in P. J. W. Debye, Collected Papers, Interscience Publ., N. Y. 1954. Other references (Book) are to Debye's volumes of which a separate list is given.

'the optical properties of an anisotropic arrangement of isotropic resonators'. This was Paul Ewald's topic and the conclusions in his thesis led Laue, who had joined Sommerfeld as his second assistant, to suggest the experiment which settled the nature of X-rays. Sommerftld was professor of theoretical physics but had insisted on having laboratory facilities: the dramatic success of this combination might well have influenced Debye but his predilection towards the experimental control of theory could already be seen in his favourite reading: Rayleigh, Maxwell, Boltzmann, Riemann, Drude. 'Debye was an avid reader of classical mathematics and physics, and in particular of Lord Rayleigh's papers and his Theory of Sound. Proper modes ofvibration had a high priority in his mind... Especially the English masters keep the two parts (theory and experiment) of Physics together.' (2) On the status acquired by Debye in these early years at Munich, Ewald must again be quoted (3) : 'Needless to say to those who know of his later development, Debye was, even then, an outstanding physicist, mathematician, and helpful friend. He was, not less than Sommerfeld himself, a centre for the senior students and graduates frequenting the Institute and the Physics Colloquium.'

Two of Debye's other publications from Munich must be mentioned. The first was a review article instigated by Sommerfeld: 'Stationary and quasi-stationary [electric] fields' (1910). Kramers has described this as a very beautiful and mathematically very elaborate ninety-page article, and other senior physicists have commented how they wish younger writers on this theme would first study Debye's article. The second was much briefer but, even in 1950, earned the special commendation of his teacher when, appropriately, presenting him with the Max Planck medal of the German Physical Society: in it Debye provided one of the first coherent deductions of Planck's radiation formula. It is entitled, 'The concept of probability in the theory of radiation' (1910).

In deducing his radiation formula (4) Planck assumed firstly that the energy can be absorbed and emitted in infinitesimal amounts. From this he found the connexion between the energy of a linear resonator and the energy density of the radiation field. In the second part of the argument, in order to establish the connexion between the equilibrium energy and the temperature, the assumption is made that the energy can be dealt with in finite energy-quanta. If this seems a remarkable procedure it must be remembered that eight weeks before presenting his deduction of the formula, Planck had advanced it on an essentially empirical basis. Debye corrected this situation.

Debye considers the Rayleigh-Jeans radiation cube and enumerates the totality of the proper modes of the electromagnetic vibrations by the usual subdivision of the cube edge. Debye ascribes hν as the energy to each of these proper modes. Accordingly, his quantum is not restricted to atomic or molecular features as with Planck but is associated with proper modes which are distributed over the whole cube. He finds the probability for the quantal distribution as does Planck and likewise the most probable state by the usual variation method. The temperature accordingly enters via the entropy. The distribution function so obtained is Planck's radiation law. In the fourth edition of his monograph [Die Theorie der warmestrahlung] Planck himself adopted Debye's method which he characterized 'as an extremely simple deduction' of his radiation law.

ZÜRICH, 1911-1912

Debye left Munich in 1911 before Friedrich and Knipping had performed their experiment (May 1912). He was chosen as Einstein's successor in the chair of theoretical physics at Zürich University (a cantonal, not federal, institute). Einstein had published three papers on the specific heats of solids: the first (1907) was one of the apparently immediately successful applications of Planck's quantum hypothesis, as it gave a good representation of the specific heat of diamond and other elements over a wide temperature range. He assumed the atoms in the lattice had a single vibrational frequency: the specific heat was deduced as the temperature derivative of Planck's energy formula. But in his third paper on specific heats, published from Prague (May 1911), Einstein showed that the single-frequency model was quite unacceptable-individual atoms would lose their vibrational energy far too rapidly and, apart from that, the thermal conductivity of the lattice would be impossibly small. He concluded the paper thus: 'The theoretical problem is so to modify the molecular mechanics as to give not only the specific heat law but also the likelihood of accounting for the simple relation for the thermal conductivity.' Debye has himself described (1964) how he proceeded: '1 considered that Planck's formula should be applicable to any kind of vibrating system. Such was in my opinion a solid or a lattice of atoms, since by using the proper coordinates, this mechanical system could be described as an agglomeration of a number of independent vibrating systems, each with its own proper frequency. The only question was to find the spectrum of frequencies. I knew about the one dimensional case of a string of atoms which had been treated by Rayleigh and started to formulate the mathematical problem for a three-dimensional cubic arrangement... This turned out to be rather complicated. Soon I realized that for my immediate purpose it would not be necessary to know too much about the details of the spectrum. At high temperatures the frequency distribution is immaterial. One has only to know the number of degrees of freedom: each gets kT. On the other hand, at low temperatures according to Planck's formula the high frequencies are going to carry less and less energy. So I decided to find out the density of distribution of proper frequencies in the limit... '

'In order to calculate the density distribution in the ionic spectrum I calculated the spectrum for a sphere of isotropic elastic material. This was a tour de force but I enjoyed it because at that time I knew so much about spherical harmonics and cylinder functions. As a matter of fact the method can be described as clumsy. Anyway, I got the right result.'

Two of the assumptions introduced in Debye's treatment were described by Sommerfeld in 1950 as 'bold'. Not only did he evaluate the total spectral distribution of the elastic modes but he equated the total number of degrees of freedom to the number of permitted proper vibrations. Secondly, he assigned to each of these an energy hν. The evaluation of the totality of elastic modes he undertook for the form of a sphere: whilst this had the advantage of a strain-free boundary, it led to difficult limiting values of the Bessel functions which Debye was only able to handle thanks to his previous work on this subject. Debye's paper on specific heats was received by the Annalen der Physik on 24 July 1912. On 12 March 1912, the Physikalische j Zeitschrift received the first of two papers by Born and Kármán, the second being received on 30 November 1912 (5). The treatments were entirely independent and remarkably different in character. Born and Kármán I accepted the point-mass structure of the crystal lattice and evaluated the vibrational spectrum by a detailed consideration of the coupled oscillations in the three-dimensional structure. This was closer to reality than the Debye model but even more intricate a problem. However, Debye himself used the Born-Kármán model in his major paper on the scattering of X-rays by a crystal lattice (lo October 1913). It is interesting and perhaps significant, that in the volume of fifty papers published with Debye's help to celebrate his seventieth birthday, the specific heat paper is the last but one in the volume in a group labelled 'Miscellaneous' (1954).

The relation he deduced for the specific heat per 9 atom at constant volume is historically the first of those to be described as *the Debye equation*:

$$C_v = \frac{9Nk}{{\nu_D}^3} \int_0^{\nu_D} \frac{e^x}{(e^x - 1)^2} x^2 \nu^2 d\nu$$

$x = (h\nu/kT)$, ν_D is the maximum frequency in the lattice.

It gives an excellent representation of the specific heat for a wide range of solids. Two features are especially noteworthy. The specific heat is a function of $x_D = \theta / T$, where θ is a temperature (the Debye temperature) characteristic of the lattice, $h\nu_D / k$. And, secondly, at the lowest temperatures C_v becomes simply proportional to T^3, a most valuable guide in approaches to absolute zero of temperature.

The year 1912, when Debye was twenty-eight, also saw the publication of another major contribution. This was a four-page paper bearing the date 1 February (1912*): Some results of a kinetic theory of insulators. 'The assumption that the interior of dielectrics contains not only elastically bound electrons but also permanent dipoles of constant electric moment enables us to explain in a completely satisfactory way the temperature dependence of the dielectric constant' [of polar compounds]. This is the scientific birthplace of molecular electric dipoles. However, the delivery process in this instance is not impressive. Rather than an a-priori deduction of the means for evaluating the molecular moments an essentially ex-cathedra statement is made: 'If the polarization, P, produced by an electric field of strength, E, is calculated, an expression with two additive terms is found. The first term of this expression measures the effect of the displaced electrons and is independent of temperature. We represent it by $(\varepsilon_0\text{-l})E$. The second term has the Curie-Langevin form $E(a/T)$, where a is a constant and T the absolute temperature.

For gases we then obtain:

$$p = E(\varepsilon_0 - l) + E(a/T).'$$

Later, the result $a = N\mu^2/9k$ is quoted: here μ is the dipole moment. This perfunctory character may well have been the result of Debye's unquestioned recognition that he was following the Langevin calculation of the mean magnetic moment of gas molecules carrying a permanent magnetic moment, although the Langevin paper (6) is not explicitly referred to and k, Boltzmann's constant, is introduced as 'the universal constant with which the natural logarithm of the probability is to be multiplied to obtain the entropy. Its value is according to Planck, 1.346 X 10-16 erg.' This is clearly an echo from the paper in which he deduces the Planck radiation formula.

The relation which has served to evaluate the electric moments (μ) of gaseous molecules and which again is always called the Debye equation is

$$P = \frac{4\pi}{3} N \left(a_0 + \frac{\mu^2}{3kT} \right)$$

P is evaluated from experiment via the Mossotti-Clausius factor:

$$P = \left(\frac{\varepsilon_0 - 1}{\varepsilon_0 + 2} \right) V,$$

where V is the molar volume. Debye has called P, the molar polarization, although it differs from the net molecular polarizability only by the universal factor $(4\pi/3)N$. This is an unfortunate misnomer but it has in no way limited the effective use of P to determine the dipole moments and thence valuable information on the stereochemical structure and bonding characteristics of many hundreds of molecules. The practical e.s.u. (c.g.s.) unit of dipole moment, 1×10^{-18} e.s.u., has universally been known as the debye (D).

In his original account Debye was obliged to forgo the application of his equation to data for gases-scarcely any existed of sufficient precision. He was obliged to test his relation on data for liquid alcohols, water and some other polar solvents. In this way he derived the first values for the molecular electric moments. The 1912 values may be quoted in debyes together with presently accepted ones: diethyl ether 1.18 (1.16): toluene 0.51 (0.34): water 0.57 (1.82): methyl alcohol 0.34 (1.71). It has long been appreciated that the simple Debye equation for p and μ is quantitatively applicable only in the molecularly dilute gaseous state. Without correction it will not give a molecularly significant μ value from data for pure liquids. Debye realized this limitation in the original paper (1912*), even to the extent of indicating that the relation predicted a critical temperature below which spontaneous polarization would appear. 'As far as I know', he wrote, 'this has not been achieved in practice.' In fact, ferroelectricity was first observed by Valasek in 1921 in crystals of Rochelle salt, and only more recently has an analogous state been surmised in ice and something similar achieved in polar liquid phases (11).

An account has been given (7) of a Debye lecture in 1912 at the University of Leiden when he presented his treatment of the electric permittivity of gases and liquids. Everyone, including the young Debye, was anxious to see how Lorentz, who was present, would take this account: it seems he somewhat drowsily nodded approval, as if in answer to the presentation directed almost personally to him.

UTRECHT, 1912-1914

Historically, it is significant that a considerable delay occurred in the development of dipole moment studies. At this early stage Debye does not appear to have initiated any measurements himself, although there are indications that his intention to do so was one reason for his leaving Zürich in 1912, where his chair was in theoretical physics: he had hopes that at Utrecht he would be able to undertake experimental dipole moment studies. However, the appointment at Utrecht was again to a professorship of mathematical physics and theoretical mechanics. His inaugural lecture was on the kinetic theory of matter and its modem development (1913). In it he expressed the opinion that 'mathematical physics is in the first place physics and it could not exist without experimental investigations'. Moreover, he was glad to express his agreement on this with his senior colleague, W. H. Julius, the professor of experimental physics. Certainly Debye developed no experimental work there. This is not surprising as during the year or so actually at Utrecht he was occupied with two major theoretical developments: the entirely new representation of dielectric dispersion and the equally far-sighted treatment of the influence of lattice vibrations upon X-ray diffraction intensities. Perhaps the somewhat sketchy nature of the 1912 dipole moment paper contributed to the delay before other scientists pursued dipole moment measurements. Debye himself promised a fuller treatment-' A comprehensive paper regarding the relation between the deviations from the law of Clausius-Mossotti and the hypothesis of fixed moments is in preparation' (1913*). Despite this 1913 statement Debye does not appear to have published such a comprehensive account before the Handbuch der Radiologie article (1925).

Again, it must be remembered that the valve, as an electrical oscillator, did not come into general laboratory use until after the 1914-1918 war. Then, a number of physicists and some chemists (Jona, Sanger, Weigt, Zahn, Smyth) made a start on measurements in gases: these were not easy. Although measurements of the permittivities of solutions had been made many years earlier, their use in evaluating dipole moments appears first to have been explored in a thesis done under Debye's direction at Gottingen by Fraulein L. Lange in 1918. When this was published in 1925, a spate of work was initiated, and was accelerated by a discussion organized by Debye at Leipzig in 1928 (Book, Ed. 1929) and by his publication of the volume Polar Molecules in 1929 (Book). By 1935 a list of dipole moments for 1100 compounds could be compiled (8).

Doubtless many reasons contributed to the brevity of the basic electric dipole moment paper. It appears probable that Debye had immediately become engaged on a considerable extension of the molecular model for dielectrics. In 1913 a masterly paper full of new insights appeared: *The theory of anomalous dispersion in the region of long-wave electromagnetic radiation*. This took the appreciation of electric permittivities (i.e. 'dielectric constants') in molecular terms from the equilibrium state, delineated by means of P, to the dynamic representation in which the rate of the dipole reorientation in the electric field is determinative. The simplicity and effectiveness of the treatment he advanced shows Debye's keen physical insight and his exceptional ability in constructing a quantitative appraisal leading to direct experimental control.

He starts from the realization that on the application (or disappearance) of an electric field the molecular dipoles will take a brief but finite time to attain their equilibrium orientation. For small molecules he estimates the time to be of the order of 10 picoseconds. This he does by applying Stokes' factor for the viscous drag on a rotating sphere to one of radius 10-8 cm. He thus defines the time-scale of the orientation and the frequency range where the permittivity would pass from its value ε_0 characteristic of static fields, to ε_∞, the value when the applied field oscillates So rapidly as not to allow any motion of the dipole: at such frequencies the material fails to show its polar character and the permittivity reverts to that of a non-polar medium. Debye evaluates the dynamics of the molecular orientation and displacement in the field by using a generalization of Einstein's equation for determining the mean square displacement of a particle having Brownian motion. This leads to an expression for the decay of polarization on removal of the orienting field which is simply exponential:

$$P(t) = P(0) \cdot e^{-kt} = P(0) \cdot e^{-t/\tau}.$$

τ is the relaxation time, i.e. the reciprocal of the first-order rate coefficient for the polarization time dependence. For spherical particles of radius r Debye originally gave

$$\tau = \frac{8\pi\eta r^3}{kT}$$

where η is the viscosity of the medium. In deducing this he restricted the rotation to a plane : the general rotation leads to a factor 4 in place of 8.

This work provided an entirely new appreciation of the frequency dependence of dielectric properties. Debye himself emphasized this: 'The hypothesis that molecules contain permanent dipole moments is the essential basis of the theory'-and that hypothesis was only a year old. Even so its most important aspect was its clear analysis of a typical molecular relaxation process, in which respect it has been a model for many later representations of rapid molecular changes. The frequency dependence of the permittivity (ε') was necessarily accompanied by an electrical energy absorption factor (ε''), these being the two experimental observables associated with the time-dependence of the polarization. Debye gave effectively the relations:

$$\varepsilon' = \varepsilon_\infty + \frac{\varepsilon_0 - \varepsilon_\infty}{1+\omega^2\tau^2}$$

$$\varepsilon'' = (\varepsilon_0 - \varepsilon_\infty)\frac{\omega\tau}{1+\omega^2\tau^2}$$

Here w is the angular frequency ($2\pi\nu$ radian s^{-1}).

That a maximum in ε'' is reached for $\omega\tau = 1$ provides a very direct means of evaluating τ. Using relations equivalent to these, Debye drew a dispersion and absorption curve for water: he took $\varepsilon_0 = 80$: $\varepsilon_\infty = 2$: and $\lambda(\varepsilon''_{max}) = 1$ cm. This was very good guesswork: at 20 °C acceptable values now are $\varepsilon_0 = 80.2$: $\varepsilon_\infty = 4.5$: $\lambda(\varepsilon''_{max}) = 1.74$ cm, and until 1947 (9) it could be said that no published data on the dielectric dispersion of water were certainly nearer to the correct values than Debye's guess of 1913. In the original paper he quotes no general numerical data but gives a reference to a recent summary (10): much of the best experimental work at that date was still due to Drude. In view of the complexity of the molecular interactions in water it is a surprising fact that the simple Debye relations, based on a single relaxation time, do fit the data almost as accurately as they can be determined. The same is true for many of the crystalline forms of ice. Although very many systems do show deviations from these Debye equations, in the sense of providing a flatter and wider absorption curve ($\varepsilon'' = f(\omega)$) than he predicted, the general character of these deviations is well understood. One way of expressing this extension of his treatment is to envisage a finite number or a continuous range of contiguous relaxation times to be present in a single molecular system.

In the original paper Debye acknowledges the cooperation of an assistant (R. Ortvay).Boltzmann's constant is now described simply and directly as such. And there is specific reference to one other aspect of molecular behaviour whose time-dependence can be treated in the same way. This is the Kerr effect, i.e. electrically induced double refraction. The general exploration of dielectric dispersion had to wait until the klystron and other v.h.f. sources became generally available, i.e. until after 1945. Debye did himself provide an introduction to experimental studies in this area in a paper at a Faraday Society Discussion (1934). The thermal assessment of the electrical loss factor ε'' which he described there was pursued by some of his associates, but it did not offer more than preliminary quantitative data.* Especially in relation to its important molecular structural aspects much of the later v.h.f. work on dielectric dispersion was developed by C.P. Smyth at Princeton: von Hippel at M.I.T. organized solid state studies and a group of French physicists at the Sorbonne also took up these problems (11). The effective examination of the frequency dispersion of the Kerr constant was not undertaken until appropriate pulsed-field methods became available in the later 1950s. For small molecules the direct observation in real time of these dipole relaxation processes is only now being achieved.

Whilst professor at Utrecht, Debye married in Munich, Matilde Alberer, a Bavarian lady, three years his junior. She was one of three daughters of the boarding house he had stayed at in Munich. On marriage she acquired Dutch

---

* Professor Sack [Cornell] recalls the important contributions of J. Errera [Brussels], a close col league of Debye's. In the stereochemical implications of molecular dipole moments (1925), in the study of dipole rotation and dispersion in solids (e.g. ice, 1924). and in dispersion measurements of the Kerr effect (1935), Errera contributed much to substantiate the Debye model of dielectric behaviour. A similar position, and friendship, was attained by S. Mizushima [Tokyo].

nationality and at present still resides at Ithaca. There were two children of the marriage, a son and a daughter. The former is now a, senior industrial physicist in the U.S.A.: for several years, both at Berlin and at Ithaca, he very substantially assisted in his father's work, making notable contributions on the experimental side. The daughter and her two sons now also live in the U.S.A.

Debye's sojourn at Utrecht was only for two years. It seems that before he had been made a member of the senate there he was probably already committed to moving to Gottingen. Certainly one reason for this move was the poor immediate prospect at Utrecht for the laboratory facilities he desired. Nevertheless, the Utrecht period was a very fruitful one as, in addition to the treatment of dielectric dispersion, it embraced his work on lattice vibrations and X-ray diffraction intensities. Three brief reports, followed by a major paper dated 10 October 1913 (1914*) originated at Utrecht: in the brief items (1913: Verh. d. Phys. Gesell.) his name appears as Debije.

The discovery of X-ray diffraction was made at Munich in May 1912 and Laue had immediately presented his theoretical interpretation of it (12). Debye was in the closest contact with developments at Munich. His own work there, the whole corpus of his interests in radiation problems and, most particularly, his recent treatment of lattice specific heats provided him with the clearest insight into the character and some of the refinements of X-ray diffraction. The paper which resulted is one of the great achievements in crystal physics. That its content should have been developed within months of the discovery of X-ray diffraction must remain for ever astonishing. It is difficult to convey briefly the depth and clarity of the approach, the coherence of the presentation, and the degree of conviction achieved by the evaluation of numerical conclusions discriminating between two models differing only in one simple but fundamentally important assumption: but some points from the various sections may be quoted.

The introduction mentions the preliminary results already published (1914*): 'The summation of the effects on the incident beam due to the atomic structure, no longer rigid as assumed by Laue, resulted in the absence of an effect on the sharpness of the interference maxima and in the existence of an intensity effect. An explanation was also found for the appearance of noticeable intensities mostly in those directions that deviate only by small angles from the direction of the incident beam, and at the same time for the fact that the so-called reflection of X-rays is observed under ordinary circumstances only if the incidence is not too steep. The computation further shows that, as a consequence of the thermal movement, the interference intensity must always be accompanied by a scattered radiation which has its maximum where the interference intensity is weakest...

'Firstly we have dropped the assumption of the mutual independence of the atoms. In this paper their movement is composed of superposed elastic waves whose wave numbers assume all values in the elastic spectrum of the body, a method that has proved successful in the theory of specific heats.

'Secondly, we thus created the opportunity to apply the quantum hypothesis in a definite manner to our case. We have not decided for or against the existence of zero-point energy... though different new articles by A. Einstein and O. Stern, H. Kamerling Onnes, W. H. Keesom, E. Oosterhuis present weighty reasons for the assumption of a zero-point energy... In fact it is clear from the very beginning that, if the fundamental concepts of the theory are correct, the mean square of the amplitude of the atomic movements itself and not its differential coefficient with respect to temperature (as in the case of specific heats) will be decisive. In view of this decision, to be made on the basis of experiments, we have developed the theory to a point where a formula is available which can be used numerically, and have appended numerical and graphical discussions.'

The subsequent sections are:

'I. Mathematical formulation of the principal problem... We therefore must establish the average of [the diffraction intensity] with respect to the displacements u, v, w,... If the probability of an arbitrary arrangement is known the mean value of this factor can be computed... this mean value constitutes the principal problem...

'II. Introducing normal coordinates. Just as in the case of the calculation of specific heats, the introduction of normal coordinates is also of great value for the representation of the atomic movement. For this purpose we might proceed from the conventional elastic differential equations for the condition of a continuously occupied space, and would thus be in a position to develop an approximate theory, as has been done for the computation of the mean heat energy. [Debye's treatment, 1912.] Instead we will here introduce the atomic structure of the body from the beginning, and therefore define our new coordinates following the exposition by Born and v. Kármán (5)... We will attempt to exploit the advantages of both methods as much as possible.

'III. Computation of the desired mean value for the case of vanishing zero-point energy.

'IV. Computation of the required mean value with the assumption of zero-point energy.

'V. General results regarding the effect of the thermal movement.

'VI. The effect of heat at low temperatures.

'VII. Approximate formula for the temperature function valid at all temperatures.

'VIII. Numerical discussion and graphical presentation.

'Summary :...

'(4) The exponent in the function just mentioned vanishes at T = 0 in the absence of zero-point energy: it remains finite and assumes a substantial value if zero-point energy exists...

'(8)... as for the specific heat of monatomic bodies... the temperature dependence is a function only of the ratio of the characteristic temperature θ to the temperature of observation.'

An addendum, added in the proofs, incorporates valuable suggestions sent to the author by Sommerfeld and Lorentz. Debye's quantitative treatment needed correction in some respects. These corrections were provided by Waller who was also one of the authors with R. W. James and D. R. Hartree of a paper which (among others) clearly established the existence of zero-point energy by a careful evaluation of the X-ray diffraction intensities in rock-salt (13).

GOTTINGEN, 1914-1920

Both Gerlach and Debye have sketched the circumstances in which Debye was invited to Gottingen. Significantly, David Hilbert took an active part, although the two established physics professorships were then occupied by Riecke and Voigt. They had been greatly impressed by Debye's contribution at a Wolfskehl discussion devoted to the kinetic theory of matter at Gottingen from 21 to 26 April 1913. The speakers at this conference and their themes were: M. Planck, 'The significance of the quantum hypothesis for the theory of gases'; P. Debye, 'The equation of state on the basis of the quantum hypothesis'; W. Nernst, 'The kinetic theory of solids' ; M. v. Smoluchowski, 'The limits of validity of the second law of thermodynamics'; A. Sommerfeld, 'Problems of free path lengths' ; H. A. Lorentz, 'The application of kinetic methods to the motion of electrons'. Debye's paper dealt with several basic aspects of the quantum theory as seen at that date, including his representation of the general relation for imposing quantization in terms of the momentum: the character of this relation he said was based 'on a remark of

Einstein' (1913). When Debye's contribution was published in the volume of the conference papers it had an addendum which included two further original developments. The first treated the anharmonic terms in lattice vibrations, which provided the explanation of the expansion coefficient in solids: the second was a treatment of heat conduction in terms of the scattering and decay of sonic (phonon) waves. On this new basis Peierls systematized heat conduction in crystals sixteen years later. Undoubtedly it was this whole chapter in the volume which he edited that so impressed David Hilbert. Advantage was taken of the fact that Voigt was prepared to relinquish responsibilities on the experimental side and it was arranged to make part of the institute available for Debye-'a very unusual thing' (Gerlach).

Manneback (13a) has pointed out another profoundly significant remark of Debye's which much preceded his transfer to Gottingen. It appears as a note on a paper by Sommerfeld and Runge (Ann. Physik, 35, 290 (1912)). Debye draws attention to the relations between the Jacobian equation of motion with partial derivatives of the first order and of second degree and the equation for linear waves of the second order. A transformation analogous to that of Ricati allows a passage from the one to the other and establishes a possible physical connexion between the two: this connexion is illustrated by the relations existing in the limit between the geometrical optics of Huygens and classical wave optics. The use that Schr6dinger made of such ideas is well known.

The Gottingen appointment was to a professorship of theoretical and experimental physics and it offered definite prospects for experimental work as new laboratories had recently been completed there. An invitation to Góttingen at this date must, in any case, have been well-nigh irresistible to a mathematically inclined physicist. It is evidence of Debye's status to recall the extraordinary brilliance of the faculty he was asked to join as a senior member. In mathematics the head of the school was Felix Klein, and associated with him were David Hilbert, Carathéodory, Landau, Toeplitz, Weil, Runge and Courant. In physics there were Voigt, Wiechert, Madelung, Prandtl, Riecke, Born, von Kármán, Tamman and Simon.

Debye must have been contemplating his move to Góttingen almost simultaneously with the publication (July 1913) of Bohr's first paper on atomic structure (14). In that paper the quantization of the energy is introduced in what now appears a remarkable way and the result is subsequently shown to be equivalent to the assumption 'that the angular momentum of the electron round the nucleus in a stationary state of the system is equal to an entire multiple of a universal value' [h/2π]. Bohr's further papers of 1913 are based on the relation mva = h/2π, as an *ad hoc* condition. Within the same year Debye presented a rationalized and more general version of the quantization condition, i.e. $\oint p \cdot dq = n\hbar$ where p is the generalized momentum and q the associated coordinate, the integral extending over a complete period of the motion (15). It is worth noting that Debye's statement of this condition (Utrecht, 10 February 1913) certainly preceded the availability to him of Bohr's first paper. This relation, introduced independently in 1915 and often associated with W. Wilson and Sommerfeld, became the accepted version of applying the quantization condition until wave-mechanics provided a more satisfactory basis in 1926. It immediately led, in Sommerfeld's model, to the introduction of the azimuthal quantum number and to the elliptic electron orbits in all atoms beyond beryllium. Such was his interest in the Bohr theory that Debye must have abandoned his intention of studying dipole moments experimentally. His first experimental work combined two other interests.

Debye took up X-ray diffraction in the hope that it might provide evidence for the electron orbits envisaged in the Bohr atomic model. A medical-type X-ray tube with a platinum anticathode was used and (hopefully) a sheet of paper taken as diffracting material. Not surprisingly, the photographs showed nothing of any hoped-for pattern: the source was capable of producing only continuous radiation. Fortunately his assistant, Paul Scherrer, was a young Swiss graduate so that the outbreak of war in 1914 caused a minimum disturbance to the work, as neither of them was of German nationality. The next step was the construction of a metal X-ray tube (in the first instance, copying the design of other workers) with a water-cooled copper target and an aluminium foil window for the escape of the rays. This copper 'home-made' type of continuously evacuated tube or, rather, modifications of it, remained in use in some laboratories until the end of the Second World War. With this new source and a cylindrical camera, Scherrer took a photograph of finely powdered lithium fluoride. In one sense the success was complete. The sharp rings of a typical powder diffraction photograph was obtained, and explained as formed by the intersection of conical beams from the randomly oriented crystals with the photographic plate. In their original paper Friedrich, Knipping and von Laue had failed to detect diffraction from a powdered crystal specimen, but later Friedrich found rings around the primary beam without appreciating their significance. The powder method was independently discovered by A. W. Hull at the General Electric Schenectady

Laboratories: his account in 1917 appeared a year after the Debye-Scherrer paper which had not reached the U.S. owing to war conditions.

The powder method was for many years a much used means of crystal analysis and although for detailed structural analysis it has long been replaced by single crystal methods, it still retains its great value in the identification of any crystalline material, be it of natural or artificial origin, of chemical, biological, mineralogical or metallurgical character. As one of its first applications, Scherrer revealed the nature of metallic colloidal solutions and showed how to determine the sizes and structures of particles far below the resolving power of the light microscope. Gerlach has described how Debye and Scherrer enjoyed showing him their laboratory in 1916. The burnt-out (Ruhmkorff) induction coils of the X-ray set showed huge swellings where the paraffin wax had come through the windings: Debye and Scherrer's amusement at their sight 'made a strange impression' on an experimental physicist trained by Paschen.

The structures of some cubic crystals were soon determined by the powder method, thanks to the clear understanding of those lattices from the early work of the Braggs. Even some crystals with complex ions, e.g. K_2PtCl_6, proved amenable to evaluation and, in this instance, provided important confirmation of the octahedral anionic structure arrived at chemically by Alfred Werner. Other lattices such as elementary boron proved too recondite for the powder method but graphite with its especially interesting layer structure was partly unravelled by Debye and Scherrer (1917). As had been realized from the sharpness of the powder photograph rings for lithium fluoride, the component ions were responsible for t4e diffraction and it was anticipated each ion would contribute intensity in relation to its total number of electrons. An evaluation of the lithium fluoride intensities led to the welcome conclusion that their total electrons were in the ratio 2 : 10. This appeared to be important confirmation for well-established anticipations but the quantitative significance of this early deduction was later criticized (16) : it was obtained by a plausible but not a reliable extrapolation. The unquestioned importance of this evaluation was its indication that significant calculations could be made from X-ray intensities.

Systematic efforts were also made to detect 'structure' in liquids by X-ray diffraction. Narrow jets of benzene and cyclohexane were used, but only diffuse interference rings were recorded: these provided at best an indication of the average separation of the molecules in the liquid. However, the problem represented by these studies recurs in various forms throughout Debye's subsequent work. For disordered condensed phases (such as a liquid) the essential feature is to define the molecular radial distribution function, whilst for the dilute gaseous state it becomes the assessment of the molecular structure from the diffraction of randomly oriented molecules. These themes grew from the early Góttingen studies and related ones were amongst the most prominent interests of Debye's last years.

At Góttingen Debye also published theoretical treatments of hydrogen-like atomic spectra. One of them gave a presentation more systematic than that achieved by Sommerfeld and was based on the Hamiltonian (1916). This method was subsequently used by Bohr in his major paper (15) in the Proceedings of the Copenhagen Academy for 1918. Another stimulating theoretical investigation was Debye's account of the Zeeman effect in terms of the quantum hypothesis (1916). He deduced the discontinuous value of the angle between the electron orbit and the direction of the magnetic field, i.e. the space quantization of the orbits which had been introduced by Sommerfeld almost simultaneously.* It is clear from the papers that Sommerfeld was the more convinced that the electron orbits were 'real' and oriented at fixed angles in the magnetic field. It was these accounts which led to the Stern-Gerlach experiment (1922). Gerlach recalls that when Otto Stern and he were telling Debye in 1921 of the difficulties and lack of success in their experiment at that date, Debye was quite explicit in his comments: they should not think that the details of the electron orbits had any physical reality, they merely arose from the mathematical model. They belonged to what Debye was accustomed to call the zoology of the quantum rules or the railway-guide for the electrons. This emphasizes the limited validity he ascribed to mathematical representation: but his quantization process was more intimately correct than he was prepared to allow. The success of the Stern-Gerlach experiment undoubtedly made a great impact on quantum-theoretical thinking: Born was to characterize it as 'perhaps the most impressive evidence we have of the fundamental difference between classical and quantum mechanics' (15). In view ofBorn's status in that field and of the fact that Debye is not always thought of as one of the architects of atomic physics, it is significant that in Born's volume of that title the only authors named

* Bohr wrote (1918): 'Subsequently Sommerfeld himself and Debye have on the same lines indicated an interpretation of the effect of a magnetic field on the hydrogen spectrum which... undoubtedly represents an important step towards a detailed understanding of this phenomenon.'

more frequently than Debye are Bohr, Dirac, Heisenberg, Planck, Rutherford, Schrödinger and Zeeman.

In molecular physics, a theoretical paper of this period was that evaluating the van der Waals cohesive forces (1920*). Keesom had emphasized that dipole-dipole interactions contributed significantly to gas molecular interactions. Debye proceeds: '[In the limit of high temperatures] it can be easily shown that two rigid electrical systems do not, in the mean, exert a force on one another... The situation is immediately and essentially changed if we consider molecules that are not completely rigid. The fact that each gas has a refractive index different from unity is proof of the mobility of the charges in the molecule. Taking this into consideration, it will be clear that a given molecule assumes an electric moment in the field F of another molecule, which moment is proportional to the field F. Thereby a mutual energy arises between two molecules which is proportional to the product of the field strength times moment, i.e. proportional to the square of F. Thus the average of the corresponding force cannot vanish. Further, it will be readily seen that the force is always one of attraction. Hence we may conclude that we have found in this force the origin of van der Waals's universal attraction.' Debye then writes the potential in the neighbourhood of any molecule consisting of a distribution of electric charges as

$$\phi = \frac{a}{r} + \frac{b}{r^2} + \frac{c}{r^3} + \ldots$$

where a arises from the net charge; b from the dipole; c from the quadrupole, etc. He determines the interaction for quadrupole (i.e. non-dipolar) molecules and evaluates it, using the molar refraction as a measure of the polarizability.

In invoking the quadrupole field Debye provides one of the first convincing statements to account for the attractive forces experienced by non-dipolar molecules, although the quadrupole field (which does not exist for the noble gas atoms) lacks the universal character of the dispersion interaction first revealed by London (1926). In a later paper (1921) Debye shows how, with a magic cross, zero field can be created along an axis about which the field's spatial variation is as large as possible. This is his suggested means of differentiation in an optical method between orientation due to a quadrupole moment interaction and that due to an induced moment (Kerr effect). The method has since been frequently used. He also offers an explanation of the universal repulsion between molecules: based on classical electrostatics only, it is not acceptable for molecular fields.

ZÜRICH, 1920-1927

Debye remained until 1920 at Góttingen: his six to seven-year stay there was the first of four such periods in European institutes. There followed Zürich (1920-1927), Leipzig (1927-1934), Berlin (1934-1940). Conditions must still have been very depressed in Germany when Debye accepted the directorship ofthe Physical Institute at the (Federal) Technische Hochschule in Zürich. He was now also editor of the Physikalische Zeitschrift. Rarely has an editor contributed so much of outstanding significance to his journal. Volume 24 (1923) contains from Debye:

pp. 161-166: a paper 'X-ray scattering and quantum theory'—received on 14 March 1923.

pp. 185-206: a paper 'On the theory of electrolytes. I. The freezing point depression and related phenomena'-submitted on 27 February 1923.

pp. 305-325: a paper 'On the theory of electrolytes. II. The limiting law for electric conductivity'-received on 19 July 1923.

The first gives an original fully quantitative account of what became known as the Compton effect, i.e. the change in X-ray wavelength on scattering by an electron. The other two, in which his assistant Dr Erich Hückel was a joint author, created a new epoch in the study of electrolytes.

The paper on X-ray scattering is a masterpiece of presentation in its conciseness, in its clarity, and in the completeness with which it covers a new discovery of major physical significance. Debye based this paper on a report by A. H. Compton of his experimental data published as a bulletin of the National Research Council (Washington) in October 1922. The Physical Review on p. 267 of Vol. 19 (1922) carried a brief account of the results and a qualitative statement of their probable significance in an undated note from Compton. On 13 December 1922, Compton's major paper, replete with quantitative theoretical appraisal, was received by the Physical Review and it appeared in the May 1923 issue. Accordingly, the simultaneity of the two independent interpretations of the Compton effect is as close as could be wished. Some quotation may succeed in suggesting the quality of the Debye paper (1923*).

'It is known that scattered X-ray radiation is polarized and varies in intensity with the direction of scattering... If θ is the angle between the primary and the secondary beam the formula $i_S \propto \dfrac{1+\cos^2\theta}{2}$ follows for the dependence of the scattered intensity i_S on the angle. At comparatively long wavelengths of the primary radiation, however, a substantial discrepancy exists... this effect is caused by the fact that the distances between the electrons in the atom are of the same order of magnitude as the wavelengths of the X-rays used for these experiments. Consequently the secondary rays originating at the different electrons of one single atom interfere with one another, and the scattered intensity at average wavelength is computed to be proportional to Z^2 for $\theta = 0$, and proportional to Z for $\theta = \pi$, where Z is the number of electrons in the atom.*

'If we now proceed to very short waves, the interference effect should, according to the calculations, contract to a progressively smaller angular region surrounding $\theta = 0$, and the intensity outside this region should be represented by Thomson's formula. All experimental evidence with short waves is contrary to this assumption. In the following I wish to present some thoughts referring exclusively to the short wave region.

'Four points seem to me to deserve particular attention:

1. The intensity of the scattered radiation is considerably higher in the direction of the primary radiation ($\theta = 0$) than in the opposite direction ($\theta = \pi$) in contradiction to the mathematically derived proportionality to ($1+\cos^2\theta$).

2. It now appears to be certain that the radiation scattered in the direction of the primary beam is harder than that scattered in the opposite direction. Thus the wavelength is changed, again in contradiction to the above-mentioned computation.

3. The total energy of the scattered radiation sinks below the limiting value corresponding to Thomson's calculation...

4. Each scattering is accompanied by electron emission. The shorter the wavelength, the more the electrons appear to be ejected in the direction of the primary beam.

'Not all the experimental results are so unequivocal that the assertions 1 to 4 can be regarded as absolutely confirmed by experiments. However, I recently gained the impression from a survey by A. H. Compton (*Bull. Nat. Res. Council*, **4**, No.20, October 1922: Nat. Acad. Sc., Washington, D.C.)

---

* It should be noted that precisely similar considerations were invoked in 1945 by Debye in treating light scattering by large polymer molecules: from these considerations he determined the effective length of the molecules.

that it is highly probable that they are correct. I will therefore hesitate no longer to present for discussion an explanation of these effects based on quantum theory which occurred to me as a possibility quite some time ago.

'*Let us assume that classical electrodynamics fails also in the computation of the energy scattered by a free electron excited by primary radiation, and must be replaced by a quantum concept.* In particular the following picture holds. The primary X-ray radiation of frequency ν_0 transfers to a free electron in an elementary process the energy $h\nu_0$. This energy is transformed quantitatively, serving, first, to generate a secondary ray of frequency ν and energy $h\nu$ and, second, to impart a velocity v to the electron.

'The secondary radiation may be considered as "needle radiation" in Einstein's interpretation. On the basis of these premises, a highly detailed picture of the processes can be secured, provided it is assumed that (a) the law of conservation of energy and (b) the principle of conservation of momentum hold also in this instance. The two theorems are sufficient without any additional hypothesis.

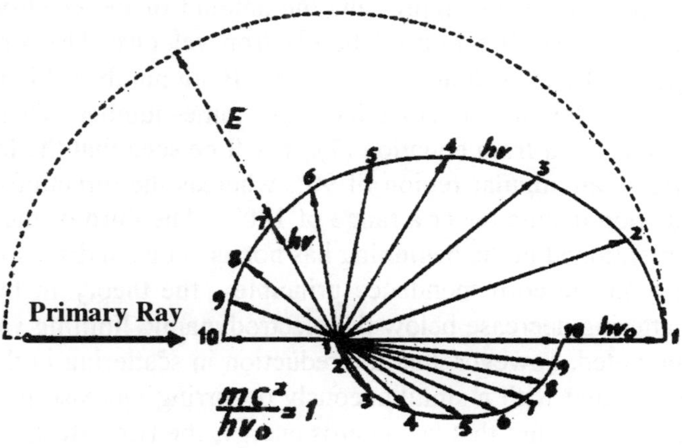

FIGURE 1

'The velocities of the electrons will be of the order of the velocity of light, c; we therefore use the relativistic formulae...'

Then follows a beautifully succinct deduction of (i) the frequency of the secondary ray; (ii) the velocity of the secondary electron; (iii) the direction in which the latter is ejected. The now well-known results are summarized diagrammatically for an incident X-ray of energy $h\nu_0$ equal to the rest energy of the electron, mc^2: i.e. $\nu_0 = 1.23\times10^{20}$ s^{-1}, or $\lambda_0 = 0.0243\times10^{-8}$ cm (Debye's values).

'The upper part of the figure shows a half-circle indicated by a dashed line with radius $h\nu_0$; further, a series of arrows bounded by a full-line curve is drawn. Their lengths indicate the magnitude of the secondary radiation quanta $h\nu$ in accordance with equation (5). The section of the radius extending between this curve and the circle measures the respective absolute value of the energy E of the electron, since it is required by the law of the conservation of energy that $h\nu$ and E add up to $h\nu_0$ (cf. eq.(6)). The lower part of the figure shows a series of arrows also bounded by a full-line curve. The length of these arrows represents the amount of the electron energy E, and they point in the direction of the electron velocity. The arrows for the radiation quanta $h\nu$ as well as the arrows for E are numbered in such a manner that those belonging together have the same number. Their directions have been calculated from equation (7). It will be seen that the lower arrows are confined to an angular region of 90°, whereas the directions of the secondary radiation quanta cover a range of 180°... The third of the experimental results mentioned in the beginning has not as yet been discussed...

'Relying on the correspondence principle... the theory in fact demands that the scattering decrease below the electrodynamic limiting value to zero. It should be noted, however, that the reduction in scattering in this representation is associated with a simultaneously occurring increase in electron energy by such an amount that, as regards energy, the two effects together correspond to the electrodynamic limiting value at any wavelength... By means of experiments which reveal characteristic deviations from this scheme, we may hope to secure deeper insight into the laws of quantum theory, particularly as regards their relation to physical optics.'

This effect in X-ray scattering was of great significance in the development of atomic physics. It emphasized the wave-particle character of the photon, i.e. the duality which also appeared in the wave nature of the electron. As A. H. Compton's experimental assessment was basic to this appreciation and as he, independently of Debye, also evaluated it theoretically, he received the Nobel Prize for Physics in 1927.

The interest in electrolytes so fully developed in the papers with Hückel was an entirely new one for Debye. He himself explained (1923*): 'The present considerations were stimulated by a lecture by E. Bauer on Ghosh's publications, held at the Physikalische Gesellschaft (Zürich). The general viewpoints taken as the basis for the computation of the freezing point depression as well as of the conductivity led me, among other things, to the limiting law involving the square root of the concentration. I could have reported on this during the winter of 1921 at the Kolloquium. With the active help of my assistant, Dr E. Hückel, a comprehensive discussion of the results and their collection took place during the winter of 1922.' This substantial period of gestation helps to explain the comprehensive nature of the two papers in 1923. In taking up the electrolyte problem Debye might be described as entering for the first time the area largely occupied by physical chemists. No one would now maintain that a border can be defined between it and physics: Debye himself is perhaps more responsible than any other single individual for its disappearance. The one functional difference involved is that a knowledge of chemical facts is needed adequately to experiment with the systems he now commenced studying.

The first paper carries a concise review of the approaches to understanding ionic solutions already taken. Arrhenius, van't Hoff, Nernst, W. Sutherland, P. Hertz, Milner, Bjerrum and Ghosh figure in this. The basic problem was that the simple laws established by van't Hoff for osmotic pressure and related equilibrium properties for solutes were seriously in error for strong electrolytes, i.e. for conventional salts and strong acids (HCl, H2SO4' etc.). Arrhenius's account of solution conductivities and ionic equilibria which was adequate for weak electrolytes (acetic acid, ammonia, etc.) was equally inapplicable.

'Recently, under the influence of Bjerrum, the impression gained strength that consideration of the electrostatic forces, exerted by the ions on one another—and of considerable importance because of the comparatively enormous size of the elementary electric charge-must supply the desired explanation. Classical theory does not discuss these forces, rather, it treats the ions as entirely independent units. A new interaction theory has to be analogous in some respects to van der Waals's generalization of the law of ideal gases to the case of real gases. However, it will have to resort to entirely different expedients, since the electrostatic forces between ions decrease only as the square of the distance and thus are essentially different from the intermolecular forces which decline much more rapidly with an increase in distance.

'Milner [in 1912 and 1913] computed the osmotic coefficient along such lines. His computation cannot be objected to as regards its outline, but it leads to mathematical difficulties which are not entirely overcome, and the final result can only be expressed in the form of a graphically determined curve for the relation between $(1-f_0)$ and the concentration [f_0 is the "osmotic coefficient"]. From the following it will further emerge that the comparison with experiment carried through by Milner, supposes the admission of his approximations for concentrations which are much too high and for which, in fact, the individual properties of the ions, not taken into account by Milner, already play an important part. In spite of this, it would be unjust to discard Milner's computation in favour of the more recent computations by Ghosh [1918, 1921] on the same subject. We shall have to revert, in the following, to the reason why we cannot agree to Ghosh's calculations, neither in their application to the conductivity nor in their more straightforward application to the osmotic pressure.'

That Debye was familiar with Milner's treatment when he carne to write this paper is not always remembered. His electrolyte model was essentially identical with Milner's and they agreed even to the extent of concurrence on a number of basic relations. However, the mathematical manipulation was markedly different in the two cases and, most significantly, Debye reduced his conclusions to quantitative forms which could be applied directly to the physicochemical data. An all-important section was the calculation of the resultant electric potential (ψ) in the neighbourhood of a positive ion in a univalent salt solution (e.g. KCl) : 'Direct calculation, as attempted by Milner, who considers each possible arrangement of ions and lets it enter into the computation with the probability corresponding to Boltzmann's principle, proved too difficult mathematically. We therefore replace it by another consideration, where the computation is, from the beginning, directed towards the average of the electric potential generated by the ions.' In a volume element dV where the potential is ψ, the average number of positive ions is $ne^{-\varepsilon\psi/kT}dV$ and of negative ions $ne^{+\varepsilon\psi/kT}dV$: ε is the electronic charge and n the average number of each kind of ion per unit volume. The net charge density is ρ:

$$\rho = n\varepsilon\left(e^{-\varepsilon\psi/kT} - e^{+\varepsilon\psi/kT}\right) = -2n\varepsilon \cdot \sinh\frac{\varepsilon\psi}{kT}$$

But Poisson's equation gives

$$\Delta\psi = -\frac{4\pi}{D}\rho$$

where D is the electric permittivity of the medium.

Except very close to the ion, $\varepsilon\psi/kT << 1$ and then

$$\Delta\psi = -\frac{8\pi n\varepsilon^2}{DkT}\psi = \kappa^2\psi$$

where

$$1/\kappa = \sqrt{\frac{DkT}{8\pi n\varepsilon^2}}$$

The length $(1/\kappa)$ introduced in this way is the basic parameter in the treatment: it replaces the average distance between ions in Ghosh's considerations. It represents the length over which the net charge density in the neighbourhood of one ion falls by a factor (1/e) and so measures the extent of the ion's 'atmosphere'. If $(1/\kappa) >> a$, where a is the mean radius of the ions in solution, the potential energy of the single positive ion with respect to its surroundings is $u = -\varepsilon^2/D\kappa$. When the ionic radii are not negligible, a factor $(1/1 + Ka)$ appears in this relation.

From these factors the calculation proceeded to evaluate the osmotic pressure, vapour pressure lowering, freezing point depression and boiling point elevation for the solutions. All these aspects are directly interrelated. As the most precise and the greatest body of data was in the form of freezing point depressions, these were calculated for typical aqueous salt solutions: $KCl : K_2SO_4 : MgSO_4 : La(NO_3)_3$. As κ was simply proportional to (concentration)$^{1/2}$, the already well-established fact that deviations from the simplest conditions in salt solutions increased as (concentration)$^{1/2}$ was immediately accounted for. The freezing point values were reproduced to within an accuracy of a few per cent up to concentrations of the order 1×10^{-2} gmol/litre—at which the deviations from the van't Hoff relations could already be more than 100 per cent. Using adjusted but plausibly acceptable mean ionic radii, Debye and Hückel extended their agreement up to concentrations approaching 0.1 gmol/litre. However, many other features apart from the ionic radius factor are now known to cause significant deviations from their original treatment. R. H. Fowler characterized the ionic radius term as an 'omnium gatherum' correction term best regarded as an empirical means of extending the fitment of experimental data to higher concentrations. Even so, the treatment was an instant success and for twenty years appraisals and refinements of it were made so that today it stands as an accepted major chapter in physical chemistry.

An appreciably more difficult task was undertaken in the second paper with Hückel: that of evaluating the concentration dependence of the ion mobilities in solution. It must be noted that the absolute values of these mobilities were not considered, merely their progressive decrease (again proportionately to [concentration $]^{1/2}$) from experimentally deduced values at 'infinite dilution'. As each ion moves under the applied electric field, the surrounding ions will also be constantly reforming the ion atmospheres. The important new feature is the relaxation time characterizing the rate of formation of the ionic atmosphere. With the ions in motion the ionic atmosphere will not attain its equilibrium distribution and the Boltzmann factor cannot now be applied to define it. Debye proceeded, not for the first nor by any means the last time, to use the equations for Brownian motion to provide a representation of the ionic distributions. For any one ion in motion its associated atmosphere will be more fully developed behind it than in the direction it moves. This provides a braking force which will increase with concentration. Another factor is the inevitable counter-movement of oppositely charged ions: as they all effectively carry some of the solvent in their immediate environs with them, any one ion will be moving through a solvent in some degree of motion in the opposite direction. This was assessed by the methods used by Helmholtz for the treatment of electrophoresis.

It will readily be appreciated that the theoretical treatment of ionic nobilities is very significantly more recondite a problem than treating the equilibrium features, and the paper presenting this material is far more impressive than the first. The evaluation of the retardation by the ion atmosphere is presented in five steps, the last representing the field and charge distribution around a moving ion. The electrophoretic forces generated by the countermovement of the ions are treated in three stages. The evaluation of the resultant conductivity and its general numerical computation for aqueous solutions form the next two sections. Then comes a detailed consideration of the available experimental data and their conformity with the theory. There is some case for suggesting that this last section is the most impressive in an impressively powerful and original paper. Debye was well versed in the variety of mathematical methods that had gone to make the theory. To find the considerable body of experimental data handled with the clarity, thoroughness, and relevance they receive, when he had done no experimental work in this field, is evidence of mastery which all can appreciate. Aqueous solutions are treated systematically in two sections, and non-aqueous solutions in a further section illustrating the roles of the permittivity and viscosity factors. Whilst most physical chemists have never seen this paper, its main features have been built into their textbooks for forty years.

Of course the Debye-Hückel papers were only the first efforts at a quantitative treatment of strong electrolytes. Much further refinement and extension has been given these considerations. Debye recurred to both aspects — equilibrium and mobility features— in a number of other papers. Especially there must be mentioned three major contributions in which H. Falkenhagen is a co-author. These treat of the frequency dependence of the conduction, an aspect clearly capable of giving critical information on the relaxation times of the ionic atmospheres, and other time dependent features in the mobility. Comparatively little has yet been done to explore these features experimentally. In the original treatment of mobilities Debye did not fully take into account the Brownian motion of the ions. This, however, initially escaped perhaps everyone's attention. He was, it seems, somewhat fond of telling the story of how, in 1925, there came uninvited to his director's office at Zürich, a timid young man smiling very nervously. Bowing to Debye over his desk he finally came out with: 'Wissen Sie dass Ihre Elektrolyt-theorie falsch ist?' It was Onsager. If this version (20) is correct, the youth of the visitor must be allowed to excuse the unqualified 'falsch'. It is always presumed that the immediate response was the offer to the visitor of a cigar, a chair and a writing pad: certainly for the first, there is a high probability.

Another item familiar to physical chemists has its origin in a Dutch paper (1923). This is an exposition by Debye of the influence of other electrolytes upon the solubility of sparingly soluble salts. No new theory is involved. The osmotic coefficient (i.e. one form of a thermodynamic activity coefficient) of the Debye-Hückel paper is used to give a quantitative account of the interionic effects. Using the data for silver sulphate as the insoluble salt, the various characteristic solubility influences are calculated and compared with experiment. This type of solubility study has proved to be a flexible and penetrating probe for many chemical equilibria. It was an unfortunate coincidence that Debye's electrolyte theory only became known in the same year as G. N. Lewis and H. M. Randall published their classic volume of chemical thermodynamics. That volume summarized with masterly clarity the experimental study over a period of thirty years of ionic and other equilibria, but it lacked the further insight made available by Debye's new treatment of electrolytes.

At Zürich, Debye's laboratory now attracted scientists from outside Europe: at the same time Debye himself spent time out of Europe. One product was a brief paper jointly with the young Linus Pauling, published from Pasadena (1925). It dealt with the electric permittivity factor in the electrolyte problem and confirmed Debye's earlier assumption of the macroscopic value for this in evaluating interionic potentials. A product of a winter term's stay at the Physics Department of M.I.T. was a more important paper (1925*) entitled *Note on the scattering of X-rays*. It reveals very clearly a formative phase in Debye's thinking on diffraction by liquids and by single molecules and it is both physically and mathematically a gem of great clarity. As these two problems justifiably occupied a great deal of Debye's later efforts let us attempt again to follow him.

'There can be no question that not only in crystals, but also in the molecules of substances in the liquid or gaseous state, the atoms occupy definite places... Therefore it should be possible to detect interference effects corresponding to the geometrical arrangement of the atoms in the molecule.

'If, for instance, the scattering of a liquid has been measured, one might possibly think that the properties of the scattering function, i.e. the scattered intensity plotted as a function of the angle between the secondary and primary X-ray, only depends on the dimensions and the form of the atomic frame constituting the molecule. From this point of view P. Debye and P. Scherrer in 1916 discussed the scattering of [liquid] benzene. In the meantime it became evident, however, that a very large number of different liquids yield diffraction-patterns which show only slight differences. Therefore, at the meeting of the German Physical Society in 1920 at Jena, Debye stated that the principal maximum of the scattering function must be due to interferences between the different molecules of the liquid... The theory of this effect seems to be as difficult as the theory of the correction for the dimensions of the molecules in the equation of state. Nevertheless, I believe that a first approximation... may be of some interest. It will be shown that even if the molecules are comparable with hard spheres and do not interact in any way... this alone is sufficient to cause a scattering function exhibiting a maximum.

'In this way it seems proved, both experimentally and theoretically, that it will only be possible, even in the case of liquids, to arrive at the scattering function characteristic of the molecule and its atomic frame, if we succeed in freeing the primary experimental result of the undesired intermolecular interferences. Further, it seems improbable that it will be possible to perform theoretical calculations which will give a reliable formula for this process in the case of dense liquids. Therefore, the only possible way to find the interferences due to the interaction of the atoms constituting the molecule, seems to be the performance of scattering experiments with gases of different densities... it should be possible to arrive at direct measurement of the atomic distances in the molecule.'

Debye then proceeds to calculate from first principles, and in the clearest possible fashion, the general form of the resultant scattering function for N structureless particles (e.g. spheres or small cubes) scattered through a volume V. With ψ = amplitude of the wave scattered by one particle: v_0 = total volume of the particles : 2a = diameter of the particles (there is confusion in the paper on this factor) : R the distance from the localized collection of particles at which the intensity (I_m) is observed: θ the angle between the secondary and primary ray, then

$$I_m = N\psi^2 / R^2 \left[1 - \frac{8v_0}{V}\phi\left(8\frac{\pi a}{\lambda}\sin\frac{\theta}{2}\right)\right]$$

$$\phi(u) = \frac{3}{u^3}(\sin u - u\cos u) \quad [\phi(0) = 1].$$

If the scattering of each particle were the same as the scattering of a single resonator, ψ^2 would be proportional to ½(1 +$\cos^2\theta$) as the primary radiation is assumed unpolarized and the scattering yields a polarization which is complete for an angle $\theta = 90°$. To visualize the effect of the scattering Debye draws the curve for the function

$$\left(\frac{1+\cos^2\theta}{2}\right)\left[1 - \frac{1}{2}\phi\left(12\pi\sin\frac{\theta}{2}\right)\right]$$

This corresponds to $(2a/\lambda) = 3$. Under these assumptions, a ring would be observed with a maximum intensity at an angle of about $\theta = 16°$.

'In general, the molecule will not act as a single resonator, but it is to be expected that there will exist an interference effect corresponding to the shape of the atomic frame. The whole scattering curve must now show a superposition of the two kinds of interference effects which may briefly be called the "inner" and the "outer" effect. Now the outer effect will always be proportional to v_0/V. It should therefore be possible to eliminate the outer effect which tends to vanish with decreasing density... then we may, in the case of nitrogen, for instance, expect as a result the direct measurement of the distance of the two N-atoms.'

Again, from the first principles, Debye modifies his previous calculation for structureless particles to one for molecules of radius a each containing two scattering centres distance I apart. He arrives at

$$I_m = 4N\frac{\psi^2}{R^2}\left\{\frac{1}{2}\left[1+\frac{\sin x}{x}\right]-\frac{8v_0}{V}\left(\frac{\sin(x/2)}{x/2}\right)^2\phi(\gamma)\right\}$$

$$x = 4\pi(l/\lambda)\sin\frac{\theta}{2}$$

$$y = 8\pi(a/\lambda)\sin\frac{\theta}{2}$$

With decreasing density the influence of the second term in the curly bracket vanishes and the scattering function finally assumes the limiting form:

$$I_m = 4N\frac{\psi^2}{R^2}\frac{1}{2}\left[1+\frac{\sin x}{x}\right]$$

agreeing with a general formula previously calculated (Debye 1915). Debye again takes $(2a/\lambda) = 3$, and $l = a$; here λ might reasonably be 1.0×10^{-8} cm. Retaining both terms in I_m (total), he 'shows how the first maximum, occurring at about $\theta = 12°$ and corresponding to intermolecular interference, disappears with decreasing density, leaving only a second maximum at $\theta =$

45° undisturbed, because this maximum corresponds to the interatomic interference of the two atoms constituting the molecule'.

Tlús model calculation must have given Debye great confidence in his persistent and, later, successful efforts to evaluate the diffraction patterns of both liquids and gases.

The list of titles of his published papers (the total exceeds two hundred—see Bibliography) emphasizes the sustained wide range of his contributions. In this account comment is restricted to the main or outstandingly original themes. It is noticeable, too, that the great majority of the papers carry only his own name: this probably results from their being still very largely theoretical and referring only to experimental data already published by other authors. Of his major interests, new aspects of electric dipole moments were delineated at Zürich. An especially valuable article appeared in the Handbuch der Radiologie, Vol. VI, 1925. In addition to a full exposition of the elements of dipole moment theory (promised in 1913), it includes an account of the Kerr effect in which Langevin's treatment is extended. This electrooptical effect is shown to offer information on the anisotropy of the polarizability of molecules-both polar and non-polar. The measurements are not easy and for forty years the systematic study of this effect has been in relatively very few hands. Laser sources and pulsed fields offer much improved sensitivity. Valuable deductions on many interesting aspects of molecular structure have been made from the Kerr effect starting from Debye's analysis (17). It was not until 1926 that Debye presented the definitive treatment of dipole-induced dipole interactions as a contributing feature in molecular cohesion. In the gas phase this is never comparable with the dipole-dipole or the dispersion force interaction, but in condensed phases it can become significant owing to the much closer proximity of the molecules and their assumption of orientations where the term may be at a maximum.

Perhaps the most original contribution published during this second period at Zürich was a brief but quite conclusive statement entitled: Some remarks on magnetization at low temperatures (1926*). This in itself would have created an international reputation for a lesser scientist. In it Debye expounds the principle of adiabatic demagnetization as a means of cooling at the lowest temperatures. It may be recalled that Langevin pointed out in his initial 1904 treatment of paramagnetism that the sudden (or, strictly, adiabatic) demagnetization of oxygen gas would necessarily lead to a cooling effect and that an observation of this kind had been made experimentally in ferromagnetic nickel (Weiss & Picard 1921). However, no serious consideration appears to have been given to its use in achieving the lowest temperatures. Debye's interest seems to have stemmed from an experimental study of gadolinium sulphate by Kamerlingh Onnes but he was also himself Weiss's successor at Zürich. He presented in quantitative form an assessment of the prospects for cooling on adiabatic demagnetization. There were two obstacles to an accurate prediction: firstly, the specific heat of gadolinium sulphate was not known at the relevant low temperatures-but he made a plausible order-of-magnitude estimate for this: secondly, the indications were that if the Langevin function for the paramagnetism held, it would result in the absolute zero being attained. 'Naturally, this does not mean that there is a possibility of reaching absolute zero with such an adiabatic magnetic process. Rather it must be concluded the Langevin function is not correct. However, the facts that on the one hand the Langevin formula is verified by experiments at 1.3° absolute and on the other that a large difference was found (above) between the possible magnetic and caloric entropy changes make it very probable that cooling considerably beyond the region of validity of the formula can be achieved. Therefore, it may be of interest to make measurements at low temperatures on the adiabatic cooling of gadolinium sulphate which can be expected when the magnetic field is switched off suddenly. *Footnote in the paper*: In addition, note that the specific heat measured at constant magnetic field CH should differ noticeably from the specific heat at zero field, Co, for from (8) it follows immediately that (CH-CO)... is already 0.15 nk. We do not dare to venture a prediction of the magnitude of the cooling which can be realized: however, it appears not to be excluded that this can be large. Only experiments can decide, and the above analysis should stimulate the carrying out of these.'

Debye's statement appeared in 1926. In 1927, quite independently, Giauque at Berkeley published a similar proposal. This proposal was certainly a stimulating one but the experiment was not then readily attempted in many laboratories. Giauque eventually reported (19 March 1933) cooling from 1.5 to 0.25 K; a group of Leiden workers achieved a very similar result almost simultaneously (6 April 1933). Subsequently, temperatures below 10-3 Khave been attained by this method.* For this and related work Giauque was awarded the 1949 Nobel Prize in Chemistry.

* Professor Bauer [Cornell] writes that in seminars (ca. 1940) Debye discussed the possibility of using nuclear spin demagnetization: this has led to temperatures of 1 Q-6K. He also emphasizes how, throughout Debye's work, one continually encounters his profound grasp and masterly deployment of classical physics.

LEIPZIG, 1927-1934

In 1927 Debye left Zürich to become professor of experimental physics and director of the Physical Institute at Leipzig University: at the same time Heisenberg was appointed to the professorship in theoretical physics. (Some of Professor Heisenberg's impressions of Debye at Leipzig are given in a letter: Appendix A). It was during his period here that Debye's orientation to chemical physics became almost complete. One characteristic feature was the organization, with the help of the Ministry of Education for Saxony, of 'small and intimate' conferences in Leipzig on specialized topics. In 1928 the subject title was Quantum Theory and Chemistry; in 1929, Dipole Moments and Electromagnetic Radiation; in 1930, Electron Diffraction; in 1931, Molecular Structure; and in 1933, Magnetism. At these symposia, leading active contributors presented up-to-date statements on the topics discussed: these were published (as five monographs) by Hirzel of Leipzig and, in English translation, by Blackie in the U.K. The volumes (edited by Debye) were highly valued additions to the scientific literature and the discussions set a pattern much copied since.

Leipzig became a Mecca for physical chemists and molecular physicists. The translation from Zürich was accompanied by a change in direction in Debye's work. The electrolyte studies were phased out and there was a resurgence of developments in the diffraction studies of molecular structure. Firstly, however, the molecular aspects of dielectric behaviour, more particularly of gases and liquids, were given a systematic presentation in what has become a *locus classicus*: the volume *Polar Molecules* also appeared in a somewhat more extended version in German: *Polare Molekeln* (Books). Largely owing to the stimulus provided by this monograph, chemists now took a lively interest in molecular dipole moments. The topic received considerable attention in the 1930s and aspects outlined by Debye, especially those dealing with dipole relaxations, are still being actively pursued.

It was at Leipzig that Debye was able to achieve for the first time clear X-ray diffraction by isolated molecules. This was an advance of major significance in the work already started with Scherrer: the possibilities in this area had been considered even before their joint paper of 1916. A notably successful experiment was made with carbon tetrachloride vapour and interference rings were photometered and evaluated to give a chlorine-chlorine intramolecular distance of 3.1 A (1929). This value was quickly improved 'to an accuracy of about 1 per cent' as 2.99 A: the later, better, value is 2.89 A.

There are two footnotes of particular interest in the definitive account of this new method (1930*) : an account, incidentally, which carries no acknowledgement of the fact that the experimental work was due to L. Bewilogua and F. Ehrhardt, although this had been made clear in an earlier note (1929). The first footnote reads: 'The investigations reported on here began in 1915 in a note entitled "Scattering of X-rays" (P. Debye, *Ann. Physik*, **46**, 809, 1915). This note contained the basic formulae used here, to demonstrate the essential point that random orientation cannot destroy the appearance of interference maxima. It is dated 15 February 1915 and is marked by the editors as "received 27 February 1915". By a strange coincidence, a paper on the same subject by P. Ehrenfest presented by H. A. Lorentz and H. Kamerlingh Onnes (*Amst. Akad.* **23**, 1132,1915) was given at the meeting of the Amsterdam Academy on Saturday, 27 February 1915. The title of that paper is: "Concerning interference phenomena to be expected when X-rays pass through a diatomic gas." The two atoms are considered as two scattering points with a fixed distance between them, and it is shown that for this special case, also, interference effects should be observable in spite of random orientation.'

A second footnote reads: 'In the meantime Messrs Mark and Wierl have succeeded in performing a beautiful experiment (*Naturwiss.* **18**, 205, 1930) in which the interferences here discussed are obtained with cathode rays, the de Broglie wavelength of which is about thirty times as small as the wavelength of the X-rays we used They give as a result of their measurements on CC14 the distance (Cl-Cl) as 3.14 A. This value, however, should be corrected as we did our earlier value of 3.1 A, with regard to the angular decrease in intensity which occurs with, β-rays just as with X-rays. Based on the Schrödinger equation, this correction can be calculated, at least for sufficiently fast electrons, in a similar manner using the Fermi distribution.

'Herr Mark mentions that Herr Bothe suggested the possibility of using β-rays instead of X-rays at the occasion of a lecture I gave at Zürich. I may add to this that Herr Pringsheim made a similar remark after an earlier lecture which I gave at the Berliner *Physikalische Gesellschaft*.'

There followed a succession of important molecular structural studies by X-ray diffraction in gases. After some simpler structures (Cl_2, O_2, CO_2, CS_2) the halogenated methanes and ethanes were examined, with notable extensions of the details of their structure and the appreciation of the limited rotational freedom in 1.2 dichloroethane, CH2Cl.CH2Cl, with its 'cis', 'trans' and 'gauche' forms. Ethylenic isomerism and benzene were successfully evaluated by 1932. In the meantime, Mark and Wierl's electron diffraction method had also developed strongly, and as the intensities were such that satisfactory diffraction patterns could be obtained with electrons in seconds where the X-rays required hours, it was the electron diffraction method which took over this aspect of molecular structural studies.

Other developments at Leipzig included the X-ray diffraction study of liquids. Following Zernicke and Prins, Debye explained how the angular dependence of the scattered intensity could be inverted, using Fourier's theorem, to give the distribution function representing the probability of finding the molecules in the liquid at particular separations. With Menke, experimental data were first obtained for mercury for which 'The curve thus shows that even in the liquid state there exists a quasi-crystalline structure which is defined quantitatively in terms of the probability curve' (1930*). Debye repeatedly approached liquid phase structure in terms of the concept of short-range order. The frequency dispersion of electrolytic conductivity and the influence of high-fields on ion mobility were also considered. For dielectric media Debye appears to have treated the non-linear field effect only in the Handbuch der Radio[ogie (1925; 2nd Ed., 1934). This is surprising, as that effect in polar liquids can be larger, more readily observed and more revealing than in electrolyte solutions. Its exploration is largely due to Piekara (18).

The dominating theme in Debye's scientific interests, the interaction of radiation with molecular systems, acquires another notable illustration in the study of the scattering and diffraction of light by ultrasonic waves. The clarity of Debye's statement again demands quotation: it is based upon his discussion in 1914 on the thermal conductivity of insulators, which marked the birth of phonon theory: developments in relation to electron conduction in metals (Bloch 1928) and thermal conductivity (Peierls 1929) had come later.

In the 1932 account with F. W. Sears, he wrote:

'In a paper published in 1922 Leon Brillouin treated the problem of light scattering. In accordance with the fact that for low temperatures Einstein's theory of specific heat has to be abandoned for Debye's theory, Brillouin attributes the thermal density fluctuations in the body (which, in his theory, as in a previous theory of Einstein's are responsible for the scattering) to a superposition of sound waves. He tries to apply his theoretical results to the explanation of X-ray scattering. We know now that this application is far from correct, as for such short waves the electron density changes due to the atomic or molecular structure are much more important than the thermal fluctuations. For light waves, however, with a wavelength much longer than molecular distances, Brillouin's analysis leads to some remarkable results. They can be stated in the following manner. Suppose the primary light travels in Fig. 1 in a direction characterized by a vector $\mathbf{S_0}$ of length unity in this direction. Let it be assumed that of the scattered light a part is observed travelling in another direction characterized by a unit vector, **S**. Then firstly, of the sound waves of all possible directions, only those are important for the scattering which are travelling in or opposite to the direction of the vector **s** = $\mathbf{S} - \mathbf{S_0}$. This can also be expressed by saying that the planes of the sound waves have to be situated such that the scattered light can be considered as optically reflected by these planes. But there is a second limitation. Of all the sound waves of direction ± **s**, only those of a definite wavelength Λ are effective. This wavelength is $\Lambda = \lambda/s$ where λ is the wavelength of the light and **s** is the length of the vector **s**, which is $2\sin(\theta/2)$, calling θ the angle between the primary and the secondary ray. This last condition can be expressed by saying that the consecutive planes of maximum density in the sound wave must be separated by a distance Λ such that the well-known relation of Bragg holds. In this case the light rays reflected by the consecutive planes will have path differences of one wavelength (of light) each and therefore the reflections will be strong. So we are left with only two sound waves, travelling with the velocity of sound q, one in the direction of + **s** and the other in the direction of – **s**. The frequency of the reflected light, according to Doppler's principle, will be changed, and instead of the primary frequency ν_0, we shall find in the scattered light the two frequencies:

$$\nu = \nu_0\left[1 \pm \frac{2nq}{c}\sin\frac{\theta}{2}\right]$$

where c is the velocity of light and n is the index of refraction of the medium.'

Debye points out that this effect had very probably been seen by Gross in 1930 and by Meyer and Ramm in 1932, but there were some divergences in the echelon examination of the scattered light. He then describes the experiment done by Sears at his suggestion during a visit to M.I.T. in April 1932. A delightfully simple arrangement revealed many significant aspects of the interaction of the two wave-trains not covered by the simple statement already quoted. 'The original purpose was to demonstrate Bragg reflection by sound waves by adjusting the reflection angle. However, another phenomenon appeared immediately. As long as the quartz crystal (generating sound waves of length a few tenths of a millimetre) did not vibrate, only the image of the slit (illuminated, e.g. by a mercury arc lamp) was visible in the focal plane of the second lens. However, as soon as the quartz crystal was activated, diffraction images appeared to the left and right of the central image. They were completely analogous to the grating spectra of an ordinary grating. Each spectrum shows colours which appear in the usual order; the spectra are equidistant. The number of visible orders depends on the intensity of the sound vibrations. It was possible to make more than twenty orders visible to the left and right of the central image. The phenomenon is very brilliant and the images can be directly projected and demonstrated in a large auditorium... The phenomenon therefore gives a very simple method for determining the speed of sound, since a simple measurement of the diffraction angle yields the ratio of the wavelength of the light to the wavelength of the sound.' Sears has commented on Debye's immediate reaction to the unexpected features in the observations: 'One thing impressed me greatly. When I had called Debye into the laboratory and showed him the experiment, he rushed back to his office and began a theoretical explanation by first writing down Maxwell's equations.' It is best, again, briefly to quote Debye.

'... a theory has been developed based on the assumption of a volume scattering, in which every volume of the liquid contributes to the total scattering in accordance with Maxwell's equations. In this way, it seems at first that one would expect only one reflection for a definite angle of incidence... Taking into account, however, that the dimensions of the illuminated volume of the liquid are finite, it can easily be shown that in our case Bragg's reflexion angle is not sharply defined and that reflexion should occur over a rather appreciable angular range. If l is the length of the path of light in the liquid, Λ the wavelength of the supersonic waves and λ the wavelength of the light, then two quantities are of importance: namely, the quotients l/Λ and Λ/λ. Only if l/Λ is large compared with Λ/λ does a sharp definition of Bragg's angle exist. Working with a frequency of 10^7 cycles, Λ is about 0.1 mm, l is of the order 10 mm and λ is about 0.5×10^{-3} mm. In this case, therefore, $l/\Lambda = 100$ and $\Lambda/\lambda = 200$, the quotient of these two quantities is ½ and cannot be considered as large. A detailed analysis shows that in such a case reflection will occur over a range of angles left and right of the critical angle which follows from Bragg's relation. Moreover, the intensity variations predicted by the theory in varying the angle continuously are just the same as described in the rather peculiar experimental results on this point.

'We are, however, still left with another difficulty. The theory predicts only the first order spectrum... but it has assumed that the variations of the refractive index are of purely harmonic character. If they are not, then we can consider the disturbances as a superposition of frequencies ν, 2ν, 3ν, etc., with the corresponding wavelengths Λ, $\Lambda/2$, $\Lambda/3$, etc.,... This departure may be due to the non-sinusoidal character of the crystal vibrations, although higher harmonics may be produced by the scattering itself, if the intensity of the supersonic waves is high enough.'

It is difficult to imagine a simpler example which would reveal so many aspects of diffraction theory. In view of his obvious appreciation of the numerous possibilities for studying liquid behaviour which this experiment provides, it is, perhaps, surprising that Debye scarcely pursued the method further.

BERLIN, 1934-1940

In 1934, it seems, Debye already found he could not fully isolate his work and the Physical Institute at Leipzig from undesirable political interference. In 1935 he moved to the directorship of the Physics Institute of the Kaiser-Wilhelm-Gesellschaft at Berlin-Dahlem. This had previously been headed by Einstein and von Laue but, then, as an institute for theoretical physics. Now, thanks to generous grants from the Rockefeller Foundation, magnificently equipped laboratories for experimental research were also available: the equipment included a very-low-temperature laboratory; high-voltage (3 MV) ; high magnetic fields; all chemical facilities; and excellent workshops. He caused the name to be changed to that of the Max-Planck Institute, a title ultimately taken over by the Gesellschaft as a whole. Simultaneously, Debye became professor of physics at the University of Berlin. These were State appointments in Germany, and as Debye had retained his Dutch nationality they necessitated his obtaining permission to accept the positions from his sovereign, Queen Wilhelmina. He quickly experienced police-state interference in the Berlin Institute, but up until his leaving, he also appears to have been treated with some deference by the Nazi authorities.

No major new themes appear in his own work at Berlin. Nevertheless, there were significant developments in many interests such as the dielectric properties and quasi-crystalline structure of liquids; paramagnetic relaxation; the achievement of the lowest temperatures by adiabatic demagnetization.

These are recognizable extensions of earlier work. The application of the method of Fourier analysis to the evaluation of electron diffraction data notably extended the power of that method, as did the testing by his son, P. P. Debye, of the sector method on the experimental side (19). Debye also provided a theoretical treatment of the Clausius thermal diffusion method for isotope separation (1939).

With the arrival of war in 1939, Debye was soon confronted with unacceptable alternatives in the Institute. The authorities decided, without his knowledge, that the Institute should be used to explore the development of a uranium reactor; this was planned as a secret operation. Debye's account in the introduction to the volume entitled Collected papers (1954*) is:

'At the time I accepted to go to Berlin I was still a Dutch citizen... At the same time the German government conceded in a letter signed by the Minister of Education, Dr Rust, that in accepting the positions offered, I did not become and would not be asked to become a German citizen. The positions, as usual, were lifetime positions. During the time the laboratory was still under construction, I received an offer from Harvard University, which I declined because I did not feel free to quit before having finished what I had undertaken to do... The war broke out and one day without previous warning I was informed by Dr Telschow from the Kaiser-Wilhelm-Gesellschaft that I could no longer enter the laboratory except by becoming a German citizen. I refused. I was advised by the Minister of Education to stay at home and occupy myself by writing a book. Instead, I was able to overcome the difficulties put in my way by different German authorities and to leave for the U.S.A., byway of Italy, in order to give the Baker Lectures at Cornell University, to which I had been invited.'

Except by omission, this statement does not tell us anything of Debye's assessment of the Nazi regime. A Dutch colleague has commented: 'Debye's strong tie to Germany and perhaps even more his notable lack of political interests, was almost fatal for him. After 1933, the year in which Niels Bohr travelled to the German universities to see what he could do for the threatened (Jewish) physicists, Debye who, as a faithful Roman Catholic and as an honest Limburger, could have nothing to do with National-Socialism, did not immediately realize that there was no further place for him in German physics' (20). Certainly for Debye to have imagined in 1940 that his position had not been radically altered by the arrival of war: to have earlier thought, if only in view of what had happened in Einstein's case, that he should not leave Nazi Germany 'before having finished what I had undertaken to do', all indicates a considerable indifference to the political situation. It is probable that he had many offers of posts in the U.S.A. As early as 1915 G. N. Lewis, head of the Chemistry School at Berkeley, had written to Rutherford on behalf of colleagues in the Physics Department requesting his opinion on two promising young scientists: Peter Debye and Niels Bohr. The reply has been published (Appendix B). By the mid-thirties Debye's status was such that his assistants claimed their salaries could best be quoted in milli-debyes. One invitation to leave Berlin for Harvard is well-established: but a variety of reasons-scientific, financial, historical attachments and family affiliations-help to explain the late date at which he left the Nazi capital. However, Debye eventually arrived in the United States for what was to be his last appointment.

Having expressed his unpreparedness to accept the German Government's wishes, in leaving Berlin in January 1940, Debye was obliged to abandon his home and all his possessions there. His son had gone to the U.S. in the summer of 1939 but his wife remained in the German capital when Debye left Berlin, ostensibly to lecture in Zürich. From there he moved to Milan and telegraphed the family in Maastricht to send him money. With that help he was quickly able to reach New York. His wife joined him later, but his daughter and her two sons did not do so until after the war.

In 1936 Debye had been awarded a Nobel Prize in Chemistry 'for his contributions to the study of molecular structure through his investigations on dipole moments and on the diffraction of X-rays and electrons in gases'. The citation provides ample grounds for the award even if it is no more inclusive of Debye's overall contributions than was Einstein's (1921, in Physics, 'for services to theoretical physics and the law of the photoelectric effect'). At this stage Debye's whole career had been as a professor of physics, but he was not the first physicist to be awarded a Chemistry prize. Rutherford was the first (1908); Marie Curie's second prize (1911) and that shared by Frédéric Joliot and Irene Joliot-Curie (1935) were other instances. Debye appears to have had no regrets in this choice. As much as anyone's, his work had served to close the gap between chemistry and physics and later he enjoyed a double status. Writing (28 December 1964) of the Lindau (Bodensee) meetings of Nobel prize-winners, he explained: 'These meetings are biennial, one year for physicists, one year for chemists. I always get invitations to both, probably because I belong to the status in between. I have only been there once as a chemist. I decided to attend a second time as a physicist' (21).

CORNELL UNIVERSITY, 1940-1966

Debye had an invitation to give the George Fischer Baker lectures in the Chemistry Department at Cornell University in 1940. These were immediately followed by his acceptance of the chairmanship of the Department. And in the charming surroundings of the Cornell campus he spent far the longest of his university stays. From 1952 he was Professor Emeritus but his work continued with unabated energy until 1966: there are some thirty three publications between 1940-1952, and fifty after that date. In 1946 he became a citizen of the U.S.A.

The disruption of war and the possibility of the U.S.A. becoming involved in it were already clear when Debye arrived in Cornell. Whilst his status as a Dutch citizen may have kept him out of some projects, he was quickly brought in as a consultant on others. The dependence of the U.S.A., as of much of the rest of the world, upon Malayan rubber had led to an intensive programme of studies for substitutes. These were necessarily polymeric materials and this determined Debye's principal new interest. Basic to the evaluation of polymer properties is the knowledge of an appropriate mean molecular weight value and, in solutions, information on the configuration of the particles and their interaction as the concentration increases. For these features Debye developed the powerful light-scattering method. He himself explained the basis of this new technique (1947): 'The problem can be approached either by making a detailed calculation of the electromagnetic field surrounding a particle in order to derive the loss of primary light energy due to its radiation (Rayleigh, 1871) or by treating the effect of the molecular inhomogeneities on the light in a second approximation as due to spontaneous fluctuations of the density and the concentration in a medium which in a first approximation is considered to be perfectly homogeneous (Einstein, 1910).' The detailed information he was able to derive from the study of light scattering provided the most practical proof of his unique grasp of the possibilities.

In 1935 two French workers, Putzeys and Brosteaux, had successfully applied Rayleigh's relations to spherical protein molecules in solution (22), but the method had not been generalized. Debye refers to this work. In one of his original presentations (1947) he deduces the turbidity of the solution (τ) using electromagnetic theory and the time average of Poynting's vector. If I is the light-beam intensity and x the direction of propagation,

$$-\frac{dI}{dx} = \tau I = \left(\frac{128\pi^5}{3} \frac{n}{\lambda^4} \frac{m^2}{F^2} \right) I$$

...'the difference in dielectric constant between the solvent and the solution is proportional to n, the number of particles per cubic centimetre, and to m/F, the electric moment characterizing the particle in a field of unit intensity. The turbidity, according to equation 2 (above) is also proportional to n, but, unlike the dielectric constant, proportional to the square of m/F'.

His working relation for small particles took the simple form:

$$H\frac{c}{\tau} = \frac{1}{M} + 2Bc\,.$$

H is a constant dependent upon l/λ^4 and the refractive index (substituted in practice for the electric permittivity) of the solution: c is the concentration (g cm^{-3}) and M the desired molecular weight. B is a factor expressing the interaction of the solute molecules in the solution and it is closely analogous to the second-virial coefficient of a gas in the form (B/V2). These are the simple essentials of the analysis. Considerable further detail can be deduced from the angular dependence of the scattered intensity, an aspect which only becomes significant when the particle (i.e. molecular) size approaches the same order of magnitude as the wavelength of the light, i.e. for particle dimensions of 1000 Å or more. The theoretical analysis and experimental study of these systems was the subject of about twenty papers (1944-1964). Some interesting practical problems associated with the small refractive index changes and the small scattered light intensities which had to be measured led to simple, novel and effective experimental methods: of these he was clearly pleased to write: 'Most of the instruments have been designed by P. P. Debye'—his son, who worked with him for a number of years.

The light scattering method is not a simple procedure for molecular weight determinations: it requires careful control in the preparation of the solutions and in the physical measurements. Its accuracy rarely exceeds a few per cent but that is ample for the polymer specimens usually measured. Most of the classical methods have to be abandoned at molecular weights above 10^4 and it is for such higher values that the light scattering method is readily applicable. The upper limit approaches 10^7 and it has been taken in special cases down to molecular weights below 10^2. This gives it a wider range than any other single method.

In 1939 Debye gave a theoretical treatment of the Clusius-Dickel thermal diffusion method of isotope separation: but in Debye's model the diffusion took place in solution. The major aspects of his analysis were confirmed experimentally. This led, in 1945-1946, to a systematic study of thermal diffusion and fractionation in solution, not only of essentially simple, medium-size molecules (e.g. toluene-chlorobenzene) but also of dissolved polymer molecules. Some striking results were achieved, supporting the general analysis. Initially, the theory did not directly relate the thermal diffusion to the molecular weight so that that aspect was of a purely experimental character.

Not surprisingly, Debye also became interested in other aspects of polymer molecule behaviour. He treated theoretically the viscosity of their solutions and produced an interpretation of the important empirical method of evaluating the molecular weights of polymers from these viscosities. This method, first introduced by Staudinger in 1926, had been the subject of much discussion and disagreement. Another aspect of such solutions to which Debye became attached was micelle formation. As. the concentration is increased in solutions of many medium molecular weight compounds, it can happen that there is a rapid, but reversible, agglomeration of the molecules to large clusters termed micelles. This process depends upon intermolecular interactions of an attractive character and were it to continue to the level of macroscopic particles it would be an example of phase separation. These features, both micelle formation and the critical conditions reached in the separation of a new phase, now received Debye's attention. They were admirably suited for light scattering studies, but other techniques including X-ray scattering, ultra-sonic measurements and high-frequency electric methods were also used. These studies were being actively pursued up to the last weeks of his never less than extremely busy life. Even after his eightieth birthday he refused to contemplate the possibility that a just normally active schedule of scientific activity would be acceptable.

A single instance of his latter-day vigour must suffice as evidence of its unusually high level. The National Bureau of Standards held a conference in Washington in the spring of 1965: the subject was 'Critical phenomena'. Over eighty years of age, Debye made a great impression on an international audience of younger scientific leaders, many of whom had not previously experienced him in action at such a meeting. Despite the intensive modern developments in the subject he was able frequently to add important and illuminating comments to the discussions; those at the end of one session led the colleagues who heard them to insist on his writing them up separately. They are, accordingly, on record (23), as are the impromptu remarks (24) he made as chairman of one session where the first speaker was unable to present a contribution so that Debye took the opportunity 'to speak two minutes to give you my conception of what he would have said. I think he would have said that if you want to look at what radiation is going to do when it falls on a medium with irregular fluctuations, you have to characterize these fluctuations first in a phenomenological way... characterizing these fluctuations by a correlation function... this correlation function has to be a correlation function in space as well as in time... from this correlation function you can calculate the intensity of the scattered radiation directly by a simple Fourier transformation... you have to look at two things in the intensity distribution. First, how is the intensity distributed over the angles? This geometrical feature gives you an opportunity to characterize a length... you can also make a spectral investigation of the scattered light, which would be characteristic for time correlations... this is characterized by a correlation time or by a relaxation time, if you want to call it that way, depending on the circumstances. This would be completely phenomenological.

'I was prepared to tell him that this is a linear theory, which you generally call a Born approximation. If you come to the neighbourhood of the critical point, the correlations become bigger and bigger and bigger, so that probably your linear theory will not work any more. If you want to tell something about the angular distribution of the frequency distribution of the scattered radiation derived from actual observations, you have to introduce the right theory. The linear theory may not be correct at the critical point, because you get such big fluctuations. So I had hoped to induce him to say something about the second approximation and about what we are going to see there.'

Professor Widom has summarized his later research activities at Cornell (25): 'The major thrust of his work at Cornell was certainly scattering from amorphous media, mostly light-scattering, applied to the determination of the sizes and shapes of the scattering centres... With his 1959 paper on critical opalescence Debye started to apply his light-scattering ideas and techniques to the study of critical phenomena, especially to critical mixing points in solutions, where from the intensity and angular dissymmetry of scattered light one obtains a measure of the magnitude of the huge concentration fluctuations that characterize a critical point and a measure of their spatial extent, or of the distance over which they are correlated. In his theoretical work on this subject he essentially re-derived (without fully realizing it) the Omstein-Zernike theory of critical opalescence...

'Debye interpreted the earlier light-scattering measurements of Zimm in terms of this theory, showing that they fit in their major features, and he went on to make measurements of his own. He really initiated the systematic measurement, by light-scattering, of correlation lengths near critical points, now a very active and important subject. Besides studying the critical mixing point of solutions of "small" molecules, he made analogous measurements near the critical solution point of polystyrene solutions, thus making contact with his earlier polymer work. Also, he managed to make contact with his early love, electrical forces, by predicting, and then measuring, the effect of an electric field on the critical solution point and on the associated critical opalescence. Indeed, he saw in this effect the possibility of determining the relaxation time of the concentration fluctuations, which was his last preoccupation before his death (Bibliography, 1967). He hoped also to see the space-and-time dependence of critical concentration fluctuations by direct visual observation (not the electric field effect) : this was the subject of the last, posthumous paper (1968). The whole question of the space-and-time dependence of fluctuations, the relation between them and their manifestation in the spectral width and shape of scattered light was the central issue in the remarks he made at the Critical Phenomena conference.'

Debye's 'retirement' from the active headship of the Department at Cornell University had included a heavy programme of lecturing and consulting up and down the U.S.A. and usually two trips per annum to Europe, on one of which he would visit his sister in Maastricht. One such trip was to be in June-July 1965 when he planned to attend meetings in Germany and the U.K. and to visit Maastricht: he was busily engaged on the details of a study week he was organizing for the Pontifical Academy of Sciences at the Vatican City on the subject of Molecular Forces, and he was due to address the IUP AC Polymer Conference at Prague. These were merely his European plans at the age of eighty-one. Most of them, however, were abandoned. He wrote in February 1965: 'A medical check-up which I had a few days ago has disturbed all that. My doctor declares that I am carrying much too heavy a load, which according to him and in his words is "killing me"... I am not an invalid by any means and may still have a birthday or two in conformity with your good wishes.' This appraisal could be taken to suggest that his judgement on personal affairs was as sound as on scientific matters. He attended the Prague meeting in September 1965 but suffered a heart attack in April 1966 at Kennedy Airport when leaving on another European trip. Even then it is recorded that within a month he had resumed activities, including travelling to lectures and consultations. Certainly he returned to active participation in his group's work at the Cornell laboratories and when finally obliged to take to hospital he at first successfully insisted on having the use of a telephone to the laboratory in his oxygen tent. 'Where your treasure is, there will your heart be also.'

HIS SCIENTIFIC STATUS

Debye's work embraces so many major contributions to present-day chemical physics that it would be a most difficult task fully to appraise it. Fortunately a detailed appraisal is unnecessary as the student of that area in science needs no more than the reminder: *Si monumentum vis, circumspice.* Were an adequate assessment to be attempted it would require a careful evaluation of what precisely had preceded Debye's work on a variety of topics and (an even larger task) what were the developments directly attributable to his contributions: only then would the true extent of its value be established. For the present a summary list of only the major items need be attempted. With an indication of any earlier contribution of significance this might be:

Specific heats of crystalline solids (Einstein).

Electric dipole moments (Langevin).

The dielectric medium and molecular dipole relaxations.

Lattice vibrations and X-ray diffraction.

The rule for quantization in terms of classical mechanics.

X-ray diffraction by powders : amorphous materials: liquids: and gaseous molecules.

The space-quantization of electron orbits (Sommerfeld).

The electron-atom energy exchange (Compton).

Equilibrium properties of strong electrolyte solutions (Milner): their conductance and its frequency and field dependence.

Adiabatic demagnetization.

Phonon theory and the diffraction of Light by sound waves.

Light scattering by solutions (Rayleigh, Einstein) : the weights and shapes of polymer molecules.

The question can at least be asked whether, in the broad area of molecular physics, any single individual since Faraday has contributed so much.

Perhaps one spot-check on the magnitude and quality of Debye's gifts may be proffered the reader. In the volume Chemistry at the Centenary (1931) Meeting of the British Association for the Advancement of Science there are three brief reports by Debye of his contributions at the London meeting (1932). Each was on a different topic of major interest; each was being pursued in his own laboratory; and each was the result of his own original contributions to molecular science. Scarcely anything in these statements needs to be modified today: more than that, within their brief compass and at their level, it would be difficult, even forty years later, to give more clearly or more correctly a better synopsis of these three topics. They provide examples for all to study of that power of insight and mastery of expression which characterized the genius that was Peter Debye.

Debye's career was of a piece with the quality of his mind and his energy. There was a frequent change of location, as of interest, and a measured description of it has the same limitations as those of a kaleidoscope: the clarity and brilliance are accompanied by repeated changes of pattern and significance. Thanks to their fundamental character and their very high degree of originality it is possible that his pre-1920 contributions outweigh in significance even the totality of his many subsequent achievements. The decade 1910-1920 provided the foundations of a new molecular physics on which we are still building. The last decades, spent essentially in physical chemistry, cannot compare with his European period in physics. Whilst it needed a Debye to evolve the light-scattering method in its powerfully complete theoretical and experimental form, it is certain that some of his many interests in polymers could equally well have been developed by others. The chemical physicist is thus left with the still profitable exercise of considering what further branches of molecular behaviour Debye might have pursued had he remained in closer contact with physics after 1945.

Debye's status was acknowledged, apart from his Nobel Prize, by the conferment of honorary degrees by sixteen universities (including Oxford) and by his election to twenty or more national or regional scientific academies. A list of the major medals presented him suggests a roll-call of the pioneers on whose work he built: with the countries making the award, these included the honours associated with Lorentz (Netherlands), Rumford and Faraday (the Royal Society and the Chemical Society, U.K.), Max Planck

(Germany), Joseph Priesdey, Benjamin Franklin and Willard Gibbs (U.S.A.). One distinction which pleased Debye perhaps before all others was the unveiling in 1939, in the Town Hall of his native Maastricht, of a shoulder-length bust. It is, in any case, certain that Debye never placed himself on a pedestal and there is likewise no chance that his place in molecular science will diminish with his passing. He belonged to a generation of academic scientists who rated themselves on the quality and continuity of their contributions to their chosen field, criteria now often diluted with other more general activities.

PERSONAL QUALITIES

It remains only to attempt a picture of the man, Peter Debye. This can fortunately be done with far greater conviction than a single individual could achieve by reference to the numerous appreciations written of him by colleagues and friends, some of whose close relations with him extended over several decades. A reader who did not himself know Debye will thus become convinced that he was an extremely pleasant, lively character who was fully as able to contribute pleasure to his friends and associates as to add to their insight on scientific matters: and that not only was he able to do this, but that he constantly enjoyed doing so.

His teacher, Arnold Sommerfeld, has already been quoted on Debye's prowess as a theoretical physicist. Forty-five years after first meeting his pupil he was to say (1): 'His motto in science and in life I would give as "It's all terribly simple"-("Das ist alles furchtbar einfach"). I have recently heard this from his lips as he retailed his present work on the long, coiling, molecular chains of fibrous materials.'

Professor R. M. Fuoss, an associate of many years in the U.S., wrote in 1954 (26): 'Debye is known to his colleagues through his published work, but better by his active participation in meetings and by the lectures he delivers on various occasions... And to hear a Debye lecture is a real treat: he has an uncanny skill in presenting seemingly complicated subjects in a fashion which gets to the nub of the problem with a penetrating clarity. Best of all, though, Debye is known as one to whom we can go for research advice. As described in the citation from Harvard when he was made an honorary alumnus, Debye is "a large-hearted physicist who gladly lends to the chemist a helping hand".'

Debye's immediate colleagues at Cornell University (Professors Sack, Widom and Bauer) have characterized his *modus cogitandi* (27): 'In science, as in art, there is style. Debye's theories, his ways of looking at physical phenomena and of expressing his understanding of them, were as uniquely Debye's as a painting is unmistakably an El Greco or a van Gogh. The essential element of his style was simplicity, which for Debye was not merely a technique: it was an earnest conviction. He knew that physical phenomena must have simple explanations; he took complexity to be lack of understanding. If a theory was not yet simple then it was not yet right—it was unfinished and imperfect. To achieve simplicity one must identify the essentials and isolate them from the irrelevancies. To recognize the essentials, to express them clearly and pictorially, and then to pursue their consequences with superb technical facility was Debye's style.

'...Though he had mathematical abilities of the highest level (one of his earliest papers contained the independent discovery of the method of steepest descents, and its application to the asymptotic behaviour of Bessel functions), he had a deep distrust of overly mathematical theories, and dismissed as "mere mathematics" any explanation of a physical phenomenon that lacked a concrete, visualizable basis.

'... Almost unique among theoreticians, he was not only vitally interested in explaining experimental results and suggesting new experiments to test a theory, but he participated actively by giving practical advice, designing new laboratory techniques, and following the day-to-day progress of his experimental co-workers... A number of his investigations actually started from industrial problems that came to his attention. It is thus not surprising that he was much sought after as a consultant.'

Reference has already been made to Debye's exceptional effectiveness as a lecturer. He possessed great facility of expression in the major European languages. His powers of clarification were a consequence of his own understanding. His mind penetrated farther than most and, equally remarkably, it seemed to possess a highly discriminating mechanism which almost automatically focused on the principal features in a hitherto unknown area. His own account of a colleague's book provides an indication of much more than the virtues he wished to praise in another's writing (28).

'The development, after the introduction of any new subject, is never presented as an example of strictly logical deduction from experimental evidence, carefully arranged Postfactum. Nor is it derived as a logical sequence of a mathematical formulation, appearing seemingly from nowhere at the very beginning of the argument.

'The sequence is described as it really happened, with historical truthfulness, and the idea is developed from its inception concurrently with its mathematical formulation from its starting point to the end. This, it seems to me, is the way to convey to the uninitiated reader that our science is essentially an art which could not live without the occasional flash of genius in the mind of some sensitive man, who, alive to the smallest of indications, knows the truth before he has the proof.'

The interplay of experiment with theory was never far from peak intensity in Debye's scientific thought. Professor Sack writes (21):

'What impressed me especially was his knack for making order of magnitude calculations that allowed a judgement of the feasibility of the experiment, his perseverance once he was convinced that meaningful results would be obtained, and his attention to details, once the design was agreed upon. Just as he was an unsurpassed master in choosing the most appropriate mathematical techniques to solve a theoretical problem, he was very ingenious in choosing or inventing the most appropriate experimental techniques too... His abilities as lecturer and teacher were particularly evident in his big introductory physics courses (e.g. to *ca.* 400 students in *Leipzig*) which he gave from 1920 to 1934 in Zürich and in Leipzig... Debye gave great attention to keeping the lecture demonstrations meaningful and up to date. Only a few months after the first observation of electron diffraction, such an experiment was shown in his lecture (using an old-fashioned CRT that had to be constantly pumped and a Wimshurst-type electrostatic generator for voltage supply, which would barely work on a hot and humid day).'

Not only did Debye successfully explain scientific ideas and principles to audiences as diverse as school children, research specialists and business executives but his presentation also communicated his own enthusiasm for yet further exploration of the subject. Naturally, his Dutch colleagues have particular memories of his ability as a lecturer. They comment approvingly on the quite unspoilt retention of his native accent and of characteristic Limburg diction. His presentation of the Hustinx Prizes in 1962 was the reason for a popular lecture at Maastricht on 'The measurement of molecules'. On this- singular occasion (he was lecturing in the theatre where his mother had been employed) he gave the largely lay audience a clear insight into quantitative aspects of the molecular world. He was able to show not only his own attachment to Maastricht and its people but to share with them his enthusiasm for exploring the physical world. Some of his introductory remarks (they were in Dutch) may be quoted:

'Ladies and Gentlemen,

'I have peculiar feelings while I stand here today.

'Firstly, I am thinking back to the day when as a youth of sixteen years, I experienced paradise in this building. That was when I attended my first opera performance—Gounod's Faust it was.

'I also recollect the occasion of thirty years ago when I had the opportunity to deliver an address in this building. Today, however, I have to think of the part I am playing in a very important occasion. This occasion is the handing over of the Hustinx Prizes.

'Finally I must also think of Maastricht. This town is well known for its hospitality, which would imply that the visitor must be treated in such a way that he feels completely at home. This would certainly mean that he must be addressed in such a way that he can understand everything. Therefore, as many of the visitors and acquaintances here today are Germans, I have decided to deliver my address in German.'

It was typical of Debye that when he attended a conference he participated fully. He would normally listen to all the contributions. Professor Long has written (29): ' At a recent conference, after he had just given his own talk in the morning session and was returning to listen to the afternoon session, one of his friends suggested that, given his eighty years and the 9000 feet altitude, he might preferably rest. Debye's reply was characteristic: "No, no, no; if I listen to the talks I may get some new ideas."'

All who knew Debye will instantly recall what a readily approachable and particularly friendly person he was. His research associates all emphasize how he insisted on enjoyment as an essential element in one's work. Professor Nauman recalls his initial instructions (30): 'Work when you wish: there is no eight to five schedule. Come when you want to come, leave when you want to leave: just get something done but most of all have fun in your work.' Despite this completely relaxed approach on detail it should not be thought that Debye could suffer fools or fooling-about gladly. He had a keen appreciation of the value of time, especially his own time. This can be illustrated by Professor Nauman's memory of a telephone conversation of Debye's with a firm for which he had done some consulting: 'Yes, I would be pleased to visit your laboratories again... Yes, I received your cheque for expenses and honorarium... No, it was not satisfactory. I would rather visit your laboratories for nothing than be insulted by being offered such a small honorarium as that you sent last time. I find your problems interesting, but your policies are insulting... Yes, I will come.'

'He was an avid trout fisherman. He caught trout in the streams near Cornell where there were few. First he stalked the fish by walking and watching the stream without his rod. Later he sought the trout whose habits he had observed. After catching them, he released them, to be caught another day' (30). This interest and ability derived from his earliest days in Maastricht when his grandfather would take him fishing. And catching a fine fish was not an acceptable reason for going home-the young boy of ten protested strongly against a halt being called.

'Debye was an affectionate husband, father and grandfather. His principal hobbies were gardening and fishing and both were done with the steady participation of his wife. In 1948, when the Debyes were well over 60, they took over the principal upbringing of two of their young grandsons, and it was a matter of great pleasure to Debye that, by the time of his death, both of the boys were successful graduate students' (29).

And finally we may quote from some remarks made at Cornell University by Professor Henri Sack, a close colleague of Professor Debye's for over forty years (31): 'I have tried to find a simple attribute-if this is possible-with which to characterize Professor Debye's multifaceted personality, and feel that I come nearest to my personal image of him by saying that he was a truly happy or lucky man. He was not only endowed with a most powerful and penetrating intellect and an unmatched ability for presenting his ideas in a most lucid way, but he also knew the art of living a full life. He greatly enjoyed his scientific endeavours, he had a deep love for his family and home life, and he had an eye for the beauties of nature and a taste for the pleasures of the out-of-doors as manifested by his hobbies such as fishing, collecting cacti, and gardening, mostly in the company of Mrs Debye. He enjoyed a good cigar and a good table, and he had affections for his students and associates and liked their company...

'At several occasions Professor Debye remarked that he did only those things he liked to do... he never was envious of other people's success; to him his own pastures were always the greenest and his greatest ambition was to do a perfect job of whatever problem he tackled. He did not worry whether he was called a physicist or a chemist... I am sure his greatest satisfaction came from the knowledge of having discovered something new, enhanced the understanding of a known phenomenon, or succeeded in communicating his ideas to others.

'... His eagerness to do any job well if it was worth doing at all was also reflected in his private life. In gardening he became a real expert, and his parties were always something special.

'His frequent comment in colloquia and meetings" Aber das ist ja so einfach" ("Look here, this is really so simple") has already become a Debye legend... For him the physical sciences were not a series of narrow specialities, but a coherent body of knowledge, where a few basic principles weaved like a red thread through the whole field. He was helped in his constant endeavour to correlate phenomena from different areas by his phenomenal memory. He may have forgotten the exact name of the author or the exact place of the publication, but he never forgot the essence of what he had read in a paper or heard at a meeting.

'At times Professor Debye could also be a stern task master. His criticism could be very articulate, especially in the early years. There could be no compromise when scientific truth was involved...

'And thus, I believe I can speak in the name of all those who had the privilege of having been closely associated with Professor Debye, in saying that we are grateful for all he has given us and that he will live in our memory as a brilliant scientist, a great teacher, a fatherly and helpful adviser, and, above all, as a happy man.'

A number of Professor Debye's colleagues and others have helped by providing information, references or permission to quote their own statements. Thanks are especially due and are gratefully offered to Professors S. H. Bauer, P. P. Ewald, F.R.S., R. M. Fuoss, W. Heisenberg, For. Mem. R.S., F. Hund, H. S. Johnston, F. A. Long, C. Manneback, R. V. Nauman, L. C. Pauling, For. Mem. R.S., A. J. Rutgers, H. S. Sack, E. J. W. Verwey and B. Widom (who also provided reprints). The Burgemeester of Maastricht, Dr A. M. I. H. Baeten and his secretary, Mr Minnis, generously provided photostats of local records and other publications, as did also Dr J. A. Poulis (Eindhoven) and Professor Debye's friend, Mr Edmond Hustinx of Maastricht, sent further material. Professor Hooykaas, Mrs Schotman and Dr Verwey helpedat Utrecht, and Mrs M. H. J. Niel, Professor Debye's niece, contributed details of the family.

Interscience Publishers Inc. (a division of John Wiley and Sons Ltd) are thanked for allowing quotations from The Collected Papers of Peter J. W. Debye.

Miss Glenys Owen is largely responsible for the bibliography; Mr Philip Davies sought literature; and other Aberystwyth colleagues, Mr Arnold Baise and Dr William Engelbrecht, helped with the Dutch sources; and Mr Huw Ceredig with proof-reading.

MANSEL DAVIES

APPENDIX A

Institut für Physik
8 München 23,
Fohringer Ring 6
16thJanuary, 1970

Dear Professor Davies,

You ask me about the time when I worked together with Professor Debye in Leipzig some forty years ago. At that time Debye was mostly interested in X-ray measurements of the structure of molecules, and frequently discussed his problems with me. He had a strong interest in the theoretical side of these problems, and occasionally he asked me to work out details for him. For instance, for the calculations of X-ray scattering by atoms he used the Thomas-Fermi model as an approximation, and he urged me to evaluate the incoherent part of X-ray scattering from this model. Occasionally he took part in our theoretical seminars, e.g. when we discussed the quantum theory of ferromagnetism. Debye had not much interest for political questions and therefore he preferred to disregard as much as possible the political riots on the streets of Leipzig. He hoped-as many Germans did-that the extreme tendencies in the national-Socialistic revolution would die down, so that Germany could return to a more or less normal state of politicallife. As a Dutchman he did not feel responsible for what happened in our country. Debye had a certain tendency to take things easy. He did not belong to that class of scientists who come to the laboratory very early in the morning and leave it not before midnight. From my room in the institute I could frequently see him walking around in his garden and watering the roses even during duty hours of the institute. But the centre of his interest was undoubt(ed)ly his science. When he saw with the beginning of war that he could not continue this quiet life of a scientist, he decided to leave Germany.

He would not have been obliged to leave Germany according to German laws or to the tendency of the university or government officials. In 1936 or 1937 Debye had moved from Leipzig to Berlin, where the Kaiser Wilhelm Society had built a new institute for him. During the war some members of his family have lived in the house connected with the institute still a few years, but-if I remember correctly-they all left not later than 1942.

With best regards,
Yours,
W. HEISENBERG.

APPENDIX B

College of Chemistry Newsletter, University of California, Berkeley. September 1969.
Letter from Professor E. Rutherford to Professor G. N. Lewis : December 1915.

Dear Lewis,

I have received your letter of November 19th asking me my opinion about Debye and Bohr, from the point of view of lecturers in Mathematical Physics.

I do not know Debye personally but I believe he is regarded as one of the best mathematicians of the day, and was recently. appointed Professor in Gottingen in succession to Professor Voigt. As you know, he is a Dutchman by birth, and is undoubtedly a man of great ability both on the Mathematical and Physical side. I am unable to express any opinion about his lecturing qualifications etc. but my experience is that most Dutchmen speak English quite well.

Bohr, as you know, holds the position of Reader in Mathematical Physics in my Laboratory. I regard Bohr as one of the coming men in Mathematical Physics, and I think he has a better grip of Physics than any of the Mathematical people I have come across. He is a man of great originality, and, as you know, his work has already attracted wide attention, and I am confident will do so even more in the future. As a matter of fact, I think Bohr is just the type of man that would fill your bill. He is thoroughly au courant of all the modern physical problems, and has an extraordinarily wide knowledge of experimental as well as of theoretical Physics. He is a pleasant fellow, speaks English quite well, and is quite a clear and interesting lecturer.

I do not, of course, know how he would regard a visit to California, especially in war time. As you probably know, he is a Dane, the son of the late Professor of Pathology in the University of Copenhagen, and I believe he is likely soon to be given a Professorship in that University. He has a very charming wife.

I am at present very much occupied with Admiralty work, and have practically no time for my own investigations. My Research School has vanished to Flanders or to the Dardenelles, but we still have a good number of women students, and men physically unfit and under age, so we have to keep all our classes going.

Yours sincerely,
E. RUTHERFORD.

REFERENCES

(1) A. Sommerfeld, 1950. Physikalische Blätter, 6, 509.

(2) P. P. Ewald, personal communication.

(3) P. P. Ewald, 1962. Editor: Fifty years of X-ray Diffraction. Oosthoek's Uitgeversmaatschappij, Utrecht.

(4) M. Planck, 1900. Verh. dt. phys. Ges. 2, 237.

(5) M. Born & Th. v. Kármán, 1912. Phys. Zeit. 13,297; 14,65 (1913).

(6) P. Langevin, 1905. J. physique (4),4, 678.

(7) H. A. Kramers, 1946. In Helden der Wetenschap, N. v. Uitgeversmaatschappij Kosmos, Amsterdam.

(8) O. Fuchs & K. L. Wolf, 1935. Hand- und Jahrbuch d. chem. Phys. 6 (IB).

(9) J. A. Saxton & J. A. Lane, 1947. The Physical and Royal Meteorological Soc. Rep. pp. 278-292.

(10) F. Eckert, 1913. Verh. dt. phys. Ges. 15,305.

(11) C. P. Smyth, 1955. Dielectric behaviour and structure. McGraw-Hill Book Co. N.Y. N. E. Hill, W. E. Vaughan, A. H. Price, M. Davies, 1969. Dielectric properties and molecular behaviour. Van Nostrand, London.

(12) M. v. Laue, 1912. Sber. kgl. bayer. Akad. Wiss. p. 303.

(13) R. W. James, I. Waller & D. R. Hartree, 1928. Proc. Roy. Soc. Lond. A, 118, 334.

(13a) C. Manneback: see Biographical or Obituary Notices.

(14) N. Bohr, 1913. Phil. Mag.26, 1,476, 857.

(15) See Physik. Zeit. 14,259 (1913) : dated Utrecht, 10 Feb. 1913. See also M. Born, Atomic physics, Blackie and Son, London, 1935, p. 105; and N. Bohr, Dansk. Vid. Selsk. Skr. natw. o. math. Afd. 8, RIV. 1, 1918; idem, The theory of spectra and atomic constitution, Camb. Univ. Press, 1922.

(16) W. L. Bragg, R. W. James & C. H. Bosanquet, 1921. Phil. Mag. 41, 309; 44, 435 (1922).

(17) C. G. Le Fevre & R.J. W. Le Fevre, 1960. Physical methods of organic chemistry (ed. A. Weissberger), Vol. I, part III, 2459-2496. Interscience Publ. N. Y. H. A. Stuart, Die Struktur der Freien Moleküls. Springer, Berlin: 2nd Ed. (1952), 3rd Ed. (1967).

(18) A. Piekara, 1939. Proc. Roy. Soc. Lond. A, 172, 360. J. Chem. Phys. 29, 1297 (1958); 36,2145 (1967).

(19) P.P. Debye, 1939. Phys. Zeit. 40,66,404. (20) E. J. W. Verwey: see Biographical or Obituary Notices.

(21) Letter to M.D.

(22) P. Putzeys & J. Brosteaux, 1935. Trans. Faraday Soc. 31, 1314.
(23) P. Debye, 1965. Phys. Rev. Lett. 14,783.
(24) P. Debye, 1966. Critical phenomena. N.B.S. Misc. Publ. 273, pp. 107,161.
(25) B. Widom, personal communication.
(26) R. M. Fuoss, in Collected Papers of Peter J. W. Debye: see Bibliography (1954).
(27) H. S. Sack, B. Widom & S. H. Bauer: see Biographical or Obituary Notices.
(28) P. Debye; preface to A. J. Rutgers, Physical chemistry. Interscience Publ. Ltd. London, 1954: p.v.
(29) F. A. Long: see Biographical or Obituary Notices.
(30) R. V. Nauman, personal communication.
(31) H. S. Sack: see Biographical or Obituary Notices.

BIOGRAPHICAL OR OBITUARY NOTICES OF P. J. W. DEBYE

Peter J. W. Debye. 1964. An interview. Science, 145, 554-559.

Debye, P. 1964. Herinneringen aan mijn schooltijd in Maastricht. Honderd Jaar Gemeentelijke V.H.M.Q. in Maastricht.

Debye, P. 1964. The early days of lattice dynamics. Wallis, R. F. (Ed.) Lattice dynamics. Pergamon Press, London.

Manneback, C. 1937. P. Debye, Prix Nobel de Chimie. Revue des Questions Scientifiques. Soc. Sci. de Broxelles.

Darrow, K. K. 1968. Rear book Amer. Philosophical Soc. pp. 123-130.

Davies, M. M. 1968. J. chem. Education, 45, 467-473.

Ewald, P. P. 1967. Acta Crystallographica, 22, 947-949.

Falkenhagen, H. 1964. Forschungen und Fortschritte, 38,91-93.

Gerlach, W. 1967. Jahrbuch d. Bayerischen Akademie d. Wirsenschaften, pp. 218-230.

Hund, F. 1968. Jahrbuch der Akademie der Wirsenschaften in Giittingen, pp. 59-64.

Long, F. A. 1967. Science, 155,979-980.

Laue, M. 1954. Zeit. Elektrochemie, 58, 151.

Mizushima. S. 1968. Pontificiae Academiae Scientarum Scripta Varia, 31, 731-750.

Sack, H. S. 1964. J. Amer. Chem. Soc. 86, 9A.

Sack, H. S. 1968. J. Amer. Chem. Soc. 90, follows p. 3000.

Sack, H. S., Widom, B. & Bauer, S. H. 1967. Cornell News, pp. 16-19.

Verwey, E.J. W. 1967. Proc. Kon. Akad. Wetens Kap. pp. 341-348.

(A list of 38 items, mostly newspaper and journal accounts of Professor Debye's career and the honours awarded him, has been prepared by the Stadsarchiefen at the Bibliotheek, Maastricht)

BIBLIOGRAPHY

1907. Wirbelstrome in Staben von techteckigem Querschnitt. Z. Math. Phys. 54, 418.

1908. Eine Bemerkung zu der Arbeit von F. A. Schulze. 'Einige neue Methoden zur Bestimmung der Schwingungszahlen hochster horbarer und unhorbarer Tone usw.' Annln. Phys. 25, 819.

1908. Das elektromagnetische Feld um ein Zylinder und die Theorie des Regenbogens. Phys. Z. 9,775; also in Verh. dt. phys. Ges. 10 (20), 741.

1909. Naherungsformeln für die Zylinder funktionen fÜr grosse Werte des Arguments und unbeschrankt veranderliche Werte des Index. Math. Annln. 67, 535.

1909. Der Lichtdruck aufKugeln von beliebigen Material. Annln. Phys. 30, 57.

1909. Das Verhalten von Lichtwellen in der Nahe eines Brennpunktes oder einer Brennlinie. Annln. Phys. 30. 755.

1910. Stationare und quasistationare Felder. Encyklopiidie der Mathematirchen Wissenschaften: vol. 5. Physik, Teil 2, 395.

1910. Zur theorie der Elektronen in Metallen. Annln. Phys. 33, 441.

1910. Semikonvergente Entwickelungen für die Zylinder-funktionen und ihre Ausdehnung ins Komplexe. Sber. bayer Akad. Wirs, no.5, 1.

1910. (With D. HONDROS.) Elektromagnetische Wellen an dielektrischen Drahten. Annln. Phys. 32, 465.

1910. Die Berechnung der Moleküldimensionen aus Radiometerbeobachtungen. Phys. Z. 11, 115.

1910. Der Wahrscheinlichkeitsbegriffin der Theorie der Strahlung. Annln. Phys. 33. 1427.

1911. Die Frage nach der atomistischen Struktur der Energie. Vschr. natuif Ges. Zürich, 56, 156.

1911. Über Abweichungen vom Curie-Langevin'schen Gesetz und ihren Zusammenhang mit der Quantenhypothese. Verh. schweiz. natuif Ges. 94,220; also in Compt. rend. Soc. Suisse de Phys. ler aout, 1911.

1912. Zur Theorie der spezifischen warmen. Annln. Phys. 39. 7B9.

1912. Einige Resultate einer kinetischen Theorie der Isolatoren. Phys. Z. 13, 97.

1913. Zur Theorie der anomalen Dispersion im Gebiete der langwelligen elektrischen Strahlung. Verh. dt. phys. Ges. 15, 777.

1913. (W. DEHLINGER.) Die kinetische Theorie der Materie in ihrer modernen Entwicklung (Ausgang aus der Utrechter Antrittsreder von Prof. P. Debye). ATChiv. ElektTOtech. 2, 167.

1913. Zustandsgleichung und Quantenhypothese. Nachr. Konigl. Ges. Wiss. Gottingen Math. Phys. Klasse, 1913, 140; also in Phys. Z. 14, 317.

1913. über den Einfluss der Warmebewegung auf der Interferenzerscheinungen bei Róntgenstrahlen. Verh. dt. phys. Ges. 15, 678.

1913. über die Intensitatsverteilung in den mit Róntgenstrahlen erzeugten Interferenzbildern. Verh. dt. phys. Ges. 15, 738.

1913. Spektrale Zerlegung des Róntgenstrahlen mittels Reflexion und Warmebewegung. Verh. dt. phys. Ges. 15, 857.

1913. (With A. SOMMERFELD.) Theorie des lichtdektrischen Effektes vom Standpunkt des Wirkungsquantums. Annln. Phys. 41, 873.

1914. (WithJ. KERN.) Über die Behandlung gekoppelter Systeme nach der Methode der Eigenschwingungen. Phys. Z. 15, 490.

1914. Zustandsgleichung und Quantenhypothese mit einem Anhang über Warmeleitung. Vortrage über die kinetische Theorie der Materie und der Elektrizitat. Math. Vorlesgn. Gottingen VI, B. G. Teubner, Leipzig.

1914. Interferenz von Róntgenstrahlen und Warmebewegung. Annln. Phys. 43,49.

1915. Die Konstitution des Wasserstoff Moleküls. Sber. bayer. Akad. Wiss. 1915, 1.

1915. Zerstreuung von Róntgenstrahlen. Annln. Phys. 46, 809; also in NachT. Ges. Wiss. Gottingen Math. Phys. Klasse, 1915, 70.

1916. (With P. SCHERRER.) Interferenzen an regellos orientierten Teilchen im Róntgenlicht. I. Phys. Z. 17, 277; also in NachT. Ges. Wiss. Gottingen Math. Phys. Klasse, 1916, 1.

1916. (With P. SCHERRER.) Interferenzen an regellos orientierten Teilchen im Róntgenlicht. II. Nachr. Ges. Wiss. Gottingen Math. Phys. Klasse, 1916, 16.

1916. Die Feinstruktur wasserstoffáhnlicher Spektren. Phys. Z. 17, 512; also in Nachr. Ges. Wiss. Gottingen Math. Phys. Klasse, 1916, 161.

1916. Quantenhypothese und Zeeman-Effekt. Phys. Z. 17,507; also in NachT. Ges. Wiss. Gottingen Math. Phys. Klasse, 1916, 142.

1917. Der erste Elektronenring der Atome. Phys. Z. 18, 276; also in NachT. Ges. Viss. Gottingen Math. Phys. Klasse, 1917, 236.

1917. (With P. SCHERRER.) über die Konstitution von Graphit und amorpher Kohle. Nachr. Ges. Wiss. Gottingen Math. Phys. Klasse, 1917, 180.

1917. Konzentrationselement und Brownsche Bewegung. Phys. Z. 18, 144.

1917. (With P. SCHERRER.) Interferenzen an regellos orientierten Teilchen im Róntgenlicht. III. Die Kohlenstoffmodifikationen. Phys. Z. 18, 291.

1917. Optische Absorptionsgrenzen. Phys. Z. 18, 428.

1917. Die Atomanordnung von Wolfram. Phys. Z. 18, 483.

1918. (With P. SCHERRER.) Atombau. Phys. Z. 19, 474; also in Nachr. Ges. Wiss. Gottingen Math. Phys. Klasse, 1918, 101.

1919. Das molekulare elektrische Feld in Gasen. Phys. Z. 20, 160.

1920. Die neuen Forschungen über den Bau der Molekule und Atome. Verh. Ges. dt. NatuTj: Á.TZte, 86, 239.

1920. Die van der Waalsschen Kohasionskrafte. Phys. Z.21, 178; also in NachT. Ges. Wiss.Gottingen Math. Phys. Klasse, 1920, 55.

1921. Adsorptie van elekenische molekulen. Physica, Eindhoven, 1, 362.

1921. Molekularkrafte und ihre Elektrische Deutung. Phys. Z. 22, 302.

1921. Moleculaire krachten van electrischen Orsprong. Handelingen v. het. xvIIr. v dld. NatuUT-en. Geneeskundg. CongTess, UtTecht.

1922. Laue-interferenzen und Atombau. Naturwissenschaften, 10, 384.

1923. Zerstreuung von Róntgenstrahlen und Quantentheorie. Phys. Z. 24, 161.

1923. (With E. HÜCKEL.) Zur Theorie der Elektrolyte I. Gefrierpunktserniedrigung und verwandte Erscheinungen. Phys. Z. 24, 185.

1923. (With E. HÜCKEL.) Zur theorie der Elektrolyte II. Das Grenzgesetz für die elektrische Leitfáhigkeit. Phys. Z. 24, 305.

1923. Kinetische Theorie der Gesetze des osmotischen Drucks bei starken Elektrolyten. Phys. Z. 24, 334; also in Recl. TTav. chim. Pays-Bas Belg. 42, 597.

1923. De modeme ontwikkeling van de elektrolyt-theorie. Handelingen v. het XVIII" Vdld. NatuuT-en Geneeskung. Congress, Maastricht.

1923. Over Ionen en Hun Activiteit. Chem. Weekbl. 20, 562.

1924. (With E. HÜCKEL.) Bemerkungen zu einem Satze über die Kataphoretische Wanderungsgeschwindigkeit suspendierter Teilchen. Phys. Z. 25, 49.

1924. Osmotische Zustandsgleichung und Aktivitat verdünnter starker Elektrolyte. Phys. Z. 25, 97.

1925. Molekulare Krafte und ihre Deutung. VeTh. schweiz. natUT.f Ges. Freeburg, 106, 128.

1925. Note on the scattering of X-rays. J. Math. Phys. 4, 133.

1925. Theorie der elektrischen und magnetischen Molekulareigenschaften. Handbuch der Radiologie. Band VI. Die Theorien der Tadiologie (Edited by Dr E. Marx), Leipzig, 597. (2nd Ed.) Band VI/2. Leipzig, 1934.

1925. (With L. PAULING.) Inter-ionic attraction theory of ionised solutes IV. The influence of variation of dielectric constant on the limiting law for small concentrations. J. Am. Chem. Soc. 47, 2129.

1925. (With J. McAULAY.) Das elektrische Feld der Ionen und- die Neutralsalzwirkung. Phys. Z. 26, 22.

1925. (With A. HUBER.) Een proef over de instelling van paramagnetische molukulen. Physica, Eindhoven, 5, 377.

1926. Die Grundgesetzeder elektrischen und magnetischen Erregung vom Standpunkte der Quantentheorie. Phys. Z. 27, 67.

1926. Molekulare Kriifte und ihre Deutung. Umschau, 30,905; also in VeTh. schweiz. natur.f Ges. 106, 128.

1926. Einige Bemerkungen zur Magnetisierung bei tiefer T emperatur. Annln. Phys. 81, 1154.

1926. Bemerkung zu einigen neuen Versuchen über einen magneto-elektrischen Richteffekt. Z. Phys. 36, 300.

1926. (With W. HARDMEIER.) Anomale Zerstreung von a-Strahlen. Phys. Z. 27, 196.

1926. (With W. HARDMEIER.) Dispersion anomale des rayons alpha. C.R. Soc. Suisse de Phys.8, 131.

1927. Wellenmechanik und Korrespondenzprinzip. Phys. Z. 28, 170.

1927. (With C. MANNEBACK.) The symmetrical top in wave mechanics. Nature, Lond. 119, 83.

1927. Report on conductivity of strong electrolytes in dilute solutions. TTans. FaTaday Soc. 23, 234.

1927. Das elektrische Ionenfeld und das Aussalzen. Z. phys. Chem. 130, 56.

1927. Über die Zerstreuung von Rijntgenstrahlen an amorphen Kijrpem. Phys. Z. 28, 135.

1927. Über elektrische Momente. Atti del Congresso Intemazionale deifisici, Como, 1927.

1927. Dielectric constants of electrolyte solutions. TTans. Am. electTochem. Soc. 51, 449.

1928. Die elektrischen Momente der Molekeln und die zwischenmolekülaren Krafte. Z. ElektTochem. angew. phys. Chem. 34,450.

1928. (With H. FALKENHAGEN.) Dispersion von Leitfähigkeit und Dielektrizitatskonstante bei starken Elektrolyten. Phys. Z. 29, 121; also in Z. ElektTochem. 34, 562.

1928. (With H. FALKENHAGEN.) Dispersion der Leitfähigkeit und der Dielektrizitats konstante starker Elektrolyte. Phys. Z. 29, 401.

1929. (With L. BEWILOGUA & F. EHRHARDT.) Zerstreuung von Rijntgenstrahlen an einzelnen Molekeln. Phys. Z. 30, 84.

1929. Interferometrische Messungen am Molekül. Phys. Z. 30, 524; also in BeT. Züricher VorlTiige, Juli 1-4,

1929. (With L. BEWILOGUA &. F. ERHARDT.) IDterferometri.sche MCSUDgen am Molekül. Ber. Verh. sachs. Akad. Wiss. 81, 29.

1929. Die zeitlichen Vorgange in Elektrolytlosüngen. In Probleme Modern Physik JS. Hirzd, Lejpzig, pp. 52-57.

1930. Röntgeninterferenzen an isomeren Molekülen. Phys. Z. 31, 142.

1930. Röntgenzerstreuung an Flüssigkeiten und Gasen. Phys. Z. 31, 348.

1930. Röntgeninterferenzen und AtomgrOsse. Phys. Z. 31, 419.

1930. (With H. MENKE.) Bestimmung der inneren Struktur von Flüssigkeiten mit..' Röntgenstrahlen. Phys. Z. 31, 797.

1930. Interferometrische Bestimmung der Struktur von Einzelmolekülen. Z. Elektrochem. angew. phys. Chem. 36, 612.

1930. Interference measurements with single molecules. Proc. Phys. Soc. Lond. 42, 340.

1931. A note on comparison of electrolytic resistance at low and radio frequencies. Indiall J. Phys. 6, 261.

1931. (Wjth H. MENKE.) Untersuchung der molekularen Ordnung in Flüssigkeiten mit Rontgenstrahlung. Ergebn. tech. Röntgenk. 2, 1.

1932. Chemistry at the Centenary (1931) Meeting of the Brit. Ass. Advmt. Science. Cambridge: Heffer.

P. Debye: The dispersion of conductivity in different solvents, pp. 32-33.

Interferometric measurement of atomic distances in molecules, pp. 204-209.

Anomalous dispersion in solids, p. 207.

1932. Polar molecules. Congres International d'Électricité, Paris. Sect. 1. Rapport No.1.

1932. De polaritare molecularum. Pontificia Acad. Sci. Novi Lyncaei Sci. Nuncius Radiophonicus; no.14, 3.

1932. (With F. W. SEARS.) On the scattering of light by supersonic waves. Proc. natn. Acad. Sci. U.S.A. 18, 409.

1932. Schallwellen als optische Gitter. Ber. Verh. sachs. Akad. Wiss,84, 125.

1932. Zerstreuung von Licht durch Schallwellen. Phys. Z. 33, 849.

1933. Die elektrische Leitfáhigkeit von Elektrolytlosüngen in starken Feldem und beihohen Frequenzen. Z. Elektrochem. angew. phys. Chem. 39, 478.

1933. (With H. SACK.) Demonstration des Hochfrequenzeffektes bei Elektrolyten. Z. Elektrochem. angew. phys. Chem. 39, 512.

1933. 15th Faraday lecture. Relations between stereochemistry and physics. J. Chem. Soc. p. 1366.

1933. A method for the determination of the mass of electrolytic ions. J. chem. Phys. 1, 13.

1933. Streuung von Rontgen- und Kathodenstrahlen. Ergebn. tech. Rontgenk. 3, 11.

1934. Rontgen und sein Entdeckung. Abh. Ber. dt. Mus. 6, 83.

1934. Die Physik der Atomkeme. Vortrag Bund. d. Freunde d. T. H. München.

1934. (With H. SACK.) Theorie der elektrischen Moleküleigenschaften. Handbuch der Radiologie, 2 Aufl. Band VI, Teil. 11, 69.

1934. Einfluss des molekularen Feldes auf den Verlaufadiabatischer Entmagnetisierungs- prozesse bei tiefsten Temperaturen. Ber. Verh. sächs. Akad. Wiss. 86, 105.

1934. Energy absorption in dielectrics with polar molecules. Trans. Faraday. Soc. 30,679.

1934. Hochfrequenzverluste und Molekülstruktur. Phys. Z.35, 101. l,

1934. Die magnetische Methode zur Erzeugung tiefster Temperaturen. Phys. Z. 35, 923; also in Z. tech. Phys. 15, 499.

1934. (With H. SACK &. F. COULON.) Expériences sur la diffraction de la lumière par des ultrasons. C.R. Acad. Sci. Paris, 198, 922.

1935. Kemphysik. Angew. Chem. 48, 381.

1935. Les propriétés diélectriques du point de vue moléculaire. Revue univlle. Mines, Liège, 11, 176.

1935. Analyse des essaies de sédimentation. Revue univlle. Mines, Liège, 11, 266.

1935. La rotation des molécules dans les liquides. Bull. Acad. roy. Belg. 21, 166.

1935. Relations entre la constitution chimique et les propriétés diélectriques. Bull. Soc. chim.Belg. 44, 167.

1935. Der Rotationszustand von Molekülen in Flüssigkeiten. Phys. Z. 36, 100; also in Bull. Sci. Acad. Roy. Belg. 21, 166.

1935. Dielektrische Sattigung und Behinderurig der freien Rotation in Flüssigkeiten. Phys. Z. 36, 193.

1936. Dielectric properties of pure liquids. Chem. Rev. 19, 171.

1936. Der Weg zum absoluten Nullpunkt. Umschau, 40, 4:1.

1936. Physik: (article in): 25 Jahre Kaiser Wilhelm-Gesellschaft. Bd. II Die Naturwissenschaften. Berlin: Springer.

1936. Die tíefsten heute erreichten Temperaturen. Forschungen und Fortschrifte, 12, 22; idem, English version: The lowest temperatures yet established. Res. Prog. 2,89.

1936. Bemerkung zu dem Artikel von E. Gehrcke: 'Wie die Energieverteilung der schwarzen Strahlung in Wirklichkeit gefunden wurde.' Phys. Z. 37, 440.

1937. Das Kaiser-Wilhelm-Institut für Physik. Naturwissenschaften, 25, 257.

1937. Johann Diderik van der Waals. Ned. Tijdrchr. Natuurk. +, 257.

1937. A contribution in: Die Welt der Strahlen (Edited by H. Woltereck). Leipzig: Verlag. Quelle und Meyer.

1937. (With H. SACK.) Constantes diélectriques, moments electriques. Tables annuelles des constantes, Nr. 2. Paris: Hermann et Cie.

1937. Structure in electrolytic solutions. J. Franklin Inst. 224, 135.

1937. (With W. RAMM.) Hochfrequenzverluste und quasikrystalline Struktur von Flüssigkeiten. Annln. Phys. 28, 28.

1937. Die Untersuchung der freien Elektronen in Metallen mit Hilfe von Rontgenstrahlen. Phys. Z.38, 161.

1937. Methoden zur Bestimmung der elektrischen und geometrischen Struktur von Molekülen. Nobelvortrage. Angew. Chem. 50, 3.

1938. A contribution in: Physique générale. Paris: Hermann et Cie.

1938. Die quasikrystalline Struktur von Flüssigkeiten. Der feste Korper. Vortrage Tagung Phys. Ges. Zürich, p. 4:2.

1938. Wege der modemen Forschung in der Physik. Stahl u. Eisen, 58, 1.

1938. (With M. H. PrRENNE.) über die Fourieranalyse von interferometrischen Messungen an freien Molekülen. Annln. Phys. 33, 617.

1938. Die Geburt des Wirkungsquantums. Z. tech. Phys. 19, 121.

1938. Abkühlung durch adiabatische Entmagnetisierung. Annln. Phys. 32, 85.

1938. (With W. RAMM.) Dispersion und Absorption polarer Substanzen. Nuovo Cimento, 15, 226.

1938. Die paramagnetische Relaxation. Phys. Z. 39,616.

1939. Die quasikrystalline Struktur von Flüssigkeiten. Z. Elektrochem. 45, 174:.

1939. über den tiefsten heute erreichbaren Temperaturen. Schr. dt. Akad. Luftforsch. No.3, p.8.

1939. Das Sektorverfahren bei der Aufnahme von Elektroneninterferenzen. Phys. Z. 40, 507.

1939. Untersuchung eines neuen Vorschlags zur Fourier-Ana)yse von Elektronenaufnahmen. Phys. Z. 40, 573.

1939. Zur Theorie des Clusiusschen Trennungsverfahrens. Annln. Phys. 36, 284:.

1941. The influence of intramolecular atomic motion on electron diffraction diagrams. J. chem. Phys. 9, 55.

194:1. Reaction rates in ionic solutions. Trans. electrochem. Soc. 82, 265.

1944. Magnetic approach to the absolute zero of temperature. Am. Scient. 32,229.

1944. Light scattering in solutions. J. appl. Phys. 15, 338.

1945. Angular dissymmetry of scattering and shape of particles. Rubber Reserve Company. Technical report no. 637.

1946. The intrinsic viscosity of polymer solutions. J. chem. Phys. 14, 636.

1946. (With R. H. EWART, C. P. ROE & J. R. McCARTNEY.) The determination of polymeric molecular weights by light scattering in solvent-precipitant systems. J. chem. Phys. 14, 687.

1947. The structure of polymers in solution. Record Chem. Progr. 8 (1/2), 1.

1947. Molecular weight determination by light scattering. J. phys. colloid Chem. 51, 18.

1948. The structure of polymers in solutions. Les grosses molécules en solution. Homage national a Paul Langevin et Jean Perrin (College de France), p. 39.

1948. Light scattering in soap solutions. J. Colloid Sci. 3, 407.

1948. (With A. M. BUECHE.) Thermal diffusion of polymer solutions. In High polymer physics (Edited by H. A. Robinson). Brooklyn: Chemical Publishing Co., p. 497.

1948. (With A. M. BUECHE.) Intrinsic viscosity, diffusion and sedimentation rate of polymers in solution. J. chem. Phys. 16, 573.

1949. Light scattering in soap solutions. Ann. N.Y. Acad. Sci. 51, 575.

1949. Light scattering in soap solutions. J. phys. colloid Chem. 53, 1.

1949. (With R. V. NAUMAN.) The scattering of light by sodium silicate solutions. J. chem. Phys. 17, 664.

1949. (With A. M. BUECHE.) Scattering by an inhomogeneous solid. J. appl. Phys. 20, 518.

1949. (With A. M. BUECHE.) Light scattering by inhomogeneous solids. India Rubb. World, 119, 613.

1949. (With W. M. CASHIN.) Determination of molecular weights and sizes by absorption. Phys. Rev. 75, 1307.

1950. (With A. M. BUECHE.) Scattering by inhomogeneous materials. Colloid Chem. 7, 33.

1950. (With A. M. BUECHE.) Light scattering by concentrated polymer solutions. J. chem. Phys. 18, 1423.

1950. Estructura de altos polímeros estudiada por métodos ópticos. Anales Real. Soc. Esp. Fis. Quim. B.46, 343.

1951. (With R. V. NAUMAN.) Light scattering investigations of carefully filtered sodium silicate solution. J. phys. colloid Chem. 55, I.

1951. (With E. W. ANACKER.) Micelle shape from dissymmetry measurements. J. phys. colloid Chem. 55, 644.

1951. (With F. BUECHE.) Dielectric constant of polystyrene solutions. J. phys. colloid Chem. 55, 235.

1951. (With C.W. TAIT, R.J. VETTER & J.M. SWANSON.) Physical characterization of cellulose xanthate in solution. J. Polym. Sci. 7, 261.

1951. (With W. M. CASHIN.) Effect of small refractive-index differences between solution and solvent on light scattering. J. chem. Phys. 19, 510.

1951. (With F. BUECHE & W. M. CASHIN.) Expressions for turbidities. J. chem. Phys. 19, 803.

1951. (With F. BUECHE.) Electric moments of polar polymers in relation to their structure. J. chem. Phys. 19, 589.

1951. (With F. BUECHE.) A study of crystallite sizes in polymers by a light scattering method. Phys. Rev. 81, 303.

1951. (With C. W. TAIT, R. J. VETTER & J. M. SWANSON.) Physical characterization of cellulose xanthate in solution. J. Polymer Sci. 7, 261.

1951. Method of and apparatus for effecting thermal diffusion. Patent U.S. 2, 567, 765. 11 Sept.

1952. (With J. O. EDWARDS.) Long-lifetime phosphorescence and the diffusion process. J. chem. Phys. 20, 236.

1952. (With F. BUECHE.) Distribution of segments in a coiling polymer molecule. J. chem. Phys. 20, 1337.

1952. (With F. BUECHE & W. M. CASHIN.) The measurement of self-diffusion in solid polymers. J. chem. Phys. 20, 1956.

1952. (With J. 0. EDWARDS.) A note on the phosphorescence of proteins. Science, 116, 143.

1953. (With P. P. DEBYE.) Proposal of a new method for determining molecular weights of polymers. Office of Synthetic Rubber Reconstruction: Finance Corporation Research Project.

1954. Equilibrium and sedimentation of uncharged particles in inhomogeneous electric fields. Ion transport across membranes. New York: Academic Press, p. 273.

1954. The structure of polymers. The present state of physics, Science.

1954. (With P. P. DEBYE, B. A. ECKSTEIN, W. A. BARBER & G.J. ARQUETTE.) Experiments on polymer solutions in inhomogeneous electrical fields. J. chem. Phys. 22, 152.

1954. (With P. P. DEBYE & B. H. ECKSTEIN.) Dielectric high frequency method for molecular weight determinations. Phys. Rev. 94, 1412.

1954. (With W. A. BARBER, P. P. DEBYE & B. H. ECKSTEIN.) A field-induced-diffraction method for molecular-weight determinations. Phys. Rev. 94, 1412.

1949. The collected papers of Peter J. w. Debye. Interscience Publ. Inc. New York.

1955. (With N. T. NOTLEY.) The extension of polystyrene chains; dependence on molecular weight and solvent. J. Polym. Sci. 17, 99.

1955. Structure of gel-catalysts by low angle X-ray scattering. Am. Chem. Soc. Dir. Petroleum Chem. General papers, No.33, p. 35.

1957. (With H. R. ANDERSON, JR. & H. BRUMBERGER.) Scattering by an inhomogeneous solid II. The correlation function and its application. J. appl. Phys. 28, 679.

1957. (With N. T. NOTLEY.) Dimensions of linear polystyrene molecules in solution: molecular weight dependence for low molecular weights. J. Polym. Sci. 24,275.

1957. (With H.BRUMBERGER.) Low-angle scattering of X-rays by glasses. J.phys.Chem. 61, 1623.

1958. (With P. DOREFUSS & N. T. NOTLEY.) Polymerization of isopropenyl-styrene. J. Polymer Sci. 28, 611.

1958. (With W. PRINS.) Micellar dispersion of a-monoglycerides in benzene and chlorobenzene. J. Colloid Sci. 13, 86.

1959. Rontgenstreuung in Korpem mit regelloser Struktur. Z. Phys. 156, 256.

1959. (With R. L. CLELAND.) Flow of liquid hydrocarbons in porous Vycor. J. appl. Phys. 30, 843.

1959. Angular dissymmetry of the critical opalescence in liquid mixtures. J. chem. Phys. 31, 680.

1959. (With L. K. H. VAN BEEK.) Effect of adsorbed water on the optical transmission properties of isotropic powders. J. chem. Phys. 31, 1595.

1959. (With J. DAEN.) Stability considerations on non-viscous jets exhibiting surface or body tension. Physics Fluids, 2, 416.

1959. Strukturbestimmung von Korpem mit regelloser Struktur mit Hilfe von Streustrahlung. Physikertagung Berlin 1959. Mosbach. Baden: Physik Verlag.

1960. Scattering of radiation by non-crystalline media. Non-crystalline solids. (Report of Coliference at Alfred University, September 3Td 1958.) New York: J. Wiley and Sons.

1960. Paul Scherrer und die StTeuung von RóntgenstTahlen. Basel-Stuttgart: Birkhauser Verlag.

1960. Die Winkelverteilung der kritischen Opalezenz und die Messung molekularer Wechselwirkung. Die Makromolek Chem. 35A, 1.

1960. (With A. PROCK & G. MCCONKEY.) Inhomogeneous field method for the study of large polarizable particles. J. chem. Phys. 32, 234.

1960. (With H. COLL & D. WOERMANN.) Critical opalescence of polystyrene solutions. J. chem. Phys. 32, 939.

1960. (With H. COLL & D. WOERMANN.) Critical opalescence of polystyrene in cyclohexane. J. chem. Phys. 33, 1746.

1960. Amold Sommerfeld und die Überlichtgeschwindigkeit. Physikalische Blatter, p. 568.

1960. (With H. COLL.) Non-ionic detergents in non-aqueous solvents. U.S. Dept. Commerce. Office of Technical Services. PB 146, 513.

1960. (With H. COLL.) Non-ionic detergents in non-aqueous solvents. II. Critical opalescence of binary liquid mixtures: the system polystyrene-cyclohexane. U. S. Dept. Commerce. Office of Technical Services. PB 149, 895.

1961. (With R. V. NAUMAN.) The slow change in turbidity of sodium silicate solutions. J. phys. Chem. 65, 5.

1961. (With R. V. NAUMAN.) The refractive indices of sodium silicate solutions. J. phys. Chem. 65, 8.

1961. (With R. V; NAUMAN.) A light scattering study of the aggregation of acidified sodium silicate solutions. J. phys. Chem. 65, 10.

1961. (With B. CHU.) Critical opalescence of Polystyrene in cyclohexane: Transmission measurements. U.S. Dept. Commerce. Office of Technical Services. AD 264, 359.

1961. (With B. CHU.) Critical opalescence of polystyrene in cyclohexane: Range of molecular forces and radius of gyration. U.S. Dept. Commerce. Office of Technical Services. AD 264, 360.

1962. Molecular forces, in International symposium on electrolytes. Trieste, 1959. Proceedings (Edited by B. Pesce). Oxford: Pergamon, p. 1.

1962. Interatomic and intermolecular forces in adhesion and cohesion in Symposium on adhesion and cohesion. Warren, Michigan, 1961. Proceedings (edited by Philip Weiss). New York, Amsterdam: Elsevier, p.1.

1962. (With D. WOERMANN & B. CHU.) Critical opalescence of polystyrene in cyclohexane: transmission measurements. J. chem. Phys. 36, 851.

1962. Critical opalescence and the range of molecular interaction. Pontificiae Academiae Scientiarum, Scripta Varia. 22, 53.

1962. (With B. CHU & D. WOERMANN.) Critical opalescence of polystyrene in cyclohexane: range of molecular forces and radius of gyration. J. chem. Phys. 36, 1803.

1962. (With B. CHU & H. KAUFMANN.) Critical opalescence of binary liquid mixtures: methanol-cyclohexane and aniline-cyclohexane. J. chem. Phys. 36, 3378.

1962. (With B. CHU.) Spectrophotometry and light scattering on supported platinum. J. phys. Chem. 66, 1021.

1962. (With H. COLL.) The association of a-monoglycerides in non-aqueous solvents. J. Colloid Sci. 17, 220.

1962. (With H. KAUFMANN, K. KLEBOTH & B. CHU.) Angular dissymmetry of critical mixtures: aniline-cyclohexane: aniline-l-hexene. Trans. Kansas Acad. Sci. 66, 260.

1962. (With B. CHU.) Critical opalescence of polystyrene in ethylcyclo-hexane. U.S. Dept. Commerce. Office of Technical Services. AD 266, 258.

1963. Structure determination by radiation scattering. Chem. Engng. News, 41, 92.

1963. (With B. CHU & D. WOERMANN.) Viscosity of critical mixtures. J. Polymer Sci. A, 1,249.

1963. (With D. WOERMANN & B. CHU.) Critical opalescence of polystyrene in ethylcyclohexane. J. Polymer Sci. A, I, 255.

1963. (With B. CHU & H. KAUFMANN.) Molecular configuration of polystyrene in benzene. J. Polym. Sci. A, I, 2387.

1963. Light scattering and molecular forces. Electromagnetic scattering. (Milton Kerker, ed.) Oxford: Pergamon Press, p. 393.

1964. The early days of lattice dynamics. Lattice dynamics. (Wallis, R. F., ed.). London: Pergamon Press. (Proceedings of Copenhagen Conference, August, 1963.)

1964. Flüssigkeiten, Gase, Makromoleküle: kritische Streuung und die Reichweite der Molekularkrafte. Z. Kristallogr. Kristallgeom. 120, 113.

1964. Light scattering as a tool. Official Digest Federation Soc. Paint Technol. 36, 518.

1964. (With D. CAULFIELD & J. BASHAW.) Critical opalescence of binary mixtures: perfluorotributylamine-isopentane. J. chem. Phys. 41, 3051.

1964. (With K. KLEBOTH.) An electrical field effect on the critical opalescence. U.S. Dept. Commerce. Office of Technical Services. AD. 604, 494.

1965. Hans Falkenhagen zum 70 Geburtstag am 13 mai 1965. Z. phys. Chem. 228, 289.

1965. Spectral width of the critical opalescence due to concentration fluctuations. Phys. Rev. Lett. 14, 783.

1965. (With K. KLEBOTH.) Electrical field effect on the critical opalescence. J. chem. Phys. 42, 3155.

1965. Static homogeneous electric field effect on critical opalescence. Cornell University Report no. TR-9. NASA N65-11285.

1965. Surface determination by X-ray scattering. Coloquio sobre Química Física de Procesos en Superficies Sólidas, pp. 1-11. Madrid: Consejo Superior de Investigaciones Científicas.

1966. Light-scattering as a tool. Pure appl. Chem. 12, 23.

1966. (With J. BASHAW, B. CHU & D. M. TANCREDI.) Critical opalescence of the polystyrene-cyclohexane system: small-angle X-ray scattering. J. chem. Phys. 44, 4302.

1966. (With C. C. GRAVATT.) The behavior of non-ionic detergents in non-polar solvents. U.S. Dept. of Commerce, Office of technical Services. Report AD-642604. N67-16150.

1966. (With C. C. GRAVATT & M. IEDA.) Electric field effect on the critical opalescence. II. Relaxation times of concentration fluctuations. U.S. Dept. of Commerce, Office of Technical Services. Report AD-642606. N67-16129.

1966. (With C. C. GRAVATT.) Behaviour of non-ionic detergents in non-Polar solvents. U.S. Dept. Commerce. Office of Technical Services. AD-642604.

1967. (With C. C. GRAVATT & M. IEDA.) Electric field effect on the critical opalescence. II. Relaxation times of concentration fluctuations. J. chem. Phys. 46, 2352.

1967. (With C. C. GRAVATT.) Measurement of relaxation times of concentration fluctuations by the electric field effect on critical opalescence. U.S. Dept. of Commerce, Office of Technical Services. Report AD-657208. N67-38297.

1968. (With R. T. JACOBSEN.) Direct visual observation of concentration fluctuations in a critical mixture. J. chem. Phys. 48, 203.

Books

1929. Polar molecules, New York: Chemical Catalog Co.

1929. Polare Molekeln. Leipzig: S. Hirzel.

1929. (Ed.) Dipolmoment und chemische Struktur. Leipzig: S. Hirzel.

1929. (Ed.) Probleme der moderne Physik. A. Sommerfeld zum 60 Geburtstage gewidmet. Leipzig: S. Hirzel.

1930. Polare Molekeln. Nachtrage I zur Tabellen elektrischen Momente. Leipzig: S. Hirzel.

1930. (Ed.) Elektronen interferenzen. Leipzig: S. Hirzel.

1931. (Ed.) Molekülstruktur. Leipzig: S. Hirzel.

1931. Polare Molekeln. Nachtrage II. Nachtrage zur Tabellenelektrischen Momente. Leipzig: S. Hirzel.

1931. (Ed.) Interference of electrons. Translated by Winifred M. Deans. London: Blackie and Son, Ltd.

1931. (Ed.) The dipole moment and chemical structure. Translated by Winifred M. Deans. London: Blackie and Son, Ltd.

1932. (Ed.) The structure of molecules. Translated by Winifred M. Deans. London: Blackie and Son, Ltd.

1933. Die Struktur der Materie. Leipzig: S. Hirzel.

1933. (Ed.) Magnetismus. Leipzig: S. Hirzel.

1934. The structure of matter: four lectures by P. Debye. Translated by F. M. Denton. Albuquerque: University of New Mexico.

1935. Kernphysik. Leipzig: S. Hirzel.

1945. Polar molecules. (Reprint of 1929 edition.) New York: Dover Publications (Re-issued, 1960).

1962. Topics in chemical physics: presented by Alfred Prock and Gladys McConkey: based on the Harvard lectures of Peter Debye. New York, Amsterdam: Elsevier.

1967. Molecular forces: presented by Benjamin Chu: based on the Baker Lectures of Peter J.W. Debye. New York: Interscience.

PART III JOHN BARDEEN

John Bardeen (1908 - 1991)

Fernando Sols

Dept. de Física Teórica de la Materia Condensada and Instituto Nicolas Cabrera. Universidad Autónoma of Madrid, E-28049 Madrid, Spain.

This year marks the tenth anniversary of the death of John Bardeen, one of the most outstanding figures in the science of the 20th century. In this article I offer a necessarily brief description of the life and work of this extraordinary physicist, whom I was lucky enough to know during my postdoctoral years at the University of Illinois with Anthony Leggett and Karl Hess. This narration will be mostly devoted to John Bardeen as an inventor and as a scientist, but some attention will also be paid to him as a consultant to industry and and advisor to government. An attempt will also be made to outline the main traits of his personality. Most of the material used to ellaborate this presentation has been taken from the account prepared by Lilian Hoddeson for the celebration of John Bardeen's eightieth birthday held at Urbana in 1988 [1], and from the articles written by Conyers Herring, Nick Holonyak Jr., J. Robert Schrieffer, George Pake, and David Pines, which were published in the April 1992 special issue of Physics Today [2].

The Inventor and the Scientist.

John Bardeen was born in Madison (Wisconsin) in 1908. From 1924 to 1929 he studied Electrical Engineering at the University of Wisconsing at Madison, where he earned the degrees of Bachelor of Science and Master of Science. There he attended lectures given by John van Vleck, one of the founding fathers of the quantum theory of magnetism. In 1930 he moved to the Gulf Research Laboratories in Pittsburgh, following one of his favorite EE professors in Madison. At Gulf Research, John Bardeen made applied geophysics research and invented a new electromagnetic method for the prospection of oil. The new scheme proved to be so useful that Gulf Research decided not to file a patent to avoid other companies learning too much about it. The approach proposed by Bardeen was not made public until thirty years after his invention. Despite his success, John Bardeen decided in 1933 that he wanted to return to academia and enrolled as a graduate student of Mathematical Physics at Princeton University. This was a bold move, considering that he was already making a very good salary at Gulf Research. His hope was to be able to work with Albert Einstein but, once in Princeton, he found out that the founder of relativity theory was not at the University but at the Institute for Advanced Study, and that he was not interested in taking students. So he settled for EugeneWigner, who proposed him to work on the theory of metal surfaces.

At that time (1933), the applications of quantum theory were undergoing a major expansion. Solid state physics was being born. Wigner and Seitz had just written a historical article on sodium showing that quantum mechanics could be used to calculate the electronic structure of simple metals. Bardeen tried to extend that work to calculate from first principles the work function at a sodium surface. He introduced several concepts that would become part of the foundations of metal theory. He introduced the *jellium model*, in which the positive charge of the ionic nuclei is assumed to be uniformly spread throughout the solid. He argued that such a simplification should still account for much of the essential physics of electrons in metals. He noted that an electric dipole layer was formed at the surface, due to the finite spread of the electronic cloud outside the confinements of the jellium background (a purely quantum mechanical effect). He argued, however, that this surface dipole distribution could not account for the entire work function, since there were two other contributions coming from many-body effects. An application of the self-consistent Hartree-Fock theory, then being popularized by Slater in the context of atoms and molecules, led to the identification of the *exchange hole*. Bardeen also noted the importance of the *correlation hole* which, he argued, had to continuously evolve towards the classical image potential as one moves to the exterior of the metal. Despite his necessarily crude calculations, his main conclusions still remain valid today.

In 1935 John Bardeen moved to Harvard University as a postdoctoral Junior Fellow. There he improved the Wigner-Seitz method to calculate the band structure of metals, making detailed comparisons with the high pressure experiments then being led by Bridgman. He also developed a first-principle calculation of the scattering of conduction electrons by phonons in metals. He included the self-consistent distortion of valence electrons during the vibrational motion of the ions. During that time he was also interested in nuclear theory. In 1938 he got married and moved to the University of Minnesota as an Assistant Professor. There he placed his thesis work on a firmer foundation by showing that the image potential of charges near metal surfaces could be derived from quantum mechanics. Motivated by the work of Fritz and Heinz London, John Bardeen developed a strong interest in superconductivity, which at that time still was a long-standing puzzle.

When World War II broke out, a large fraction of scientists abandoned academic research to work in defense related problems. Although he was an obvious candidate for the Manhattan project, John Bardeen moved to the Naval Ordnance Laboratory in Washington DC, because of his professional and personal links to the world of geophysics. There he worked on magnetic mines and torpedoes. At the end of the War, Bardeen was made an attractive offer to work at Bell Telephone Laboratories which he accepted.

The historical context was a very special one. War research projects had shown the importance of well funded basic research, and the value of coordinated team work involving theorists and experimentalists. Moreover, a new generation of solid state theorists with a strong background in quantum theory, had reached full maturity. This group includes the names of Herring, Seitz, Bardeen, Shockley, and Bethe, with Wigner and Slater as the historical leaders. Among their main colleagues in Europe were Jones and Mott in Great Britain, and Schottky in Germany. Already in the thirties, substantial advances had been made on the theory of semiconductors. Research on radar and other electronic equipment during the war had made it clear that there was a strong need to replace the vacuum tube amplifier in favor of more sophisticated semiconductor devices. This awareness had arisen in the context of collaborative work by several academic institutions (Purdue, Pennsylvania, Oxford) and industrial companies (Westinghouse, General Electric, Du Pont, Thompson). It was clear that after the War a fierce competition would develop in the race for more sophisticated amplifiers.

In that context, Marvin Kelly, the visionary Vice-President of Bell Telephone at that time, decided to foster research on solid-state electronic devices and asked William Shockley to recruit the best possible people for that enterprise. John Bardeen, who was already recognized as one of the outstanding solid theorists in America, was offered to join the company, at a time (1945) that was ripe for full investment in semiconductor research. During his stay at Bell Laboratories, several concepts were introduced that would play a key role in the theory of semiconductor devices, an in whose formulation John Bardeen was heavily involved. The new ideas dealt with the formation of a charge accumulation layer, the role of surface states, the existence of an inversion layer, and, most importantly, the realization that injection of minority carriers was essential in determining the transport behavior of metal-semiconductor interfaces. This latter discovery was an almost single handed contribution of John Bardeen's extraordinary physical intuition. Those two years of close collaboration with the experimentalist Walter Brattain led to the successful amplification of an electronic signal by the "point-contact transistor", in the morning of 16 December 1947. One week later, Brattain and Bardeen demonstrated the amplification of an audio signal to the company executives. The word transistor was coined by combining the popular use of words like thermistor, varistor, and resistor, which named electric devices, and the observation that transconductance and therefore transresistance were essential concepts in the new device. The discovery was held secret for several months, until it was released to the press on June 30, 1948. The invention was heralded by the press as the possible beginning of a revolution in electronics. This time they were right.

The point-contact transistor was later eclipsed by a whole family of solid-state amplification devices which eventually gave rise to a new industry. These included the field-effect transistor developed by Shockley, who would share with Bardeen and Brattain the 1956 Nobel Prize for Physics. It must be emphasized, however, that the bipolar transistor of Bardeen and Brattain was the first semiconductor device involving electronic injection. According to Holonyak [2], the date of 16 December 1947 must be viewed as the beginning of modern electronics. To emphasize that the point-contact transistor operated on fundamentally new principles, Bardeen introduced the new terminology of emitter, collector, and base, different from the terms source, drain, and gate used for analogous devices.

Later in his life, he was once asked by a Sharp executive who were the theoreticians who had most influenced him in the invention of the transistor. Rather mildly, Bardeen told him that he had chosen to work with experimenters, see the facts, sort out the data and make suggestions accordingly. Quite characteristically, he did not tell him openly that no theoretician had had much to do with his thinking on semiconductors.

In 1950 John Bardeen moved to the University of Illinois at Urbana-Champaign, where he settled for the rest of his life. Bardeen's decision was largely motivated by his desire to have greater freedom to pursue basic research on superconductivity theory. He realized that the atmosphere of independent and creative thinking which he felt he needed, would be difficult to attain at Bell Laboratories, where the brilliant but slightly oppressive figure of Shockley was pervasive, and where his proposal to form an independent theory group had been turned down.

Once in Urbana, John Bardeen started in parallel a major research program on semiconductors. As a fruit of that effort, Nick Holonyak Jr., his first graduate student, eventually became a pioneer of semiconductor lasers. But it was in superconductivity theory where Bardeen was going to write another major chapter in the history of science. His first step was the writing of a review on superconductivity for *Handbuch der Physik*. He formulated several requirements that would have to be satisfied by a hypothetical theory of superconductivity. Among them, he noted that phonons had to be involved (to account for the isotope effect discovered by Serin in Rutgers in 1950), superconductivity had to be a Fermi surface effect, and an energy gap had to exist. He also pointed out that the tiny scale of the condensation energy (eight orders of magnitude below typical electron energies) made it necessary the use of the Landau theory of Fermi liquids, which was considered to account for all the physics not essential to superconductivity.

The Physics Department in Urbana was hectic at that time. Lee, Low, and Pines showed that a retarded interaction mediated by phonons could be *attractive*, overcoming the residual Coulomb repulsion. John Bardeen felt that the then recently developed techniques of quantum field theory could be useful in the search for a superconductivity theory. He phoned C. N. Yang, then at the Institute for Advanced in Princeton, and asked him whether he knew an expert in those techniques that might be willing to do theoretical work on superconductivity. Yang recommended Leon Cooper, who moved to Urbana and joined the tandem of John Bardeen and his student Bob Schrieffer. Soon it was clear that a perturbative approach was not promising. Instead Cooper made the important discovery that two electrons above a Fermi surface experiencing mutual attractive interaction would form a bound state. The concept of *Cooper pair* was born.

Bardeen, Cooper and Schrieffer (BCS) realized that the next step was to generalize Cooper's result to the whole Fermi sea, treating all electrons on the same footing, but they did not know how. Discouraged, in December 1956 Schrieffer asked Bardeen to give him another research topic. Bardeen suggested him to try further for a few more months, since he knew they were on the right track. Inspired by recent work on pion emission by Tomonaga, and realizing that some type of field theory should be adequate (Bardeen had noted that the radius of the Cooper pair had to be much larger that the electron spacing), Schrieffer tried a new wave function and showed it to Bardeen. It was essentially a generalization of bosonic coherent states to Fermion pairs. Bardeen immediately realized that it was what they needed. The BCS trio divided tasks and started a frenzy activity. In two weeks they were ready to submit a letter to the Physical Review. This was in February 1957. In the following months, BCS polished their calculation and, in July 1957, they submitted their famous long paper on the theory of superconductivity.

The BCS theory gained quick acceptance among experimentalists, since it was clearly successful in explaining and predicting experiments. However, the theoretical community was more reluctant to accept it. The theory was not manifestly gauge invariant; electric charge was not conserved. Work by Anderson and Nambu helped to clarify the issue. They noted that it was the first instance of broken gauge symmetry. The two-electron phase within the solid had become rigid and BCS had just chosen one possible global phase for mathematical convenience. Anderson noted that the concept of gauge symmetry breaking could have bearing in elementary particle physics, where it could explain the generation of massive particles from massless fields (the Anderson-Higgs mechanism). Based on a generalization of this new concept to nonabelian gauge symmetries, Higgs predicted the existence of a scalar boson and the electroweak theory was eventually formulated. Nowadays, the concept of broken gauge symmetry permeates the theory of elementary particles. The BCS theory has also been influential in nuclear physics, where the nucleon pairing model proposed by Bohr, Mottelson and Pines has played a fundamental role ever since.

Soon after the development of the BCS theory, a heated debate arose on the role of an insulator tunneling barrier (or a segment of normal metal) sandwiched between two superconductors. On the other side of the Atlantic, a young British theoretician named Brian Josephson challenged John Bardeen's respected view that superconducting coherence could not survive under a tunneling barrier and that therefore coherent tunneling of Cooper pairs between two superconductors was not possible [3]. The two theoretical standpoints seemed irreconcilable. Agreement was reached only after experiments performed at Bell Labs by Rowell and Anderson unambiguously observed the predictions made by Josephson. The story of this confrontation reminds us that even a scientist of Bardeen's stature can occasionally be wrong, and certainly speaks of the greatness of science.

Their work on superconductivity theory earned Bardeen, Cooper, and Schrieffer the 1972 Nobel Prize in Physics (together with Giaever and Esaki, Josephson won the same prize one year later for his work on tunneling in superconductors). John Bardeen remains to date the only person ever to win two Nobel Prizes in the same discipline.

His double ground in applied and fundamental physics was surely one of John Bardeen's greatest assets. Combined with his exceptional talent, it allowed him to excel both as an inventor and as a scientist. He remained an active researcher throughout the rest of his life, playing the role of a world leader in solid state theory. He retired in 1975. A few years before his death, he was excited by the discovery of high temperature superconductors and still had time to make some notable contributions.

The Consultant and the Advisor.

In parallel to his role as an applied and basic scientist, John Bardeen was an efficient consultant to many industrial companies and an advisor to the Government. He was clearly aware of the value of research and of the knowledge it generated. Addressing a Chicago-area electronics industry seminar he once said: "It is those who carry on advanced research programs on their own who are in best position to profit from new discoveries in science wherever they may occur." He was referring both to national governments and industrial companies.

His continued advice to Xerox Corporation was instrumental in the development of the photocopying machine. He also consulted for General Electric, Supertex, Sunnyvale, Texas Instrument, and Sony Corporation in Japan. He was an expert witness for TI when this company had to fight in court for a patent on aspects of Jack Kilby's integrated circuit invention. His relation with with Sony Corporation was very close since the early fifties. In 1989, Sony endowed the John Bardeen Chair to the University of Illinois.

John Bardeen was also and advisor to many American Presidents, from Eisenhower to Reagan. In 1983, he resigned from the White House Scientific

Council because of his strong opposition to the Strategic Defense Initiative (Stars War). In 1986, he wrote about SDI: "At a time when our civilian economy needs all the help it can get to remain competitive in world markets, the best scientific and technical brains in the country may be drawn off to work on a project of dubious value."

The Man.

Since an early age John Bardeen showed an exceptional talent for Mathematics. At the age of nine, his parents and teachers decided to shift him from the 3rd to the 7th grade, a move that would be impossible in many countries. Unlike other child prodigies, John Bardeen does not seem to have been shaped by the experience. There are several circumstances that may have prevented him from being spoiled. The son of a Professor of Medicine, he lived in a university environment where high intellectual achievements are somewhat less rare. On the other hand, being the second of a family of five children probably helped him not to be the continued center of attention. In addition, his parents made sure that his after-school life was totally normal; as soon as he left school he would resume activities with children of their age. Finally, he was always interested in his sports, particularly in swimming and golf, which probably helped him not to become eccentric. Throughout his life, John Bardeen was a pleasant family man who enjoyed spending time with children and relatives.

He was deep and silent. He expressed his thoughts, always in a low voice, only after he had meditated them carefully. (His quiet nature seemed to mutate in the golf course, where reportedly he could shout and show more excitement than he ever displayed for any of his two Nobel Prize discoveries.) As a scientist, he was generous, always ready to give credit to others. He would rarely give credit to himself. When he had to, he used to find ways to convey the message indirectly.

He was aware that he had always built on the work of great scientists that had preceded him. He was particularly thankful to Fritz London for having pointed out the macroscopic quantum nature of superconductivity, a well-aimed indication that set the scientific community on the right track. With his share of the 1972 Nobel Prize, he generously endowed the Fritz London Award, presently the most prestigious prize in the field of Low Temperature Physics.

Out of admiration and sorrow for the loss of that exceptional figure, Bob Schrieffer wrote after his death on 31 January 1991 [2]: "To me, John

Bardeen represents the best that mankind has to offer. We all miss this man for all seasons."

Bibliography

[1] L. Hoddeson, *John Bardeen and the discoveries of the transistor and the BCS theory of superconductivity*, in "A Collection of Professor John Bardeen's Publication on Semiconductors and Superconductivity", University of Illinois internal publication.

[2] Physics Today, April 1992, p. 23.

[3] Physics Today, July 2001, p. 46.

Semiconductor Research Leading to the Point Contact Transistor

John Bardeen

Nobel Lecture, December 11, 1956

INTRODUCTION

In this lecture we shall attempt to describe the ideas and experiments which led to the discovery of the transistor effect as embodied in the point-contact transistor. Some of the important research done subsequent to the discovery will be described in the following lectures by Shockley and Brattain. As we shall see, the discovery was but a step along the road of semiconductor research to which a great many people in different countries have contributed.

It was dependent both on the sound theoretical foundation largely built up during the thirties and on improvement and purification of materials, particularly of germanium and silicon, in the forties. About half of the lecture will be devoted to an outline of concepts concerning electrical conduction in semiconductors and rectification at metal-semiconductor contacts as they were known at the start of our research program.

The discovery of the transistor effect occurred in the course of a fundamental research program on semiconductors initiated at the Bell Telephone Laboratories in early 1946. Semiconductors was one of several areas selected under a broad program of solid-state research, of which S. O. Morgan and W. Shockley were co-heads. In the initial semiconductor group, under the general direction of Shockley, were W. H. Brattain, concerned mainly with surface properties and rectification, G. L. Pearson, concerned with bulk properties, and the writer, interested in theoretical aspects of both. Later a physical chemist, R. B. Gibney, and a circuit expert, H. R. Moore, joined the group and made important contributions, particularly to chemical and instrumentation problems, respectively.

It is interesting to note that although Brattain and Pearson had had considerable experience in the field prior to the war, none of us had worked on semiconductors during the war years. We were able to take advantage of the important advances made in that period in connection with the development of silicon and germanium detectors and at the same time have a fresh look at the problems. Considerable help was obtained from other groups in the Laboratories which were concerned more directly with wartime developments. Particular mention should be made of J. H. Scaff, H. C. Theuerer and R. S. Ohl.

The general aim of the program was to obtain as complete an understanding as possible of semiconductor phenomena, not in empirical terms, but on the basis of atomic theory. A sound theoretical foundation was available from work done during the thirties:

(1) Wilson's quantum mechanical theory[1], based on the energy band model, and describing conduction in terms of excess electrons and holes. It is fundamental to all subsequent developments. The theory shows how the concentration of carriers depends on the temperature and on impurities.

(2) Frenkel's theories of certain photoconductive phenomena[2] (change of contact potential with illumination and the photomagneto electric effect) in which general equations were introduced which describe current flow when non-equilibrium concentrations of both holes and conduction electrons are present. He recognized that flow may occur by diffusion in a concentration gradient as well as by an electric field.

(3) Independent and parallel developments of theories of contact rectification by Mott[3], Schottky[4] and Davydov[5]. The most complete mathematical theories were worked out by Schottky and his co-worker, Spenke.

Of great importance for our research program was the development during and since the war of methods of purification and control of the electrical properties of germanium and silicon. These materials were chosen for most of our work because they are well-suited to fundamental investigations with the desired close coordination of theory and experiment.

Depending on the nature of the chemical impurities present, they can be made to conduct by either excess electrons or holes.

Largely because of commercial importance in rectifiers, most experimental work in the thirties was done on copper oxide (Cu_2O) and selenium. Both have complex structures and conductivities which are difficult to control.

While the theories provided a good qualitative understanding of many semiconductor phenomena, they had not been subjected to really convincing quantitative checks. In some cases, particularly in rectification, discrepancies between experiment and theory were quite large. It was not certain whether the difficulties were caused by something missing in the theories or by the fact that the materials used to check the theories were far from ideal.

In the U.S.A., research on germanium and silicon was carried out during the war by a number of university, government and industrial laboratories in connection with the development of point-contact or « cat's whisker » detectors for radar. Particular mention should be made of the study of germanium by a group at Purdue University working under the direction of K. Lark-Horovitz and of silicon by a group at the Bell Telephone Laboratories. The latter study was initiated by R. S. Ohl before the war and carried out subsequently by him and by a group under J. H. Scaff. By 1946 it was possible to produce relatively pure polycrystalline materials and to control the electrical properties by introducing appropriate amounts of donor and acceptor impurities. Some of the earliest work (1915) on the electrical properties of germanium and silicon was done in Sweden by Prof. C. Benedicks.

Aside from intrinsic scientific interest, an important reason for choosing semiconductors as a promising field in which to work, was the many and increasing applications in electronic devices, which, in 1945, included diodes, varistors and thermistors. There had long been the hope of making a triode, or an amplifying device with a semiconductor. Two possibilities had been suggested. One followed from the analogy between a metal semiconductor rectifying contact and a vacuum-tube diode. If one could somehow insert a grid in the space-charge layer at the contact, one should be able to control the flow of electrons across the contact. A major practical difficulty is that the width of the space-charge layer is typically only about 10-4 cm. That the principle is a sound one was demonstrated by Hilsch and Pohl[6], who built a triode in an alkali-halide crystal in which the width of the spacecharge layer was of the order of one centimeter. Because amplification was limited to frequencies of less than one cycle per second, it was not practical for electronic applications.

The second suggestion was to control the conductance of a thin film or slab of semiconductor by application of a transverse electric field (called the *field effect).* In a simple form, the slab forms one plate of a parallel plate condenser, the control electrode being the other plate. When a voltage is applied across the condenser, charges are induced in the slab. If the induced charges are mobile carriers, the conductance should change with changes of voltage on the control electrode. This form was suggested by Shockley; his calculations indicated that, with suitable geometry and materials, the effect should be large enough to produce amplification of an a.c. signal[7].

Point-contact and junction transistors operate on a different principle than either of these two suggestions, one not anticipated at the start of the program. The transistor principle, in which both electrons and holes play a role, was discovered in the course of a basic research program on surface properties.

Shockley's field-effect proposal, although initially unsuccessful, had an important bearing on directing the research program toward a study of surface phenomena and surface states. Several tests which Shockley carried out at various times with J. R. Haynes, H. J. McSkimin, W. A. Yager and R. S. Ohl, using evaporated films of germanium and silicon, all gave negative results. In analyzing the reasons for this failure, it was suggested[8] that there were states for electrons localized at the surface, and that a large fraction of the induced charge was immobilized in these states. Surface states also accounted for a number of hitherto puzzling features of germanium and silicon point-contact diodes.

In addition to the possibility of practical applications, research on surface properties appeared quite promising from the viewpoint of fundamental science.

Although surface states had been predicted as a theoretical possibility, little was known about them from experiment. The decision was made, therefore, to stress research in this area.. The study of surfaces initiated at that time (1946) has been continued at the Bell Laboratories and is now being carried out by many other groups as well[9].

It is interesting to note that the field effect, originally suggested for possible value for a device, has been an extremely fruitful tool for the fundamental investigation of surface states. Further, with improvements in semiconductor technology, it is now possible to make electronic amplifiers with high gain which operate on the field-effect principle.

Before discussing the research program, we shall give first some general background material on conduction in semiconductors and metal-semiconductor rectifying contacts.

NATURE OF CONDUCTION IN SEMICONDUCTORS

An electronic semiconductor is typically a valence crystal whose conductivity depends markedly on temperature and on the presence of minute amounts of foreign impurities. The ideal crystal at the absolute zero is an insulator. When the valence bonds are completely occupied and there are no extra electrons in the crystal, there is no possibility for current to flow.

Charges can be transferred only when imperfections are present in the electronic structure, and these can be of two types: *excess electrons* which do not fit into the valence bonds and can move through the crystal, and *holes,* places from which electrons are missing in the bonds, which also behave as mobile carriers. While the excess electrons have the normal negative electronic charge *-e,* holes have a positive charge, *+e*. It is a case of two negatives making a positive; a missing negative charge is a positive defect in the electron structure.

The bulk of a semiconductor is electrically neutral; there are as many positive charges as negative. In an intrinsic semiconductor, in which current carriers are created by thermal excitation, there are approximately equal numbers of excess electrons and holes. Conductivity in an *extrinsic* semiconductor results from impurity ions in the lattice. In n-type material, the negative charge of the excess electrons is balanced by a net positive space charge of impurity ions. In p-type, the *positive* charge of the holes is balanced by negatively charged impurities. Foreign atoms which can become positively charged on introduction to the lattice are called *donors*; atoms which become negatively ionized are called *acceptors.* Thus donors make a semiconductor n-type, acceptors p-type. When both donors and acceptors are present, the conductivity type depends on which is in excess. Mobile carriers then balance the *net* space charge of the impurity ions. Terminology used is listed in the table below:

Designation of *conductivity* type		Majority carrier	Dominant *impurity* ion
n-type	*excess*	electron (n/cm^3)	donor
p-type	defect	hole (p/cm^3)	acceptor

Table 1.

These ideas can be illustrated quite simply for silicon and germanium, which, like carbon, have a valence of four and lie below carbon in the Periodic Table. Both crystallize in the diamond structure in which each atom is surrounded tetrahedrally by four others with which it forms bonds. Carbon in the form of a diamond is normally an insulator; the bond structure is complete and there are no excess electrons. If ultraviolet light falls on diamond, electrons can be ejected from the bond positions by the photoelectric effect. Excess electrons and holes so formed can conduct electricity; the crystal becomes photoconductive.

The energy required to free an electron from a bond position so that *it* and the hole left behind can move the crystal, is much less in silicon and germanium than for diamond. Appreciable numbers are released by thermal excitations at high temperatures; this gives intrinsic conductivity.

Impurity atoms in germanium and silicon with more than four valence electrons are usually donors, those with less than four acceptors. For example, Group V elements are donors, Group III elements acceptors. When an arsenic atom, a Group V element, substitutes for germanium in the crystal, only four of its valence electrons are required to form the bonds. The fifth is only weakly held by the forces of Coulomb attraction, greatly reduced by the high dielectric constant of the crystal. The energy required to free the extra electron is so small that the arsenic atoms are completely ionized at room temperature. Gallium, a typical Group III acceptor, has only three valence electrons. In order to fill the four bonds, Ga picks up another electron and enters the crystal in the form of a negative ion, Ga-. The charge is balanced by a free hole.

While some of the general notions of excess and defect conductivity, donors and acceptors, go back earlier, Wilson1 was the first to formalize an adequate mathematical theory in terms of the band picture of solids. The band picture itself, first applied to metals, is a consequence of an application of quantum mechanics to the motion of electrons in the periodic potential field of a crystal lattice. Energy levels of electrons in atoms are discrete.

When the atoms are combined to form a crystal, the allowed levels form continuous bands. When a band is completely occupied, the net current of all of the electrons in the band is zero. Metals have incompletely filled bands. In insulators and semiconductors, there is an energy gap between the highest filled band and the next higher allowed band of levels, normally unoccupied.

The relations are most simply illustrated in terms of an energy-level diagram of the crystal. In Fig. 1 is shown a schematic energy-level diagram of an intrinsic semiconductor. Electrons taking part in the chemical bonds form a continuous band of levels called the valence band. Above these is an energy gap in which there are no allowed levels in the ideal crystal, and then another continuous band of levels called the conduction band. The energy gap, *EG,* is the energy required to free an electron from the valence bonds.

Excess, or conduction, electrons have energies in the lower part of the conduction band. The very lowest state in this band, E_C, corresponds to an electron at rest, the higher states to electrons moving through the crystal with additional energy of motion. Holes correspond to states near the top of the valence band, E_V, from which electrons are missing. In an intrinsic semiconductor, electrons and holes are created in equal numbers by thermal excitation of electrons from the valence to the conduction band, and they are distributed at random through the crystal.

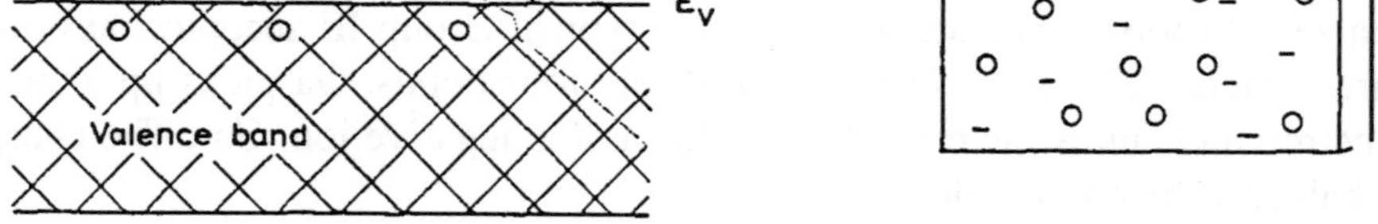

Figure 1. Energy-level diagram of an intrinsic semiconductor. There is a random distribution of electrons and holes in equal numbers.

In an n-type semiconductor, as illustrated in Fig. 2a, there is a large number of electrons in the conduction band and very few holes in the valence band. Energy levels corresponding to electrons localized around Group V donor impurity atoms are typically in the forbidden gap and a little below the conduction band. This means that only a small energy is required to ionize the donor and place the electron removed in the conduction band.

The charge of the electrons in the conduction band is compensated by the positive space charge of the donor ions. Levels of Group III acceptors (Fig. 2b) are a little above the valence band. When occupied by thermal excitation of electrons from the valence band, they become negatively charged. The space charge of the holes so created is compensated by that of the negative acceptor ions.

Occupancy of the levels is given by the position of the Fermi level, E_F.

The probability, f, that a level of energy E is occupied by an electron is given by the Fermi-Dirac function:

$$f = \frac{I}{1 + \exp[(E - E_F)/kT]}$$

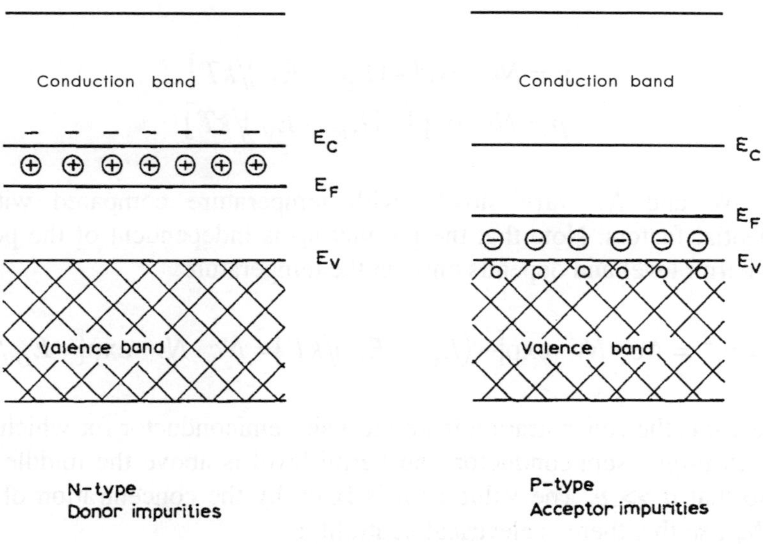

Figure 2. Energy-level diagrams for n- and p- type semiconductors.

The energy gap in a semiconductor is usually large compared with thermal energy, kT (~ *0.025* eV at room temperature), so that for levels well above E_F one can use the approximation

$$f \approx \exp[-(E - E_F)/kT]$$

For levels below E_F, it is often more convenient to give the probability

$$f_p = 1 - f = \frac{1}{1 + \exp[[E_F - E]/kT]}$$

that the level is unoccupied, or «occupied by a hole». Again, for levels well below E_F,

$$f_p \approx \exp[-(E_F - E)/kT]$$

The expressions for the total electron and hole concentrations (number per unit volume), designated by the symbols n and p respectively, are of the form

$$n = Nc \cdot \exp[-(E_C - E_F)/kT]$$
$$p = Nv \cdot \exp[-(E_F - E_V)/kT]$$

where Nc and Nv vary slowly with temperature compared with the exponential factors. Note that the product np is independent of the position of the Fermi level and depends only on the temperature:

$$np = n_i^2 = Nc \cdot Nv \cdot \exp[-(E_C - E_V)/kT] = Nc \cdot Nv \cdot \exp[-E_G/kT]$$

Here n is the concentration in an intrinsic semiconductor for which $n = p$.

In an n-type semiconductor, the Fermi level is above the middle of the gap, so that $n >> p$. The value of n is fixed by the concentration of donor ions, N_d^+, so that there is electrical neutrality:

$$n - p = N_d^+$$

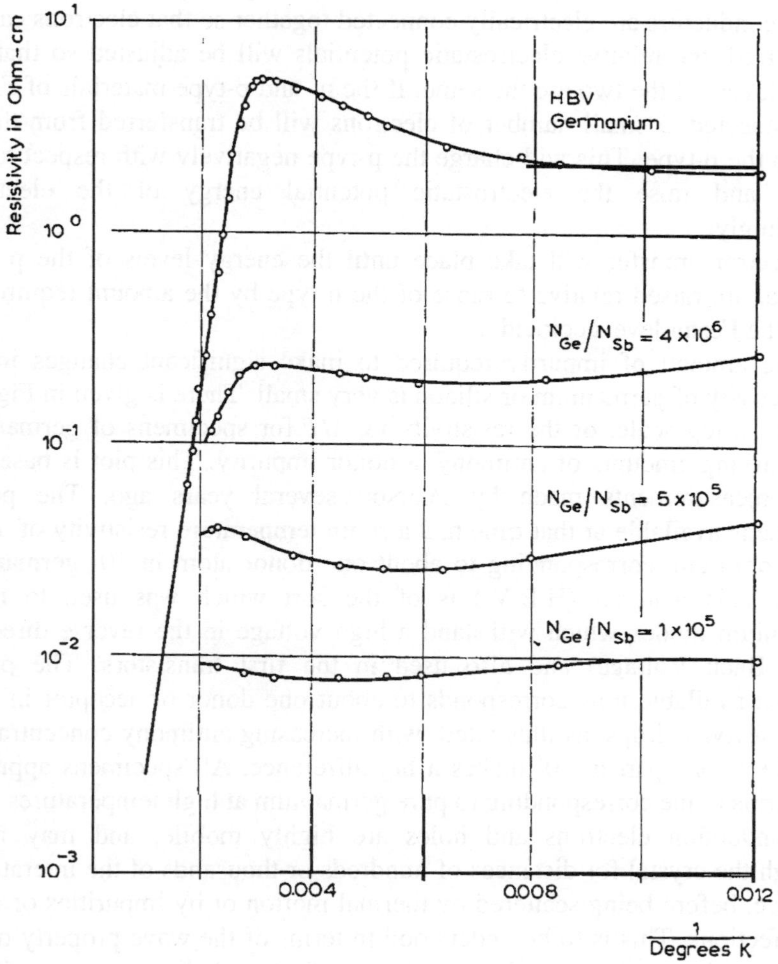

Figure 3. Conductivity vs. $1/T$ for germanium with antimony added as a donor impurity

The minority carrier concentration, p, increases rapidly with temperature and eventually a temperature will be reached above which n and p are both large compared with N_d^+ and the conduction is essentially intrinsic. Correspondingly in a p-type semiconductor, in which there are acceptor ions, $p >> n$, and the Fermi level is below the center of the gap.

The Fermi level is equivalent to the chemical potential of the electrons. If two conductors are electrically connected together so that electrons can be transferred, the relative electrostatic potentials will be adjusted so that the Fermi levels of the two are the same. If the n- and p-type materials of Fig. 2 are connected, a small number of electrons will be transferred from the n-type to the p-type. This will charge the p-type negatively with respect to the n-type and raise the electrostatic potential energy of the electrons accordingly.

Electron transfer will take place until the energy levels of the p type material are raised relative to those of the n-type by the amount required to make the Fermi levels coincide.

The amount of impurity required to make significant changes in the conductivity of germanium or silicon is very small. There is given in Fig. 3 a plot, on a log scale, of the resistivity *vs.* $1/T$ for specimens of germanium with varying amounts of antimony, a donor impurity. This plot is based on some measurements made by Pearson several years ago. The purest specimens available at that time had a room temperature resistivity of about 10-20 ohm cm, corresponding to about one donor atom in 10^8 germanium atoms. This material (H.B.V.) is of the sort which was used to make germanium diodes which withstand a high voltage in the reverse direction (High Back Voltage) and also used in the first transistors. The purest material available now corresponds to about one donor or acceptor in 10^{10}. The resistivity drops, as illustrated, with increasing antimony concentration; as little as one part in 10^7 makes a big difference. All specimens approach the intrinsic line corresponding to pure germanium at high temperatures.

Conduction electrons and holes are highly mobile, and may move through the crystal for distances of hundreds or thousands of the interatomic distance, before being scattered by thermal motion or by impurities or other imperfections. This is to be understood in terms of the wave property of the electron; a wave can travel through a perfect periodic structure without attenuation. In treating acceleration in electric or magnetic fields, the wave aspect can often be disregarded, and electrons and holes thought of as classical particles with an effective mass of the same order, but differing from the ordinary electron mass. The effective mass is often anisotropic, and different for different directions of motion in the crystal. This same effective mass picture can be used to estimate the thermal motion of the gas of electrons and holes. Average thermal velocities at room temperature are of the order of 10^7 cm/sec.

Scattering can be described in terms of a mean free path for the electrons and holes. In relatively pure crystals at ordinary temperatures, scattering occurs mainly by interaction with the thermal vibrations of the atoms of the crystal. In less pure crystals, or in relatively pure crystals at low temperatures, the mean free path may be determined by scattering by impurity atoms. Because of the scattering, the carriers are not uniformly accelerated by an electric field, but attain an average drift velocity proportional to the field. Ordinarily the drift velocity is much smaller than the average thermal velocity. Drift velocities may be expressed in terms of the mobilities, μ_n and μ_p of the electrons and holes respectively*.

In an electric field *E*,

$$(V_d)_n = -\mu_n E$$
$$(V_d)_p = \mu_p E$$

Because of their negative charge, conduction electrons drift oppositely to the field. Values for pure germanium at room temperature are μ_n = 3,800 cm^2/volt set; μ_p = 1,800 cm^2/volt sec. This means that holes attain a drift velocity of 1,800 cm/sec in a field of one volt/cm.

Expressions for the conductivity are:

n-type : $\sigma_n = ne\mu_n$

p-type : $\sigma_p = pe\mu_p$

intrinsic : $\sigma = ne\mu_n + pe\mu_p$

It is not possible to determine *n* and μ_n separately from measurements of the conductivity alone. There are several methods to determine the mobility; one which has been widely used is to measure the Hall coefficient in addition to the conductivity. As part of the research program at the Bell Laboratories, Pearson and Hall made resistivity measurements over a wide range of temperatures of silicon containing varying amounts of boron (a Group III acceptor) and of phosphorus (a Group V donor). Analysis of the data[10] gave additional confirmation of the theory we have outlined.

* A subscript *n* (referring to negative charge) is used for conduction electrons, *p* (*pos*itive) for holes.

Similar measurements on germanium were made about the same time by Lark-Horovitz and co-workers, and more recently more complete measurements on both materials have been made by other groups. The result of a large amount of experimental and theoretic work has been to confirm the Wilson model in quantitative detail.

Carriers move not only under the influence of an electric field, but also by diffusion; the diffusion current is proportional to the concentration gradient. Expressions for the particle current densities of holes and electrons, respectively, are

$$j_p = p\mu_p E - D_p \operatorname{grad} p$$

$$j_n = n\mu_n E - D_n \operatorname{grad} n$$

Einstein has shown that mobilities and diffusion coefficients are related:

$$\mu = \frac{e}{kT} D$$

where k is Boltzmann's constant. Diffusion and conduction currents both play an important role in the transistor.

The diffusion term was first considered by Wagner in his theory of oxidation of metals. The equations were worked out more completely by Frenkel[2] in an analysis of the diffusive flow which occurs when light is absorbed near one face of a slab, as shown schematically in Fig. 4. The light quanta raise electrons from the valence to the conduction bands, creating conduction electrons and holes in equal numbers. These diffuse toward the interior of the slab. Because of recombination of conduction electron and hole pairs, the concentration drops as the diffusion occurs. Frenkel gave the general equations of flow when electrons and holes are present in attempting to account for the Dember effect (change in contact potential with light) and the photomagnetoelectric (PME) effect. The latter is a voltage analogous to a Hall voltage observed between the ends of a slab in a transverse magnetic field (perpendicular to the paper in the diagram). The Dember voltage was presumed to result from a difference of mobility, and thus of diffusion coefficient, between electrons and holes. Electrical neutrality requires that the concentrations and thus the concentration gradients be the same. Further, under steady-state conditions the flow of electrons to the interior must equal the flow of holes, so that there is no net electrical current. However, if *Dn* is greater than *Dp,* the diffusive flow of electrons would be greater than that of holes. What happens is that an electric field, *E,* is introduced which aids holes and retards the electrons so as to equalize the flows. The integral of *E* gives a voltage difference between the surface and the interior, and thus a change in contact potential. As we will mention later, much larger changes in contact potential with light may come from surface barrier effects.

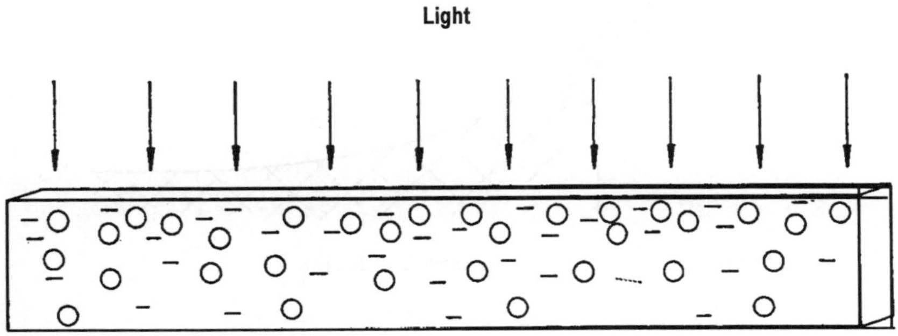

Figure 4. Schematic diagram of diffusive flow of electrons and holes created near the surface by absorption of light.

CONTACT RECTIFIERS

In order to understand how a point-contact transistor operates, it is necessary to know some of the features of a rectifying contact between a metal and semiconductor. Common examples are copper-oxide and selenium rectifiers and germanium and silicon point-contact diodes which pass current much more readily for one direction of applied voltage than the opposite. We shall follow Schottky's picture[4], and use as an illustration a contact to an ntype semiconductor. Similar arguments apply to p-type rectifiers with appropriate changes of sign of the potentials and charges. It is most convenient to make use of an energy-level diagram in which the changes in energy bands resulting from changes in electrostatic potential are plotted along a line perpendicular to the contact, as in Fig. 5. Rectification results from the potential energy barrier at the interface which impedes the flow of electrons across the contact.

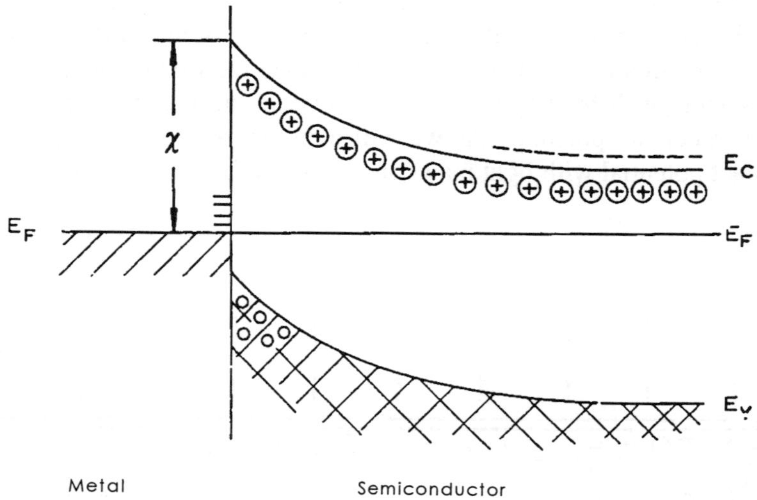

Figure 5. Equilibrium energy-level diagram for a metal-semiconductor rectifying contact along a line perpendicular to the interface. Variations in the energy bands of the semiconductor result from changes in electrostatic potential due to the layer of uncompensated space-charge. The overall change in potential from the surgace to the interior is such as to bring the Fermi level in the interior of the semiconductor into coincidence with that of the metal. In this example, there is an inversion from n-type conductance in the bulk to p-type at the surface.

The Fermi level of the metal is close to the highest of the normally occupied levels of the conduction band. Because of the nature of the metal-semiconductor interface layers, a relatively large energy, χ, perhaps of the order of 0.5 eV, is required to take an electron from the Fermi level of the metal and place it in the conduction band in the semiconductor. In the interior of the semiconductor, which is electrically neutral, the position of the Fermi level relative to the energy bands is determined by the concentration of conduction electrons, and thus of donors. In equilibrium, with no voltage applied, the Fermi levels of the metal and semiconductor must be the same.

This is accomplished by a region of space charge adjacent to the metal in which there is a variation of electrostatic potential, and thus of potential energy of the electron, as illustrated.

In the bulk of the semiconductor there is a balance between conduction electrons and positive donors. In the barrier region which is one of high potential energy for electrons, there are few electrons in the conduction band.

The uncompensated space charge of the donors is balanced by a negative charge at the immediate interface. It is these charges, in turn, which produce the potential barrier. The width of the space-charge region is typically of the order of 10^{-5} to 10^{-4} cm.

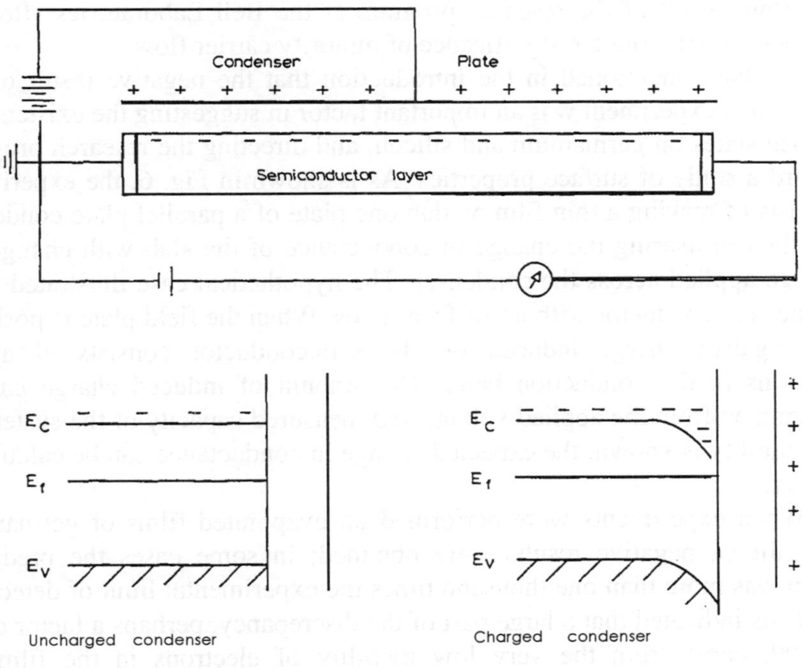

Figure 6. Schematic diagram of a field-effect experiment for an n-type semiconductor with no surface states.

When a voltage is applied, most of the drop occurs across the barrier layer. The direction of easy flow is that in which the semiconductor is negative relative to the metal. The bands are raised, the barrier becomes narrower, and electrons can flow more easily from the semiconductor to the metal. In the high resistance direction, the semiconductor is positive, the bands are lowered relative to the metal, and the barrier is broadened. The current of electrons flowing from the metal is limited by the energy barrier, c, which must be surmounted by thermal excitation.

If χ is sufficiently large, the Fermi level at the interface may be close to the valence band, implying an inversion from n-type conductivity in the bulk to p-type near the contact. The region of hole conduction is called, following Schottky, an inversion layer. An appreciable part of the current flow to the contact may then consist of minority carriers, in this case holes. An important result of the research program at the Bell Laboratories after the war was to point out the significance of minority carrier flow.

We have mentioned in the introduction that the negative result of the field effect experiment was an important factor in suggesting the existence of surface states on germanium and silicon, and directing the research program toward a study of surface properties. As is shown in Fig. 6, the experiment consists of making a thin film or slab one plate of a parallel plate condenser and then measuring the change in conductance of the slab with changes in voltage applied across the condenser. The hypothetical case illustrated is an n-type semiconductor with no surface states. When the field plate is positive, the negative charge induced on the semiconductor consists of added electrons in the conduction band. The amount of induced charge can be determined from the applied voltage and measured capacity of the system. If the mobility is known, the expected change in conductance can be calculated readily.

When experiments were performed on evaporated films of germanium and silicon, negative results were obtained; in some cases the predicted effect was more than one thousand times the experimental limit of detection. Analysis indicated that a large part of the discrepancy, perhaps a factor of 50 to 100, came from the very low mobility of electrons in the films as compared with bulk material. The remaining was attributed to shielding by surface states.

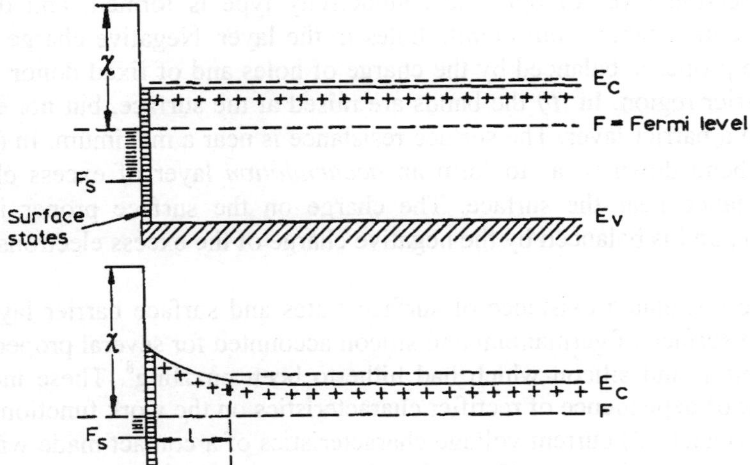

Figure 7. Formation of a space-charge barrier layer at the free surface of a semiconductor *Experiments on surface states*

It was predicted that if surface states exist, a barrier layer of the type found at a metal contact might be found at the free surface of a semiconductor. The formation of such a layer is illustrated schematically in Fig. 7. Occupancy of the surface levels is determined by the position of the Fermi level at the surface. In the illustration, it is presumed that the distribution of surface states is such that the states themselves would be electrically neutral if the Fermi level crossed at the position *FS* relative to the bands. If there is no surface barrier, so that the Fermi level crosses the surface above *FS*, there are excess electrons and a net negative charge in the surface states. When the surface as whole is neutral, a barrier layer is formed such that the positive charge in the layer is compensated by the negative surface states charge. If the density of surface states is reasonably high, sufficient negative charge is obtained with the Fermi level crossing only slightly above *FS*.

Types of barriers which may exist at the surface of an n-type semiconductor are illustrated in Fig. 8. On the left (a) the energy bands are raised at the surface so as to bring the valence band close to the Fermi level. An inversion layer of opposite conductivity type is formed, and there is excess conductance from mobile holes in the layer. Negative charge on the surface proper is balanced by the charge of holes and of fixed donor ions in the barrier region. In *(b)* the bands are raised at the surface, but not enough to form a barrier layer. The surface resistance is near a maximum. In (c), the bands bend down so as to form an *accumulation* layer of excess electron conductance near the surface. The charge on the surface proper is now positive, and is balanced by the negative charge of the excess electrons in the layer.

The postulated existence of surface states and surface barrier layers on the free surface of germanium and silicon accounted for several properties of germanium and silicon which had hitherto been puzzling[8]. These included (1) lack of dependence of rectifier characteristics on the work function of the metal contact, (2) current voltage characteristics of a contact made with two pieces of germanium, and (3) the fact that there was found little or no contact potential difference between n- and p-type germanium and between n- and p-type silicon.

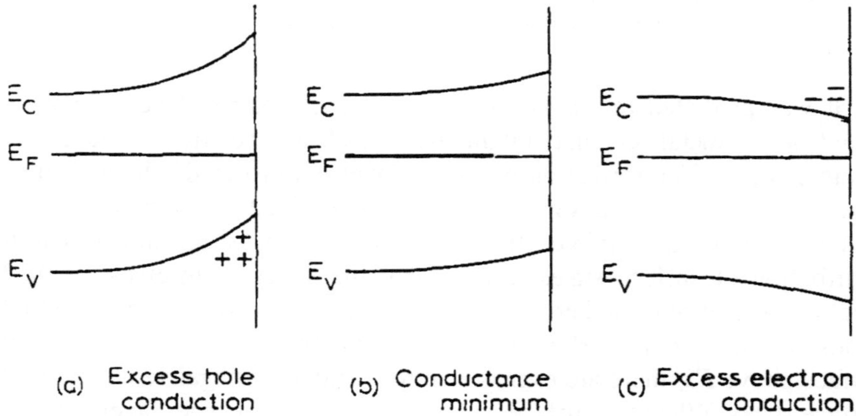

Figure 8. Types of barrier layers which may exist at the free surface of an n-type semiconductor: (a) excess conductance from an inversion layer of p-type conductivity; (b) near the minimum surface conductance; (c) excess conductance from an accumulation layer of electrons.

While available evidence for surface states was fairly convincing, it was all of an indirect nature. Further, none of the effects gave any evidence about the height of the surface barrier and of the distribution of surface states. A number of further experiments which might yield more concrete evidence about the surface barrier was suggested by Shockley, Brattain and myself.

Shockley predicted that a difference in contact potential would be found between n- and p-type specimens with large impurity concentration. A systematic study of Brattain and Shockley[12] using silicon specimens with varying amounts of donor and acceptor impurities showed that this was true, and an estimate was obtained for the density of surface states. Another experiment which indicated the presence of a surface barrier was a measurement of the change in contact potential with illumination of the surface. This is just the Dember effect, which Frenkel had attempted to account for by the difference in mobilities of the electrons and holes generated by the light and diffusing to the interior. It was found's that the change is usually much larger and often of the opposite sign than predicted by Frenkel's theory, which did not take into account a surface barrier.

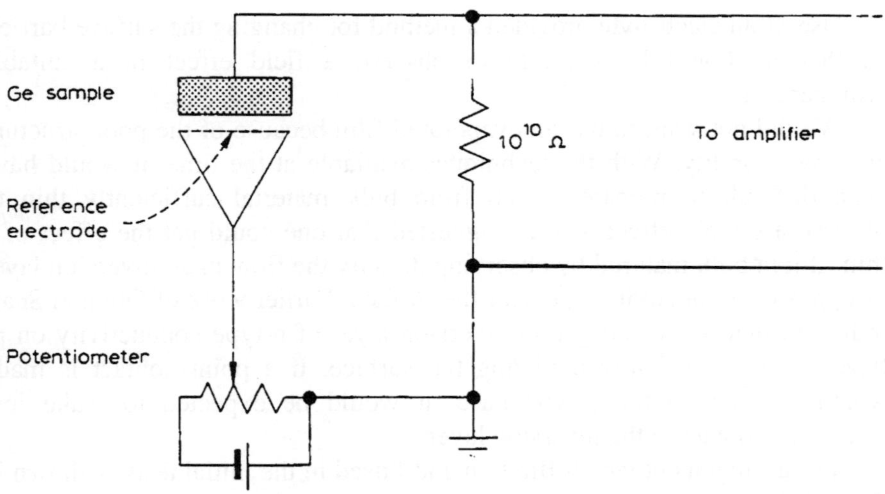

Figure 9. Schematic diagram of apparatus used by Brattain to measure contact potential and change of contact potential with light.

Some rather difficult experiments which at the time gave negative results have been carried out successfully much later by improved techniques, as will be described by Dr. Brattain in his talk.

Apparatus used by Brattain to measure contact potential and change in contact potential with illumination is shown in Fig. 9. The reference electrode, generally platinum, is in the form of a screen so that light can pass through it. By vibrating the electrode, the contact potential itself can be measured by the Kelvin method. If light chopped at an appropriate frequency falls on the surface and the electrode is held fixed, the change with illumination can be measured from the alternating voltage developed across the condenser. In the course of the study, Brattain tried several ambient atmospheres and different temperatures. He observed a large effect when a liquid dielectric filled the space between the electrode and semiconductor surface. He and Gibney then introduced electrolytes, and observed effects attributed to large changes in the surface barrier with voltage applied across the electrolyte. Evidently ions piling up at the surface created a very large field which penetrated through the surface states.

EXPERIMENTS ON INVERSION LAYERS

Use of an electrolyte provided a method for changing the surface barrier, so that it should be possible to observe a field effect in a suitable arrangement.

We did not want to use an evaporated film because of the poor structure and low mobility. With the techniques available at the time, it would have been difficult to prepare a slab from bulk material sufficiently thin to observe a sizable effect. It was suggested that one could get the effect of a thin film in bulk material by observing directly the flow in an inversion layer of opposite conductivity type near the surface. Earlier work of Ohl and Scaff indicated that one could get an inversion layer of n-type conductivity on p-type silicon by suitably oxidizing the surface. If a point contact is made which rectifies to the p-type base, it would be expected to make low resistance contact to the inversion layer.

The arrangement which Brattain and I used in the initial tests is shown in Fig. 10. The point contact was surrounded by, but insulated from, a drop of electrolyte. An electrode in the electrolyte could be used to apply a strong field at the semiconductor surface in the vicinity of the contact. The reverse, or high resistance direction is that in which point is positive relative to the block. Part of the reverse current consists of electrons flowing through the n-type inversion layer to the contact. It was found that the magnitude of this current could be changed by applying a voltage on the electrolyte probe, and thus, by the field effect, changing the conductance of the inversion layer.

Since under static conditions only a very small current flowed through the electrolyte, the set-up could be used as an amplifier. In the initial tests, current and power amplification, but not voltage amplification, was observed.

As predicted from the expected decrease in number of electrons in the inversion layer, a negative voltage applied to the probe was found to decrease the current flowing in the reverse direction to the contact.

It was next decided to try a similar arrangement with a block of n-type germanium. Although we had no prior knowledge of a p-type inversion layer on the surface, the experiments showed definitely that a large part of the reverse current consisted of holes flowing in an inversion layer near the surface. A positive change in voltage on the probe decreased the reverse current. Considerable voltage as well as current and power amplification was observed.

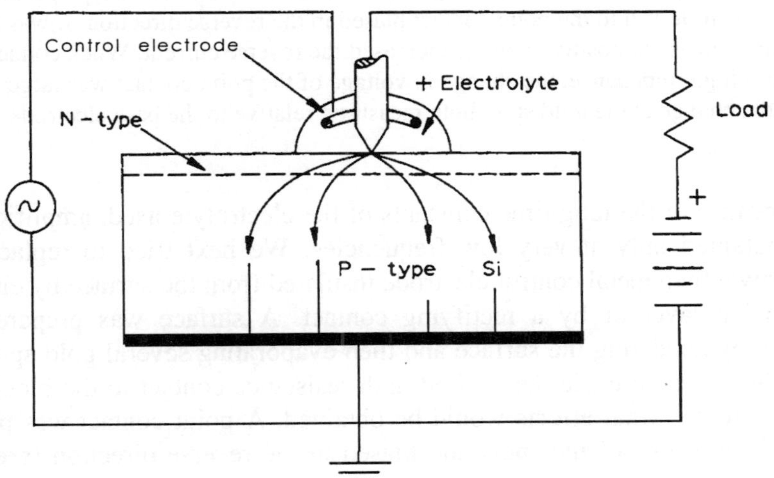

Figure 10. Diagram of experiment used to observe effect of the field produced by an electrolyte on an inversion layer of n-type conductance at the surface of a p-type silicon block. Negative potential applied to the probe in the electrolyte decreases the number of electrons in the inversion layer and thus the current of electrons flowing to the point contact biased in the reverse direction. Arrows indicate the conventional direction of current flow; electrons move in the opposite direction.

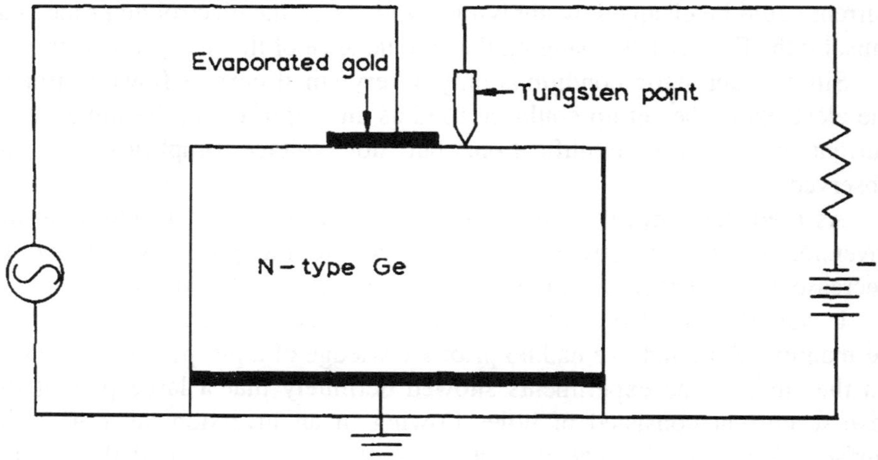

Figure 11. Diagram of experiment in which the transistor effect was first observed. Positive voltage applied to the gold spot introduced holes into the n-type germanium block which flowed to the point contact biased in the reverse direction. It was found that an increase in positive voltage increased the reverse current. When connected across a high impedance, the change in voltage of the point contact was larger than the change at the gold spot, both measured relative to the base electrode.

Because of the long time constants of the electrolyte used, amplification was obtained only at very low frequencies. We next tried to replace the electrolyte by a metal control electrode insulated from the surface by either a thin oxide layer or by a rectifying contact. A surface was prepared by Gibney by anodizing the surface and then evaporating several gold spots on it. Although none made the desired high resistance contact to the block, we decided to see what effects would be obtained. A point contact was placed very close to one of the spots and biased in the reverse direction (see Fig. 11).

A small effect on the reverse current was observed when the spot was biased positively, but of *opposite* direction to that observed with the electrolyte. An increase in positive bias *increased* rather than decreased the reverse current to the point contact. The effect was large enough to give some voltage, but no power amplification. This experiment suggested that holes were flowing into the germanium surface from the gold spot, and that the holes introduced in this way flowed into the point contact to enhance the reverse current.

This was the first indication of the transistor effect. It was estimated that power amplification could be obtained if the metal contacts were spaced at distances of the order of 0.005 cm. In the first attempt, which was successful, contacts were made by evaporating gold on a wedge, and then separating the gold at the point of the wedge with a razor blade to make two closely spaced contacts. After further experimentation, it appeared that the easiest way to achieve the desired close separation was to use two appropriately shaped point contacts placed very close together.

Success was achieved in the first trials; the point-contact transistor was born[13]. It was evident from the experiments that a large part of both the forward and reverse currents from a germanium point contact is carried by minority carriers, in this case holes. If this fact had been recognized earlier, the transistor might have come sooner.

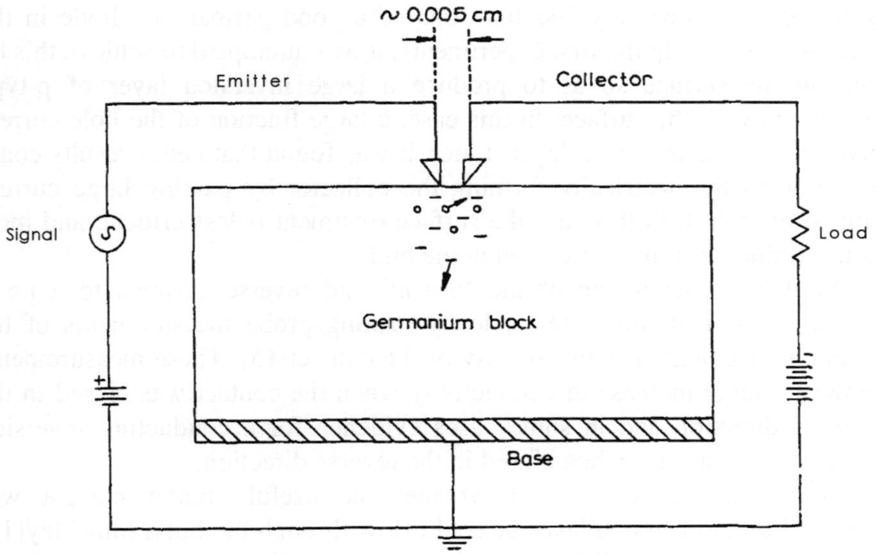

Figure 12. Schematic diagram of point-contact transistor.

Operation of a point-contact transistor is illustrated in Fig. 12. When operated as an amplifier, one contact, the emitter, is biased with a d.c. voltage in the forward direction, the second, the collector, in the negative or high resistance direction. A third contact, the base electrode, makes a low resistance contact to the block. A large part of the forward current consists of holes flowing into the block. Current from the collector consists in part of electrons flowing from the contact and in part of holes flowing toward the contact. The collector current produces an electric field in the block which is in such a direction as to attract holes introduced at the emitter. A large part of the emitter current, introduced at low impedance, flows in the collector circuit. Biased in the reverse direction, the collector has high impedance and can be matched to a high impedance load. There is thus a large voltage amplification of an input signal. It is found[14] that there is some current amplification as well, giving an overall power gain of 20 db. or more.

An increase in hole current at the collector affects the barrier there in such a way as to enhance the current of electrons flowing from the contact.

The collector current must be sufficiently large to provide an electric field to attract the holes from the emitter. The optimum impendance of the collector is considerably less than that of a good germanium diode in the reverse direction. In the first experiments, it was attempted to achieve this by treating the surface so as to produce a large inversion layer of p-type conductivity on the surface. In this case, a large fraction of the hole current may flow in the inversion layer. Later, it was found that better results could be obtained by electrically forming the collector by passing large current pulses through it. In this case the surface treatment is less critical, and most of the emitter current flows through the bulk.

Studies of the nature of the forward and reverse currents to a point contact to germanium were made by making probe measurements of the variation of potential in the vicinity of the contact[15]. These measurements showed a large increase in conductivity when the contact was biased in the forward direction and in some cases evidence for a conducting inversion layer near the surface when biased in the reverse direction.

Before it was established whether the useful emitter current was confined to an inversion layer or could flow through the bulk, Shockley[16] proposed a radically different design for a transistor based on the latter possibility. This is the junction transistor design in which added minority carriers from the emitter diffuse through a thin base layer to the collector. Independently of this suggestion, Shive[17] made a point-contact transistor in which the emitter and collector were on opposite faces of a thin slab of germanium. This showed definitely that injected minority carriers could flow for small distances through bulk material. While transistors can be made to operate either way, designs which make use of flow through bulk material have been most successful. Junction transistors have superseded point-contact transistors for most applications.

Following the discovery of the transistor effect, a large part of research at the Bell Laboratories was devoted to a study of flow on injected minority carriers in bulk material. Much of this research was instigated by Shockley, and will be described by him in the following talk. Research on surface properties of germanium and silicon, suspended for some time after 1948 because of the pressure of other work, was resumed later on by Brattain and others, and is now a flourishing field of activity with implications to a number of scientific fields other than semiconductors such as adsorption, catalysis, and photoconductivity. This research program will be described by Dr. Brattain in his talk.

It is evident that many years of research by a great many people, both before and after the discovery of the transistor effect, has been required to bring our knowledge of semiconductors to its present development. We were fortunate enough to be involved at a particularly opportune time and *to* add another small step in the control of Nature for the benefit of mankind.

In addition to my colleagues and to others mentioned in the lecture, I would like to express my deep gratitude to Drs. M. J. Kelly and Ralph Bown for the inspired leadership of the Laboratories when this work was done.

BIBLIOGRAPHY

[1] A. H. Wilson, *Proc. Roy. Soc. London,* A 133 (1931) 458; A 134 *(1931)* 277; A 136 (1932) 487.
[2] J. Frenkel, *Physik. Z. Sowjetunion, 8 (1935)* 185.
[3] N. F. Mott, *Proc. Roy. Soc. London,* A 171 (1939) 27.
[4] W. Schottky, Z. *Physik,* 113 (1939) 367; 118 (1942) 539.
[5] B. Davydov, *J. Tech. Phys. U.S.S.R., 5* (1938) 87.
[6] R. Hilsch and R. W. Pohl, Z. *Physik,* III (1938) 399.
[7] Amplifiers based on the field-effect principle had been suggested earlier in the patent literature (R. Lillienfeld and others), but apparently were not successful. Shockley's contribution was to show that it should be possible according to existing semiconductor theory to make such a device. An early successful experiment is that of W. Shockley and G. L. Pearson, *Phys. Rev.,* 74 (1948) 232.

[8] J. Bardeen, *Phys. Rev.*, 71 (1947) 717.
[9] A review is given in the lecture of Dr. Brattain, this volume, p. 337.
[10] G. L. Pearson and J. Bardeen, *Phys. Rev.*, 75 (1949) 865.
[11] *See* K. Lark-Horovitz, *Elec. Eng.*, 68 (1949) 1047.
[12] W. H. Brattain and W. Shockley, *Phys. Rev.*, 72 (1947) 345.
[13] W. H. Brattain, *Phys. Rev.*, 71 (1947) 345.
[14] J. Bardeen and W. H. Brattain, *Phys. Rev.*, 74 (1948) 230; 75 (1949) 1208.
[15] W. H. Brattain and J. Bardeen, *Phys. Rev.*, 74 (1948) 231.
[16] W. Shockley, *Electrons and Holes in Semiconductors,* D. Van Nostrand Co., Inc., New York, 1950, p. 86.
[17] J. N. Shive, *Phys. Rev.*, 75 (1949) 689.

Electron-Phonon Interactions and Superconductivity

John Bardeen

Nobel Lecture, December 11, 1972

INTRODUCTION

Our present understanding of superconductivity has arisen from a close interplay of theory and experiment. It would have been very difficult to have arrived at the theory by purely deductive reasoning from the basic equations of quantum mechanics. Even if someone had done so, no one would have believed that such remarkable properties would really occur in nature. But, as you well know, that is not the way it happened, a great deal had been learned about the experimental properties of superconductors and phenomenological equations had been given to describe many aspects before the microscopic theory was developed. Some of these have been discussed by Schrieffer and by Cooper in their talks.

My first introduction to superconductivity came in the 1930's and I greatly profited from reading David Shoenberg's little book on superconductivity, [I] which gave an excellent summary of the experimental findings and of the phenomenological theories that had been developed. At that time it was known that superconductivity results from a phase change of the electronic structure and the Meissner effect showed that thermodynamics could be applied successfully to the superconductive equilibrium state. The two fluid Gorter-Casimir model was used to describe the thermal properties and the London brothers had given their famous phenomenological theory of the electrodynamic properties. Most impressive were Fritz London's speculations, given in 1935 at a meeting of the Royal Society in London, [2] that superconductivity is a quantum phenomenon on a macroscopic scale. He also gave what may be the first indication of an energy gap when he stated that "the electrons be coupled by some form of interaction in such a way that the lowest state may be separated by a finite interval from the excited ones."

He strongly urged that, based on the Meissner effect, the diamagnetic aspects of superconductivity are the really basic property.

My first abortive attempt to construct a theory, [3] in 1940, was strongly influenced by London's ideas and the key idea was small energy gaps at the Fermi surface arising from small lattice displacements. However, this work was interrupted by several years of wartime research, and then after the war I joined the group at the Bell Telephone Laboratories where my work turned to semiconductors. It was not until 1950, as a result of the discovery of the isotope effect, that I again began to become interested in superconductivity, and shortly after moved to the University of Illinois.

The year 1950 was notable in several respects for superconductivity theory.

The experimental discovery of the isotope effect [4, 5] and the independent prediction of H. Fröhlich [6] that superconductivity arises from interaction between the electrons and phonons (the quanta of the lattice vibrations) gave the first clear indication of the directions along which a microscopic theory might be sought. Also in the same year appeared the phenomenological Ginzburg-Landau equations which give an excellent description of superconductivity near Tc, in terms of a complex order parameter, as mentioned by Schrieffer in his talk. Finally, it was in 1950 that Fritz London's book [7] on superconductivity appeared. This book included very perceptive comments about the nature of the microscopic theory that have turned out to be remarkably accurate. He suggested that superconductivity requires "a kind of solidification or condensation of the average momentum distribution." He also predicted the phenomenon of flux quantization, which was not observed for another dozen years.

The field of superconductivity is a vast one with many ramifications. Even in a series of three talks, it is possible to touch on only a few highlights. In this talk, I thought that it might be interesting to trace the development of the role of electron-phonon interactions in superconductivity from its beginnings in 1950 up to the present day, both before and after the development of the microscopic theory in 1957. By concentrating on this one area, I hope to give some impression of the great progress that has been made in depth

of understanding of the phenomena of superconductivity. Through developments by many people, [8] electron-phonon interactions have grown from a qualitative concept to such an extent that measurements on superconductors are now used to derive detailed quantitative information about the interaction and its energy dependence. Further, for many of the simpler metals and alloys, it is possible to derive the interaction from first principles and calculate the transition temperature and other superconducting properties.

The theoretical methods used make use of the methods of quantum field theory as adopted to the many-body problem, including Green's functions, Feynman diagrams, Dyson equations and renormalization concepts. Following Matsubara, temperature plays the role of an imaginary time. Even if you are not familiar with diagrammatic methods, I hope that you will be able to follow the physical arguments involved.

In 1950, diagrammatic methods were just being introduced into quantum field theory to account for the interaction of electrons with the field of photons.

It was several years before they were developed with full power for application to the quantum statistical mechanics of many interacting particles. Following Matsubara, those prominent in the development of the theoretical methods include Kubo, Martin and Schwinger, and particularly the Soviet physicists, Migdal, Galitski, Abrikosov, Dzyaloshinski, and Gor'kov. The methods were first introduced to superconductivity theory by Gor'kov [9] and a little later in a somewhat different form by Kadanoff and Martin. [10] Problems of superconductivity have provided many applications for the powerful Green's function methods of many-body theory and these applications have helped to further develop the theory.

Diagrammatic methods were first applied to discuss electron-phonon interactions in normal metals by Migdal [11] and his method was extended to superconductors by Eliashberg. [12] A similar approach was given by Nambu. [13] The theories are accurate to terms of order *(m/M)1/2,* where m is the mass of the electron and *M* the mass of the ion, and so give quite accurate quantitative accounts of the properties of both normal metals and superconductors.

We will first give a brief discussion of the electron-phonon interactions as applied to superconductivity theory from 1950 to 1957, when the pairing theory was introduced, then discuss the Migdal theory as applied to normal metals, and finally discuss Eliashberg's extension to superconductors and subsequent developments. We will close by saying a few words about applications of the pairing theory to systems other than those involving electron-phonon interactions in metals.

DEVELOPMENTS FROM 1950-1957

The isotope effect was discovered in the spring of 1950 by Reynolds, Serin, et al, [4] at Rutgers University and by E. Maxwell [5] at the U. S. National Bureau of Standards. Both groups measured the transition temperatures of separated mercury isotopes and found a positive result that could be interpreted a s $TcM^{1/2} \cong$ constant, where M is the isotopic mass. If the mass of the ions is important, their motion and thus the lattice vibrations must be involved.

Independently, Fröhlich, [6] who was then spending the spring term at Purdue University, attempted to develop a theory of superconductivity based on the self-energy of the electrons in the field of phonons. He heard about the isotope effect in mid-May, shortly before he submitted his paper for publication and was delighted to find very strong experimental confirmation of his ideas. He used a Hamiltonian, now called the Fröhlich Hamiltonian, in which interactions between electrons and phonons are included but Coulomb interactions are omitted except as they can be included in the energies of the individual electrons and phonons. Fröhlich used a perturbation theory approach and found an instability of the Fermi surface if the electron-phonon interaction were sufficiently strong.

When I heard about the isotope effect in early May in a telephone call from Serin, I attempted to revive my earlier theory of energy gaps at the Fermi surface, with the gaps now arising from dynamic interactions with the phonons rather than from small static lattice displacements. [14] I used a variational method rather than a perturbation approach but the theory was also based on the electron self-energy in the field of phonons. While we were very hopeful at the time, it soon was found that both theories had grave difficulties, not easy to overcome. [15] It became evident that nearly all of the self-energy is included in the normal state and is little changed in the transition. A theory involving a true many-body interaction between the electrons seemed to be required to account for superconductivity. Schafroth

[16] showed that starting with the Fröhlich Hamiltonian, one cannot derive the Meissner effect in any order of perturbation theory. Migdal's theory, [II] supposedly correct to terms of order *(m/M)1/2,* gave no gap or instability at the Fermi surface and no indication of superconductivity.

Of course Coulomb interactions really are present. The effective direct Coulomb interaction between electrons is shielded by the other electrons and the electrons also shield the ions involved in the vibrational motion. Pines and I derived an effective electron-electron interaction starting from a Hamiltonian in which phonon and Coulomb terms are included from the start. [17] As is the case for the Fröhlich Hamiltonian, the matrix element for scattering of a pair of electrons near the Fermi surface from exchange of virtual phonons is negative (attractive) if the energy difference between the electron states involved is less than the phonon energy. As discussed by Schrieffer, the attractive nature of the interaction was a key factor in the development of the microscopic theory. In addition to the phonon induced interaction, there is the repulsive screened Coulomb interaction, and the criterion for superconductivity is that the attractive phonon interaction dominate the Coulomb interaction for states near the Fermi surface. [18]

During the early 1950's there was increasing evidence for an energy gap at the Fermi surface. [19] Also very important was Pippard's proposed non-local modification [20] of the London electrodynamics which introduced a new length the coherence distance, to, into the theory. In 1955 I wrote a review article [17] on the theory of superconductivity for the Handbuch der Physik, which was published in 1956. The central theme of the article was the energy gap, and it was shown that Pippard's version of the electrodynamics would likely follow from an energy gap model. Also included was a review of electron-phonon interactions. It was pointed out that the evidence suggested that all phonons are involved in the transition, not just the long wave length phonons, and that their frequencies are changed very little in the normal-superconducting transition. Thus one should be able to use the effective interaction between electrons as a basis for a true many-body theory of the superconducting state.

Schrieffer and Cooper described in their talks how we were eventually able to accomplish this goal.

GREEN'S FUNCTION METHOD FOR NORMAL METALS

By use of Green's function methods, Migdal[11] derived a solution of Fröhlich's Hamiltonian $H = H_{el} + H_{ph} + H_{el\text{-}ph}$, for normal metals valid for arbitrarily strong coupling and which involves errors only of order $(m / M)^{1/2}$. The Green's functions are defined by thermal average of time ordered operators for the electrons and phonons, respectively

$$G = -\mathrm{i}\langle T\psi(1)\psi^{+}(2)\rangle \tag{1a}$$

$$D = -\mathrm{i}\langle T\theta(1)\theta^{+}(2)\rangle \tag{1b}$$

Here $\psi(r,t)$ is the wave field operator for electron quasi-particles and $\phi(r,t)$ for the phonons, the symbols 1 and 2 represent the space-time points (r_1,t_1) and (r_2,t_2) and the brackets represent thermal averages over an ensemble.

Fourier transforms of the Green's functions for $H_0 = H_{el}+H_{ph}$ for non-interacting electrons and phonons are

$$G_0(P) = \frac{1}{\omega_n - \varepsilon_0(k) + \mathrm{i}\,\delta_k} \tag{2a}$$

$$D_0(Q) = \left\{\frac{1}{\nu_n - \omega_0(q) + \mathrm{i}\,\delta} - \frac{1}{\nu_n + \omega_0(q) - \mathrm{i}\,\delta}\right\}, \tag{2b}$$

where $P = (k, \omega_n)$ and $Q = (q,\nu_n)$ are four vectors, $\varepsilon_0(k)$ is the bare electron quasiparticle energy referred to the Fermi surface, $\omega_o(q)$ the bare phonon frequency and ω_n and ν_n the Matsubara frequencies

$$\omega_n = (2n+1)\pi\,\mathrm{i}\,k_B T\,;\ \ \nu_n = 2n\pi\,\mathrm{i}\,k_B T \tag{3}$$

for Fermi and Bose particles, respectively.

As a result of the electron-phonon interaction, $H_{el\text{-}ph}$, both electron and phonon energies are renormalized. The renormalized propagators, G and D, can be given by a sum over Feynman diagrams, each of which represents a term in the perturbation expansion. We shall use light lines to represent the bare propagators, G_o and D_o, heavy lines for the renormalized propagators, G and D, straight lines for the electrons and curly lines for the phonons.

The electron-phonon interaction is described by the vertex

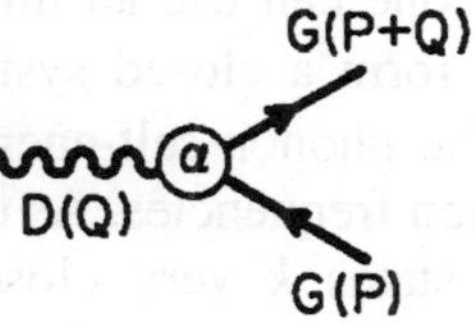

which represents scattering of an electron or hole by emission or absorption of a phonon or creation of an electron and hole by absorption of a phonon by an electron in the Fermi sea. Migdal showed that renormalization of the vertex represents only a small correction, of order $(m/M\ )1/2$, a result in accord with the Born-Oppenheimer adiabatic-approximation. If terms of this order are neglected, the electron and phonon self-energy corrections are given by the lowest order diagrams provided that fully renormalized propagators are used in these diagrams.

The electron self-energy $\Sigma(P)$ in the Dyson equation:

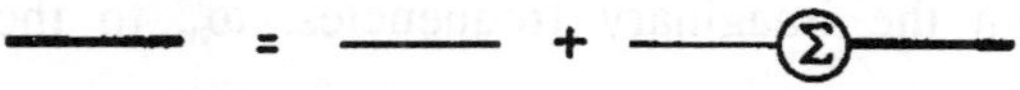

$$G(P)=G_0(P)+G_0(P)\Sigma(P)G(P) \tag{4}$$

is given by the diagram

(5)

The phonon self-energy, $\pi(Q)$, defined by

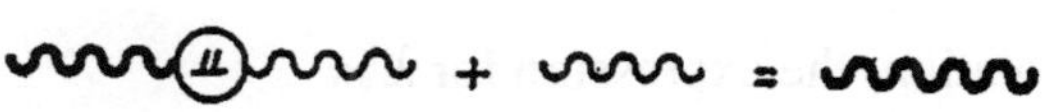
(6)

is given by

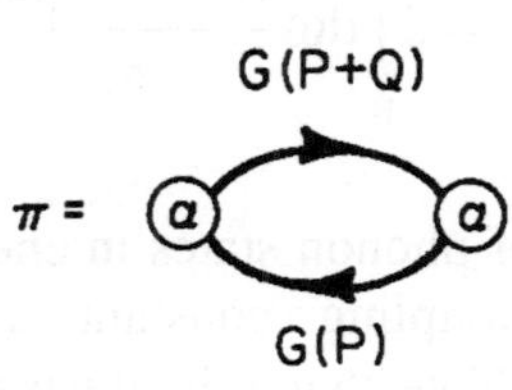

Since to order $(m/M)^{1/2}$ one can use an unrenormalized vertex function $a=a_0$, the Dyson equations form a closed system such that both $\Sigma(P)$ and $\pi(Q)$ can be determined. The phonon self-energy, $\pi(Q)$, gives only a small renormalization of the phonon frequencies. As to the electrons, Migdal noted that we are interested in states k very close to k_F, so that to a close approximation $\Sigma(k,\omega)$ depends only on the frequency. For an isotropic system,

$$\Sigma(k,\omega)\approx\Sigma(k_F,\omega)\equiv\Sigma(\omega) \tag{7}$$

The renormalized electron quasi-particle energy, ω_k, is then given by a root of

$$\varepsilon(k)=\omega_k=\varepsilon_0(k)+\Sigma(\omega_k) \tag{8}$$

In the thermal Green's function formalism, one may make an analytic continuation from the imaginary frequencies, ω_n, to the real ω axis to determine $\Sigma(\omega)$.

Although $\Sigma(\omega)$ is small compared with the Fermi energy, E_F, it changes rapidly with energy and so can affect the density of states at the Fermi surface and thus the low temperature electronic specific heat. The mass renormalization factor m^*/m, at the Fermi surface may be expressed in terms of a parameter λ:

$$m^*/m=Z(k_F)=1+\lambda=(d\varepsilon_0/dk)_F/(d\varepsilon/dk)_F \tag{9}$$

In modern notation, the expression for λ is

$$\lambda=2\int_0^\infty d\omega\frac{\alpha^2(\omega)F(\omega)}{\omega} \tag{10}$$

where $F(\omega)$ is the density of phonon states in energy and $a^2(\omega)$ is the square of the electron-phonon coupling constant averaged over polarization directions of the phonons. Note that λ is always positive so that the Fermi surface is stable if the lattice is stable. Values of λ for various metals range from about 0.5 to 1.5. The parameter λ corresponds roughtly to the $N(0)V_{\text{phonon}}$ of the BCS theory.

NAMBU-ELIASHBERG THEORY FOR SUPERCONDUCTORS

Migdal's theory has important consequences that have been verified experimentally for normal metals, but gave no clue as to the origin of superconductivity.

Following the introduction of the BCS theory, Gor'kov showed that pairing could be introduced through the anomalous Green's function

$$F(P) = \mathrm{i}\langle T\psi_\uparrow \psi_\downarrow \rangle, \tag{11}$$

Nambu showed that both types of Green's functions can be conveniently included with use of a spinor notation

$$\psi = \begin{pmatrix} \psi_\uparrow(r,t) \\ \psi_\downarrow{}^+(r,t) \end{pmatrix} \tag{12}$$

where $\psi_\uparrow$ and $\psi_\downarrow$ are wave field operators for up and down spin electrons and a matrix Green's function with components

$$\tilde{G}_{\alpha\beta} = -\mathrm{i}\langle T\psi_\alpha \psi_\beta{}^+ \rangle \tag{13}$$

Thus G_{11} and G_{22} are the single particle Green's functions for up and down spin particles and $G_{12} = G_{21}{}^* = F(P)$ is the anomalous Green's function of Gor'kov.

There are two self-energies, Σ_1 and Σ_2, defined by the matrix

$$\tilde{\Sigma} = \begin{pmatrix} \Sigma_1 & \Sigma_2 \\ \Sigma_2 & \Sigma_1 \end{pmatrix} \tag{14}$$

Eliashberg noted that one can describe superconductors to the same accuracy as normal metals if one calculates the self-energies with the same diagrams that Migdal used, but with Nambu matrix propagators in place of the usual normal state Green's functions. The matrix equation for $\tilde{G}$ is

$$\tilde{G} = \tilde{G} + \tilde{G}_0 \tilde{\Sigma} \tilde{G} \tag{15}$$

The matrix equation for $\tilde{\Sigma}$ yields a pair of coupled integral equations for Σ_1 and Σ_2. Again Σ_1 and Σ_2 depend mainly on the frequency and are essentially independent of the momentum variables. Following Nambu, [13] one may define a renormalization factor $Z_s(\omega)$ and a pair potential, $\Delta(\omega)$, for isotropic systems through the equations:

$$\omega Z_s(\omega) = \omega + \Sigma_1(\omega) \tag{16}$$

$$\Delta(\omega) = \Sigma_2(\omega) / Z(\omega) \tag{17}$$

Both Z_S and Δ can be complex and include quasi-particle life-time effects. Eliashberg derived coupled non-linear integral equations for $Z_S(\omega)$ and $\Delta(\omega)$ which involve the electron-phonon interaction in the function $a^2(\omega)F(\omega)$.

The Eliashberg equations have been used with great success to calculate the properties of strongly coupled superconductors for which the frequency dependence of Z and Δ is important. They reduce to the BCS theory and to the nearly equivalent theory of Bogoliubov [21] based on the principle of "compensation of dangerous diagrams" when the coupling is weak. By weak coupling is meant that the significant phonon frequencies are very large compared with k_BT_c, so that $\Delta(\omega)$ can be regarded as a constant independent of frequency in the important range of energies extending to at most a few k_BT_c. In weak coupling one may also neglect the difference in quasi-particle energy renormalization and assume that $Z_S = Z_n$.

The first solutions of the Eliashberg equations were obtained by Morel and Anderson [22] for an Einstein frequency spectrum. Coulomb interactions were included, following Bogoliubov, by introducing a parameter μ^* which renormalizes the screened Coulomb interaction to the same energy range as the phonon interaction, In weak coupling, $N(0)V = \lambda - \mu^*$. They estimated λ from electronic specific heat data and μ^* from the electron density and thus the transition temperatures, Tc, for a number of metals. Order-of-magnitude agreement with experiment was found. Later work, based in large part on tunneling data, has yielded precise information on the electron-phonon interaction for both weak and strongly-coupled superconductors.

ANALYSIS OF TUNNELING DATA

From the voltage dependence of the tunneling current between a normal metal and a superconductor one can derive A(w) and thus get direct information about the Green's function for electrons in the superconductor. It is possible to go further and derive empirically from tunneling data the electron-phonon coupling, $a^2(\omega)F(\omega)$, as a function of energy. That electron tunneling should provide a powerful method for investigating the energy gap in superconductors was suggested by I. Giaever, [23] and he first observed the effect in the spring of 1960.

The principle of the method is illustrated in Fig. 1. At very low temperatures, the derivative of the tunneling current with respect to voltage is proportional to the density of states in energy in the superconductor. Thus the ratio of the density of states in the metal in the superconducting phase, N_s, to that of the same metal in the normal phase, N_n, at an energy eV above the Fermi surface is given by

$$\frac{N_S(eV)}{N_n}=\frac{(\mathrm{d}I/\mathrm{d}V)_{ns}}{(\mathrm{d}I/\mathrm{d}V)_{nn}} \tag{18}$$

The normal density is essentially independent of energy in the range involved (a few meV). In weak coupling superconductors, for a voltage V and energy $\omega = eV$,

$$\frac{N_S(\omega)}{N_n}=\frac{\omega}{\sqrt{\omega^2-\Delta^2}} \tag{19}$$

As $T \to 0$ K, no current flows between the normal metal and the superconductor until the applied voltage reaches Δ/e, when there is a sharp rise in $\mathrm{d}I/\mathrm{d}V$ followed by a drop. This is illustrated in Fig. 2 for the case of Pb.

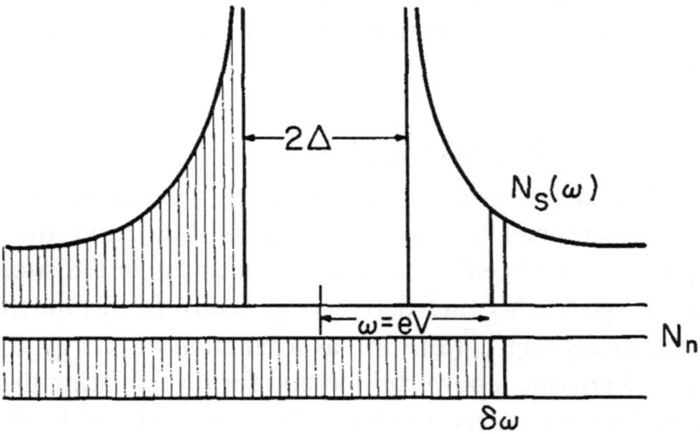

$$\left(\frac{dI}{dV}\right)_{ns} \sim N_s(\omega) \sim \frac{\omega}{\sqrt{\omega^2-\Delta^2}}$$

Tunneling from a normal metal into a superconductor

Figure 1. Schematic diagram illustrating tunneling from a normal metal into a superconductor near $T = 0°K$. Shown in the lower part of the diagram is the uniform density of states in energy of electrons in the normal metal, with the occupied states shifted by an energy eV from an applied voltage V across the junction. The upper part of the diagram shows the density of states in energy in the superconductor, with an energy gap 2Δ . The effect of an increment of voltage δV giving an energy change $\delta\omega$ is to allow tunneling from states in the range 60. Since the tunneling probability is proportional to density of states Ns (ω), the increment in current δI is proportional to $N_s(\omega)\delta V$.

The first experiments of Giaever were on aluminum, which is a weak coupling superconductor. Good agreement was found between theory and experiment. In later measurements on tunneling into Pb, a strongly coupled superconductor, Giaever, Hart and Megerle [24] observed anomalies in the density of states that appeared to be associated with phonons, as shown in Fig. 2. These results were confirmed by more complete and accurate tunneling data on Pb by J. M. Rowe1l et al. [25]

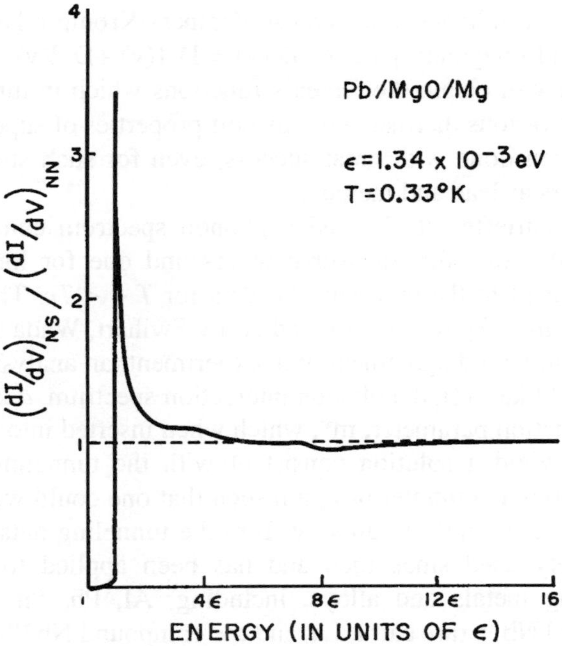

Figure 2. Conductance of a Pb-Mg junction as a function of applied voltage (from reference 24).

In the meantime, in the summer of 1961, Schrieffer had derived numerical solutions of the Eliashberg equations working with a group engaged in developing methods for computer control using graphical display methods. [26] He and co-workers calculated the complex $\Delta(\omega)$ for a Debye frequency spectrum. Later, at the University of Pennsylvania, he together with J. W. Wilkins and D. J. Scalapino [27] continued work on the problem with a view to explaining the observed anomalies on Pb. They showed that for the general case of a complex $\Delta(\omega)$

$$\frac{(\mathrm{d}I/\mathrm{d}V)_{ns}}{(\mathrm{d}I/\mathrm{d}V)_{nn}} = \frac{N_S(\omega)}{N_n} = \mathrm{Re}\left\{\frac{\omega}{\sqrt{\omega^2 - \Delta^2(\omega)}}\right\} \tag{20}$$

where Re represents the real part. From measurements of the ratio over the complete range of voltages, one can use Kramers-Kronig relations to obtain both the real and imaginary parts of D (v) = D 1(v) +D 2(v). From analysis of the data, one can obtain the Green's functions which in turn can be used to calculate the various thermal and transport properties of superconductors.

This has been done with great success, even for such strongly-coupled super conductors as lead and mercury.

For lead, Schrieffer et al, used a phonon spectrum consisting of two Lorentzian peaks, one for transverse waves and one for longitudinal and obtained a good fit to the experimental data for $T << Tc$. The calculations were extended up to *Tc* for Pb, Hg, and Al by Swihart, Wada and Scalapino, [28] again finding good agreement with experiment. an analysis of tunneling data, one would like to find a phonon interaction spectrum, *a 2(*v*)F(*v*),* and a Coulomb interaction parameter, m*, which when inserted into the Eliashberg equations will yield a solution consistent with the tunneling data. W. L. McMillan devised a computer program such that one could work backwards and derive a2(v)F(v) and m* directly from the tunneling data. His program has been widely used since then and has been applied to a number of superconducting metals and alloys, including, Al, Pb, Sn, the transition elements Ta and Nb, a rare earth, La, and the compound Nb3Sn.

In all cases it has been found that the phonon mechanism is dominant with reasonable values of m *. Peaks in the phonon spectrum agree with peaks in the phonon density of states as found from neutron scattering data, as shown in Fig. 3 for the case of Pb. In Fig. 4 is shown the real and imaginary parts of D(w) for Pb as derived from tunneling data.

One can go further and calculate the various thermodynamic and other properties. Good agreement with experiment is found for strongly coupled superconductors even when there are significant deviations from the weak coupling limits. For example, the weak-coupling BCS expression for the condensation energy at $T = 0$ K is

$$E_{BCS} = \frac{1}{2} N(0) Z_n \Delta_0^{\ 2} \tag{21}$$

where $N(0)Z_n$ is the phonon enhanced density of states and Δ_0 is the gap parameter at $T = 0$ K. The theoretical expression with $Z_s(\omega)$ and $\Delta(\omega)$ derived from tunneling data, again for the case of Pb, gives [29, 30, 31]

$$E_{theor} = 0.78 \cdot E_{BCS} \tag{22}$$

in excellent agreement with the experimental value

$$E_{\exp} = (0.76 + 0.02) E_{BCS} \,. \tag{23}$$

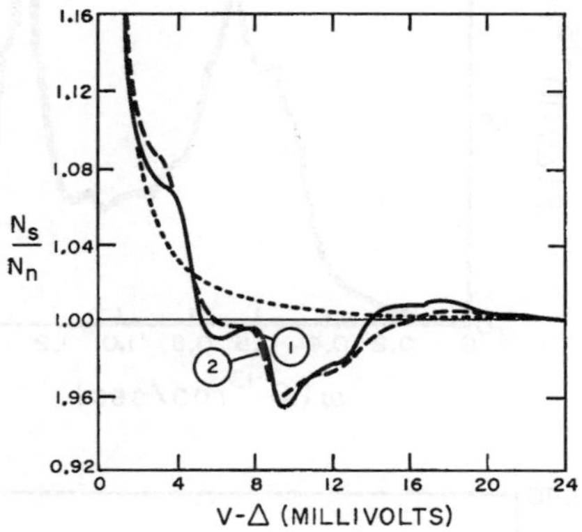

Figure 3. Density of states versus energy for Pb. Solid line, calculated by Schrieffer et al; long dashed line, observed from tunneling; short dashed line, BCS weak coupling theory.

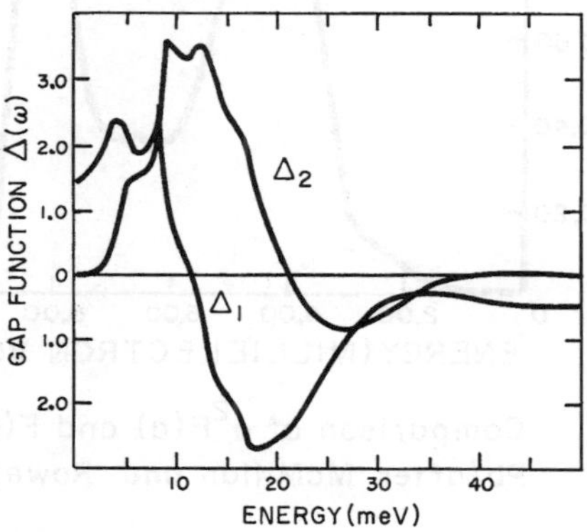

Figure 4. Real and imaginary parts of $\Delta(\omega) = \Delta_1(\omega) + i\Delta_2(\omega)$ versus energy for Pb. (After McMillan & Rowell).

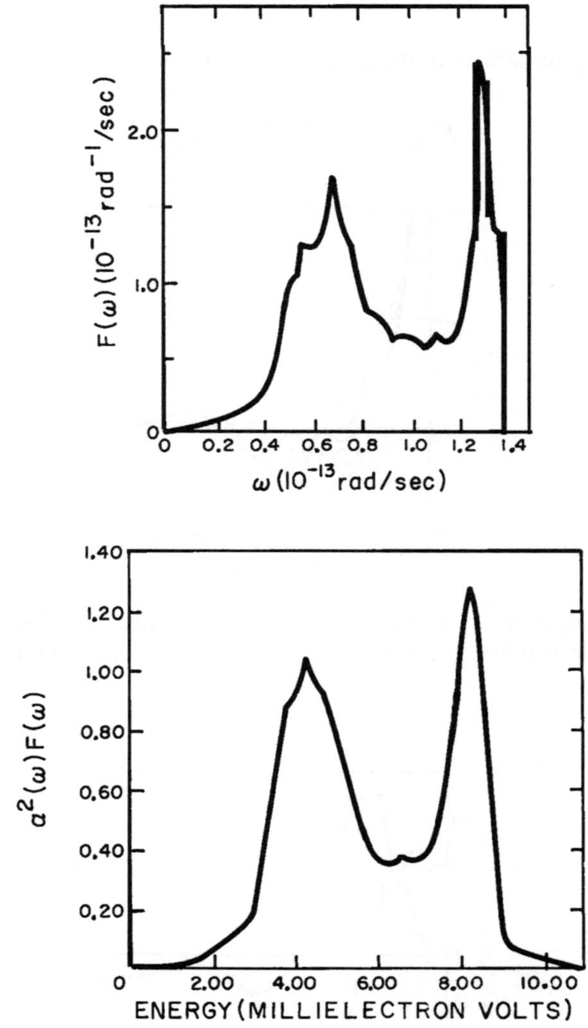

Comparison of $a^2F(a)$ and $F(\omega)$ for Pb (after McMillan and Rowell)

Figure 5. Comparison of a^2F for Pb derived from tunneling data with phonon density of states from neutron scattering data of Stedman et al. [8].

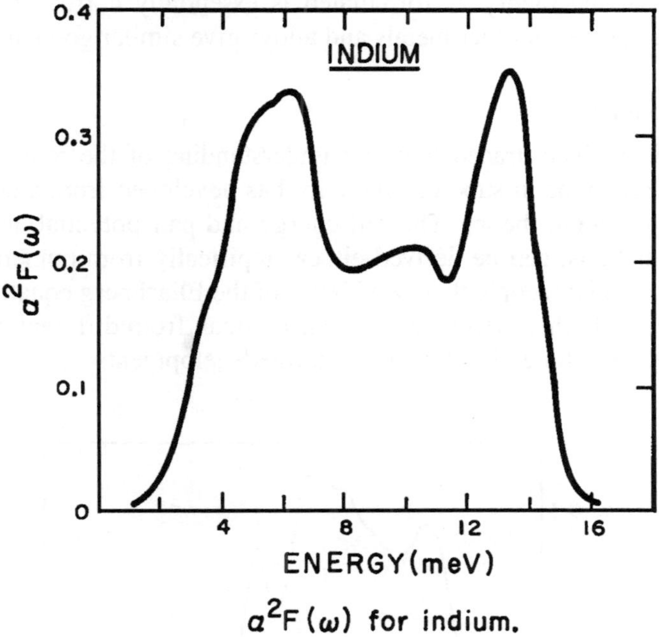

Figure 6. a^2F for In (after McMillan and Rowell).

In Figs. 5, 6, 7, and 8 are shown other examples of *a2(w)F(w)* derived from tunneling data for Pb, In, [31] La, [32] and Nb3Sn. [33] In all cases the results are completely consistent with the phonon mechanism. Coulomb interactions play only a minor role, with μ* varying only slowly from one metal to another, and generally in the range 0.1-02.

As a further check, it is possible to derive the phonon density of states, F(v) from neutron scattering data and use pseudo-potential theory to calculate the electron-phonon interaction parameter *a9(w).* From these values, one can use the Eliashberg equations to calculate &s(w) and **D** (v) and the various superconducting properties, including the transition temperature, Tc. Extensive calculations of this sort have been made by J. P. Carbotte and co-workers [34] for several of the simpler metals and alloys. For example, for the gap edge, D 0, in Al at $T = 0$ K they find 0.19 meV as compared with an experimental value of 0.17. The corresponding values for Pb are 1.49 meV from theory as compared with 1.35 meV from experiment.

These are essentially first principles calculations and give convincing evidence that the theory as formulated is essentially correct. Calculations made for a number of other metals and alloys give similar good agreement.

CONCLUSIONS

In this talk we have traced how our understanding of the role of electron-phonon interactions in superconductivity has developed from a concept to a precise quantitative theory. The self-energy and pair potential, and thus the Green's functions, can be derived either empirically from tunneling data or directly from microscopic theory with use of the Eliashberg equations.

Physicists, both experimental and theoretical, from different parts of the world have contributed importantly to these developments.

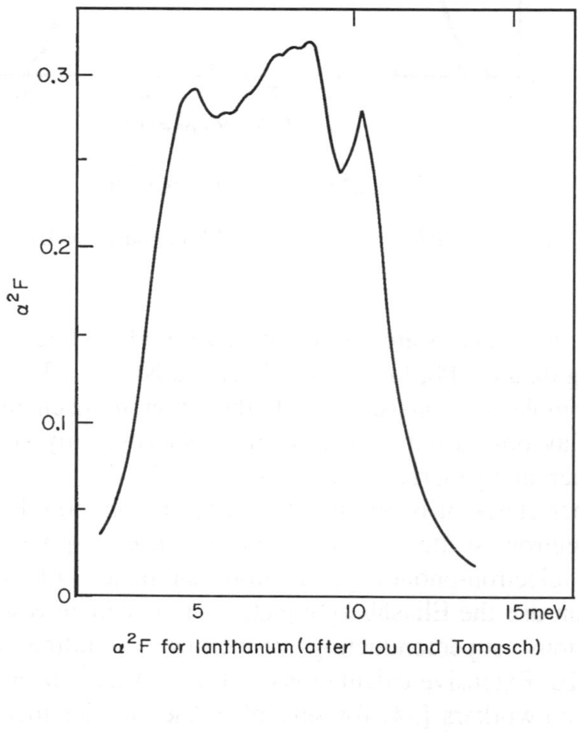

Figure 7. a^2F for La (after Lou and Tomasch).

All evidence indicates that the electron-phonon interaction is the dominant mechanism in the cases studied so far, which include many simple metals, transition metals, a rare earth, and various alloys and compounds. Except possibly for the metallic form of hydrogen, [35] which is presumed to exist at very high pressures, it is unlikely that the phonon mechanism will yield substantially higher transition temperatures than the present maximum of about 21 K for a compound of Nb, Al and Ge.

Other mechanisms have been suggested for obtaining higher transition temperatures. One of these is to get an effective attractive interaction between electrons from exchange of virtual excitons, or electron-hole pairs. This requires a semiconductor in close proximity to the metal in a layer or sandwich structure. At present, one can not say whether or not such structures are feasible and in no case has the exciton mechanism been shown to exist. As Ginzburg has emphasized, this problem (as well as other proposed mechanisms) deserves study until a definite answer can be found. [36]

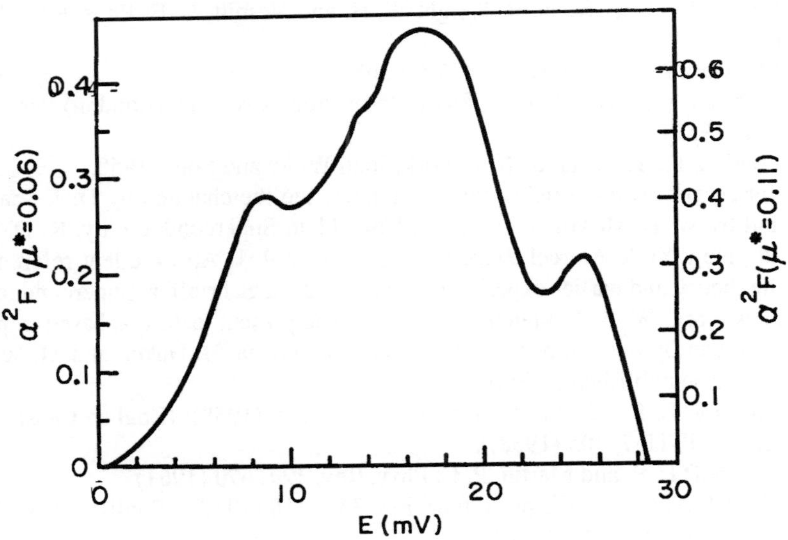

$\alpha^2 F$ for Nb_3Sn(after L.Y.L. Shen)

Figure 8. a^2F for Nb_3Sn (after Y. L. Y. Shen).

The pairing theory has had wide application to Fermi systems other than electrons in metals. For example, the theory has been used to account for many aspects of nuclear structure. It is thought the nuclear matter in neutron stars is superfluid. Very recently, evidence has been found for a possible pairing transition in liquid He3 at very low temperatures [37]. Some of the concepts, such as that of a degenerate vacuum, have been used in the theory of elementary particles. Thus pairing seems to be a general phenomenon in Fermi systems.

The field of superconductivity is still a very active one in both basic science and applications. I hope that these lectures have given you some feeling for the accomplishments and the methods used.

BIBLIOGRAPHY

[1] Shoenberg, D. Superconductivity, Cambridge Univ. Press, Cambridge (1938). Second edition, 1951.

[2] London, F. Proc. Roy. Soc. (London) *152A,* 24 (1935).

[3] Bardeen, J. Phys. Rev. 59, 928A (1941).

[4] Reynolds, C. A., Serin, B. Wright W. H. and Nesbitt, L. B. Phys. Rev. 78, 487 (1950).

[5] Maxwell, E., Phys. Rev. 78, 477 (1950).

[6] Frohlich, H., Phys. Rev. 79, 845 (1950); Proc. Roy. Soc. (London) Ser. A 213, 291 (1952).

[7] London, F., Superfluids, New York, John Wiley and Sons, 1950.

[8] For recent review articles with references, see the chapters by D. J. Scalapino and by W. L. McMillan and J. M. Rowe11 in Superconductivity, R. D. Parks, ed., New York, Marcel Bekker, Inc., 1969, Vol. 1. An excellent reference for the theory and earlier experimental work is J. R. Schrieffer, Superconductivity, New York, W. A. Benjamin, Inc., 1964. The present lecture is based in part on a chapter by the author in Cooperative Phenomena, H. Haken and M. Wagner, eds. to be published by Springer.

[9] Gor'kov, L. P., Zh. Eksper i. teor. Fiz. 34, 735 (1958). (English transl. Soviet Phys. – JETP 7, 505 (1958)).

[10] Kadanoff L. P. and Martin, P. C. Phys. Rev. 124, 670 (1961).

[11] Migdal, A. B., Zh. Eksper i. teor. Fiz. 34, 1438 (1958). (English transl. Soviet Phys. - JETP 7, 996 (1958)).

[12] Eliashberg, G. M., Zh. Eksper i. teor. Fiz. 38, 966 (1960). Soviet Phys. – JETP ***11,*** 696 (1960).

[13] Nambu, Y., Phys. Rev. 117, 648 (1960).

[14] Bardeen, J., Phys. Rev. 79, 167 (1950); 80, 567 (1950); 81 829 (1951).

[15] Bardeen, J., Rev. Mod. Phys. 23, 261 (1951).

[16] Schafroth, M. R., Helv. Phys. Acta 24, 645 (1951); Nuovo Cimento, 9, 291 (1952).
[17] For a review see Bardeen, J., Encyclopedia of Physics, S. Flugge, ed., Berlin, Springer-Verlag, (1956) Vol. XV, p. 274.
[18] Bardeen, J., L. N. Cooper and J. R. Schrieffer, Phys. Rev. 108, 1175 (1957).
[19] For references, see the review article of M. A. Biondi, A. T. Forrester, M. B. Garfunkel and C. B. Satterthwaite, Rev. Mod. Phys. 30, 1109 (1958).
[20] Pippard, A. B., Proc. Roy. Soc. (London) ***A216, 547*** (1954).
[21] See N. N. Bogoliubov, V. V. Tolmachev and D. V. Shirkov, A New Method in the Theory of Superconductivity, New York, Consultants Bureau, Inc., 1959.
[22] Morel, P. and Anderson, P. W., Phys. Rev. ***125,*** 1263 (1962).
[23] Giaever, I., Phys. Rev. Letters, 5, 147; 5, 464 (1960).
[24] Giaever, I., Hart H. R., and Megerle K., Phys. Rev. 126, 941 (1962).
[25] Rowell, J. M., Anderson P. W. and Thomas D. E., Phys. Rev. Letters, ***10, 334*** (1963).
[26] Culler, G. J., Fried, B. D., Huff, R. W. and Schrieffer, J. R., Phys. Rev. Letters 8, 339 (1962).
[27] Schrieffer, J. R., Scalapino, D. J. and Wilkins, J. W., Phys. Rev. Letters ***10, 336*** (1963); D. J. Scalapino, J. R. Schrieffer, and J. W. Wilkins, Phys. Rev. 148, 263 (1966).
[28] Scalapino, D. J., Wada, Y. and Swihart, J. C., Phys. Rev. Letters, 14, 102 ***(1965);14, 106*** (1965).
[29] Eliashberg, G. M., Zh. Eksper i. teor. Fiz. 43, 1005 (1962). English transl. Soviet Phys. – JETP ***16, 780*** (1963).
[30] Bardeen, J. and Stephen, M., Phys. Rev. 136, Al485 (1964).
[31] McMillan, W. L. and Rowell, J. M. in Reference 8.
[32] Lou, L. F. and Tomasch, W. J., Phys. Rev. Lett. 29, 858 (1972).
[33] Shen, L. Y. L., Phys. Rev. Lett. 29, 1082 (1972).
[34] Carbotte, J. P., Superconductivity, P. R. Wallace, ed., New York, Gordon and Breach, 1969, Vol. 1, p. 491; J. P. Carbotte and R. C. Dynes, Phys. Rev. ***172, 476*** (1968); C. R. Leavens and J. P. Carbotte, Can. Journ. Phys. 49, 724 (1971).
[35] Ashcroft, N. W., Phys. Rev. Letters, 21, 1748 (1968).
[36] See V. L. Ginzburg, "The Problem of High Temperature Superconductivity," *Annual Review of Materials Science,* Vol. 2, p. 663 (1972).
[37] Osheroff D. D., Gully W. J., Richardson R. C. and Lee, D. M., Phys. Rev. Lett. 29, 1621 (1972).

PART IV LEV DAVIDOVICH LANDAU

About L.D. Landau — the Great Physicist*#

V.L. Ginzburg

P.N. Lebedev Physical Institute, Russian Academy of Sciences, 117924 Moscow, Russia

If not all, then very much can be known from comparison. Every time, when we call somebody: "great person", "a man of genius" —we make a comparison with other people. Although each person is unrepeatable it is clear, that some people differ from the average very much. At the same time all words undergo depreciation to some extent; we can observe the inflation of the epithets. Besides making an appearance we tend to speak well of everybody. This was understood even by ancients - they used to say that about people, who had passed away one should speak "either well or nothing". Surely this is the reason, why reading numerous memoirs we have the impression, that there are so many distinguished people. Indeed there are not so many.

That's we can say about remembrances. Now I want to mention two items from the Statute of the Soviet Union Academy of Sciences. The 16th item: "The real member of the USSR Academy of Sciences can be the scientist, who contributed to the science with his work of first-class scientific importance". The 17th item: "The correspondent-member of the USSR Academy of Sciences can be the scientist, who contributed to the science

---

* The presentation during the evening of Landau's remembrance in the Technical Museum (Moscow, January the 20th, 1978).

\# The authorised translation from the original Russian was made by Dr. Agata Kaminska (Inst. of Physics. Polish Academy of Science. Warszawa).

with his outstanding scientific work". Now look at the announcement of the present evening put in the poster and your invitations. We meet here to have a session concerning the "outstanding" soviet physicist L. D. Landau, who in this way appeared to be on the level of the correspondent-member of the USSR Academy of Sciences. By no means I am intending not to accuse the organisers of this evening. All we know, that we should understand the epithets literally. Butter for instance can be first quality, better quality or "extra". The same concerns scientific and academic degrees. Using the similar terminology I can say, that Landau was the "super-extra quality" physicist. He was the entirely unique physicist. I always wonder at one thing. In the USSR there are thousands of young physicists graduating the high schools every year. They are very often good physicists indeed. Although during the years there was nobody as talented as Landau.

Landau was a man with the completely exceptional personality. Considering all the people I have ever known I can compare him with R. Feynman, the famous physicist especially because of his books. Naturally in our century lived many brilliant physicists: Einstein, Bohr, Planck, Schrödinger, Heisenberg, Dirac. Landau of course did not exceed them with his achievements. He appreciated himself quite correct regarding them as having much greater achievements. He assigned himself more modest position. But in my opinion to the high appreciation of Landau contribute many different factors. The first is that his scientific achievements are of first-class. They are the quantum theory of liquids (especially the theory of superfluidity of helium), the theory of phase transitions and many others. The second is that he was the very versatile physicist and he knew well all the physics. And the third is that he was the Teacher with the capital T, the born Teacher. The product of these three factors is extremely big.

I want to remark that Landau was not a wondrous child in the common sense – he never didn't played piano at three years nor solved any mathematical exercises. Although he graduated from a school at thirteen, started his study at the university at sixteen and published his first paper at eighteen years. Surely this fast progress is a mark of a great talent. For instance Pauli wrote his famous relativity theory book when he was eighteen.

Landau liked all the kinds of calculations. Ones he told me: "I am thirteen years older than you, because my first paper was published in 1926th and yours in 1939th." Thus he calculated that he was thirteen years older than

I. According to such other "indications" he was naturally much more older than I.

Landau was an unusual man also, let's say, from the "biological" point of view. I remember my astonishment that he could not pick up more than ten kilos. Unfortunately this physical weakness, which was of no consequence in common circumstances, played a tragic part in the crash, which has completely changed his life. During collision of two cars, eggs lying in the Landau's car stayed untouched, whereas he became entirely smashed. But of course, if we talk about Landau's exceptionality, we first of all think about his talent as a physicist. Existence of such people like Landau induces us to think about the limits of human possibilities and tremendous reserves of man's brain.

Physics is a many-sided science. It is difficult to measure its capabilities. For instance memory can be measured, although man's memory changes in an enormous range. Some years ago was published a book in which the author inquired about a man, who had a phenomenal memory. The limit of his memory could not be determined with the use of any available tests and he was not the inhabitant of the other planet, but as man as you and me. This is the proof of the fact that the man's brain has tremendous reserves. Landau's capabilities as a physicist demonstrate it as well.

Landau stopped his work sixteen years ago. I can't remember without pain this Landau, who he became after the catastrophe till his death in 1968th. Although sixteen years is a quite long time, there is no doubt that Landau is still alive in the sense, which we always have in mind speaking about the people, who passed away. The same we can say about his books and actually about the course "Theoretical physics" written by L. D. Landau and E. M. Lifshitz. The books of this course are the handbooks in the literal sense. For instance I don't use any other books more often. Each physicist in the USSR and abroad has them in his room. They are excellent books, a kind of encyclopaedia and in connection with this – the magnificent monument of L. D. Landau.

I want to mention here the role of E. M. Lifshitz. The course "Theoretical physics" could not be created without Landau, but it also could not be created without Lifshitz. And, what was excellent, Lifshitz continued this work*. The books were published all the time, fully elaborated, worked out. There were published even these books, which had been only decided when Landau had lived. I feel very grateful for this to E. M. Lifshitz. There

* E. M. Lifshitz lived till 1985 .

is difficult to imagine the better monument in Landau's honour or the better prolongation of his journey through the physics.

Now I want to tell you about a little episode. Recently I saw the Landau's paper, which had been written in 1933rd when he had been twenty-five years old. This was the paper concerning the superconductivity. The origin of the superconductivity had not been understood yet (this was done twenty-four years later – in 1957th). One of the hypotheses proposed and developed in the paper mentioned consisted in the assumption of the spontaneous currents existing in the superconductors. Afterwards there occurred that superconductivity in some determined events could be explained on the quite different ground. This was the reason that this paper was not taken into consideration as a Landau's achievement. Indeed, in this paper were no mistakes. Nowadays it appears that there exist materials showing spontaneous currents*. Today after years this article is still astonishing with its clearness and precision. It is still alive and helpful.

I am not intending to make the memory of Landau looking better as it is. Sometimes he was violent, sometimes he did not wanted to hear or even offended somebody. But he never was lordly or boorish. I must admit however, that he two time's humiliated me in public. The first time it happened in 1943rd in Kazan. He then abused me public being extremely irritated. He was the master yet, whereas I was still inexperienced youth, although I had defended my doctor's thesis (I can say here that the defence of doctor's thesis only is not the evidence of anything). The second time he roughly offended me in front of other people about 1960, after that I had been regarded as a prominent scientist, according to the statement of the statute of the USSR Academy of Sciences (i. e. I was a correspondent-member of the Academy of Sciences). But the point was that at that time we were friends yet. In both cases Landau unconditionally disobeyed the generally accepted social norms, so I had all the reasons to bear a grudge against him and I bore. Although I understood that it was not the insult of the principal, who wanted to disdain me. He simply disregarded some behaviour rules and some of them he didn't understand. There were the people, to whom this circumstance was an obstacle to make a friendship with Landau. I am very glad that it was not the case with me. Besides there is one essential remark, that in both cases mentioned when Landau severely criticised me, actually he was right, not me.

---

* This and the previous sentences are not entirely exact and at any rate they need some comments (*the author's note to the present edition*).

Sometimes people wonder if Landau made any serious mistakes, in particular evaluating some new theories and ideas. But it goes without saying that the only one who slips does not make any mistakes. Landau of course made mistakes. But I think that he misled more rarely than many other people and, what is important, he misled in the interesting manner. I described it in details in the article devoted to 60-th Landau's anniversary, which unfortunately turned out to be the obituary. I'll restrict myself to the remark that Landau as an analyst and a man of a broad knowledge of physics was particularly aware of the weakness and difficulties of new hypotheses, theories, etc. Together with the style of severe expression of his opinions without thinking about the form of this, without caution, etc. this made an impression, that Landau always criticised new ideas and in general was conservative. In addition he emphasised himself not to be "an inventor". In my opinion we have to understand it in the sense only, that Landau was especially mighty in solving the difficult problems, in analysing and criticising, but not in the field of generating new vague hypotheses or inventing some devices, measuring methods, etc. Landau's critical remarks occurred to be sometimes false, but finally they were useful. This critique was the necessary element in the creation process of the school and Landau was creating and he created the school. Formally I don't belong to this school, as Landau was not my teacher and I didn't passed any theory exam (apropos Landau emphasised many times that I lost a lot because I hadn't passed this exam and he was quite right). But this just happened and I was lucky that I could learn and be on friendly terms with my highly respected and beloved teacher I. E. Tamm and at the same time I worked together with L. D. Landau. And I must say that no "organising problem" arose in connection with this fact. Both I. E. Tamm and L. D. Landau created the schools and they considered as natural the co-operation between them, mutually attending their seminars, common discussions, etc.

Landau's name became a legend and in the legends the reality is very often changed or mixed with myths. I hope that the present evening will help to separate the truth from fables and to draw the correct image of the excellent physicist L. D. Landau.

ADDITION

During rather monotonous rowing—I was fishing on Ladoga before my journey—my recollections about Landau began to glide. And it occurred that I remembered something, that was not inserted into the note, placed above and also to the article, devoted to Landau's 60-th anniversary, which was written early (УФН. – 1968. – V. 94. – P.181.). Nothing of special importance or interest. But nevertheless I decided to put it down. And this is an excuse, of course. Now the raw of collected reminiscences about prominent Soviet physicists has already printed or is being prepared for publication. Undoubtedly, such collection about Landau must be published some day.

Meanwhile I. J. Pomeranchuk and A. S. Kompaniejec—"the pupils of the first appeal" had died, V. B. Berestecky has died also. Now, when these lines are being written I am 64. It is 10 years more than Dau was by the time of the accident (November 7, 1962)*. So I think there is no point to postpone and decided to prepare this "Addition", which together with the note will be possible to insert in the collection.

1. Fishing is my only "hobby", and even it Dau considered to be a useless activity, or may be even worse. He said laughing: "Ha, ha, fisherman, the warm is from the one side, the fool is from the other side, as Voltaire said". And I invariably answered: "Dau, I don't fish with a warm, but with the minnow". But this didn't work. Dau, as required, held his ground. This repetition was typically for him, as the same disk was playing; may be it can be called stereotype. This touched the science too. I must admit that we used this. It was known that it was possible to turn Dau on asking him certain question; it always worked. I don't know why, but I kept in my mind only one example—the question about Lorentz—Lorentz equation**. Mention this equation produced his anger (what was done for effect, of course) and the stream of invectives or caustic remarks. Unfortunately I don't remember exactly the words he said but their sense was: there is no such an equation, it is semiempirical relation. He was right. Lorentz-Lorentz equation works only for the simple models of optical and

---

* As it is known L.D. Landau has lived more than 6 years after the accident (he died on April, 1,1968). But he was infirm, may be even better to say –the other person. I was among those people who were watching by the patient's bedside in the hospital and also visited him the further years. But I'm not going to touch that period of time, though, may be psychologist or psychoanalyst "studying" the ill Landau could do medical certificate for him when he was healthy. But refuse doing such an analysis and actually couldn't do it.

** The Lorentz-Lorentz equation in fact coincides with the Clasius-Mosotti formula and determines dependence of refraction coefficient or permitivity on liquid density (optical isotropic substance). This equation is obtained when we consider the electric field E which acts on the molecule (doublet) to be equal to $E+(1\pi/3)P$, with E—average macroscopic field and P—medium polarization.

isotropic medium, and the main supposition lies in identifying molecule with the point doublet. Certainly, in the liquid and in generally in the condensed medium, where the distance between molecules is equal to their size, dipole approach can not be allowable a priori. Nevertheless for the number of liquids the Lorentz-Lorentz equation suits experimental data. This obviously led to the exaggeration of equation value, to its wide adaptation (the simplicity of this equation assists the last). Probably, Landau faced the misunderstanding the role and true sense of this equation and "record" his disapproval on the disc in his mind. And than, with accordance to his marked manner he played over this disc. I'm sure, Dau understood that everybody knew his position, but he gave the show, perform the scene of righteous anger.

2. It was not easy to argue with Landau. Sometimes he didn't want to listen, sometimes he stung, the other time he evaded the question, saying: "Think about it by himself". But now Landau is disputing some opinion, declaring the opposite one to be delirium and so on. And he understands that he's wrong. And than, Landau stands the new position with the same assurance, i.e. that one, which he called delirium before. Actually, I don't see anything bad in it. Rather widespread statement, which says that it is very bad to change the opinion, seems to be nonsense. It is much worse when you don't change your mind because of stubbornness or misunderstanding even after cogent arguments or new facts. I don't like very much when a person occurred to be wrong and than begins to prove that he was taken in the wrong way and denies his former statements and so on. There were not such cases with Landau. But occurred to be wrong he usually didn't say "I was mistaken" or something like that but went on to the new point as it was natural thing. As long as I was not right very often in disputes with Landau, I was not satisfied with that his behaviour. It was not possible to be delighted with the pleasure of victory over the Teacher. That's why we began to sign our opinions. Now I remember only two such receipts, which are lost, unfortunately. In one case Landau asserted that long-living mesons (heavier than muon) exist, and I insisted on the reverse. Actually it was not scientific dispute, because there were not trustful theoretical arguments "for" and "against" any of those opinions and they don't exist till now. The question was that we believed the different groups of experimentalists. I won that dispute. The second case was when Landau asserted that in the solids (it seems to me he mentioned metal) there are no plasmons. As usually when

matter touched physics Landau's opinion had reasonable foundation. In that case he considered damping factor of plasma waves in metal must be the same order as their frequency ratio. It's clear that there is no any sense to talk about quantum plasma waves—plasmons, if they damp already on the one wavelength. My opinion about plasmons existence was based on the experimental data, and it's known that damping of longwave plasmons, at least in the number of cases, is not very high and conception about them has sense.

3. If in physics Landau's opinions usually were sober and deep, so this can not be said about him in the other fields. I don't want it sounds like a reproach; I only establish the facts. There were and are a lot of people with different exceptional abilities (for example, Leonardo da Vinci). But, certainly, one bright talent is revealed only in one field. Landau belongs to this category; he possessed the great gift as a physicist. But he didn't paint the pictures; he was not sculpture or poet. Frankly I'll notice that I'm even glad that it is not so. If Dau, for example, had painted the bad pictures and considered them to be in earnest (it happens so), we, probably, might have only regretted about it. At the same time, Landau had the wide range of interests; he was well-educated person and knew languages, what was not typical for the people of his generation, whose childhood and youth were passing the stormy epoch of the breaking the old school and so on. All this was important and the alive person can be divided in some sharply bounded elements. I wanted (though I probably do it rather awkward) only emphasise here that Landau felt his superiority only in the field of physics. If I, let say, didn't share some of his literature opinions (for example, he appreciated Dreiser very highly, whom I don't appreciate at all), I did not find anything extraordinary in it.

Here I allow myself to do one digression, although it is linked with the former paragraph and concerns the theme. Here and there "marked people"-researchers*, writers, etc .are interviewed about their opinion concerned those topics that are not connected with their profession. Ok, the interest to the prominent people can be understood and is not contra-indicated. But it is necessary to know and remember that behind the frames of his profession even eminent person does not have any right to aspire to (at any rate to aspire without supplemented reasons) some special authority. The variation of this theme is the question about behaviour in the society. It is a charisma

* L.D. Landau didn't like the word "scientist " very much.

to possess the big talent in some field, but this doesn't give the right to break the common norms. Theoretically, probably, everybody will agree with it. The real life is difficult. A very talented person understands very early that he is superior to many elder people who have already taken high positions. Like the method of self-affirmation the young man of talent begins to kick and shock people. Conflicts begin. Landau passed through this too. Some years later when he got the recognition his behaviour was changed very much. But his known behaviour eccentricity did not leave him. This can explain attitude of those people who were brought up in different surroundings toward Landau, those people who couldn't manage to know him closer.

In the light of said above, as we can think, Landau had luck. Fortunately, there are no duels nowadays. But there are a lot of methods with the help of which it is possible if not get out of the way, but at least vex and traumatise a young man. Landau was early recognised and got what he deserved. Here I don't touch that fact that he had a lot of troubles for the whole year. I want to emphasise here the merit of the older generation of Soviet physicists (in comparison with Landau). They showed their worth towards Landau from their good side (one of such examples I call the election Landau at the age of 38 for the real member of USSR AS, escaping the usual preliminary being at the post of corresponding member).

But I come back to that topic that Landau could be mistaken in the fields outside physics as well as in the evaluation of physicists. I will give you example of such his evaluation of physicists experimentalists X and Y. In the evacuation In Kazan (1941—1943) Landau strongly declared several times: "X and Y are the best physicists experimentalists in the Soviet Union". Why? "I judge by their faces". Of course he didn't: X and Y talked "smoothly", they had good reputation, and by the way, they admit willingly Landau to be the best theorist physicist in the USSR. Years passed and everyone saw that "the best physicists experimentalists" are bad experimentalists. In fact the definitive opinion about X I didn't hear from Landau, but one day (approx. 1960) while talking to me, I don't remember the question I asked, but he said: "Y is not physicist at all". I was even taken aback and asked him rather stupid question: "So why are you dealing with him?" And the answer followed: Y is a clever man, I ask advice from him in everyday questions. I allow myself to doubt that "everyday questions" by Y was good for Landau, but this is another business.

Landau was mistaken in other people too, but everyone does. Unfortunately he excused behaviour which was absolutely inadmissible for me.

4. Landau treated his physicists colleagues critically and scolded some of them. But it occurs often. It was important for me that he didn't scold those people who I liked and respected. In particular, Dau treated Tamm highly. Although I think he undervalued him as a physicist. May be it was because of the different style of working. Somehow or other it's everyone's right to appreciate the achievements of his colleagues and there are no absolutely objective tests.

In some connection with that said above I'll touch the history of advancing for the Nobel Prize the discovering and explanation Vavilov—Cherenkov effect. In early 50-th (but after 1953) somebody (I don't know who) of us decided to start the Nobel club, what meant to begin to advance candidate for the Nobel Prizes. So, I.V. Kurchatov asked E.K. Zavoysky and me to prepare the presentation of I.E Tamm, I.M. Frank, P.A. Cherenkov (Vavilov had already died by that time and not more than three person could be awarded the Nobel Prize and only during their life). Of course we prepared that. I know that others were preparing the presentation of P.L. Kapica and L.D. Landau for their works in the field of helium II superfluidity. The time passed and we found out suddenly that someone somewhere decided to advance only Cherenkov and Kapica. It seems to me that presentation had been done but here it doesn't matter. I don't know exactly all details but here it doesn't matter. It is important that we decided not to allow such injustice take place. In the USSR only the members of the Academy of Science in corresponding fields* can get the proposal to advance for the Nobel Prize. So it was decided that academicians physicists must write a letter for the Nobel committee. People from the Institute of Physical Problems did it about Landau. I don't remember who wrote the letter. E.L. Feinberg wrote the letter where we reported to the Nobel committee about Tamm's and Frank's roles, we applied the prints and affirmed that those three men must be awarded a Prize. Than we had to collect signatures. I remember when I went to the one of the leading academician who agreed

* I don't know why not just every academician gets such proposals, but people outside the Academy get them also. The details of it are not known for me (advancing is considered to be confidential, and it is noted in the letter of Nobel committee. I began to get the proposals only after I was elected for the academician in 1966.

absolutely with the letter but refused signing it. If it was decided "on high" to advance only Cherenkov, so how he could report different opinion. Than I went to Landau. He said me that he didn't appreciate Vavilov—Cherenkov effect very much. (I knew it earlier, and Landau said it not for excuse for not signing the letter). He was ready to sign it if we would write "if awarded" instead of "must be awarded". So we did it. In addition N.N. Andreev and A.I. Alihanov signed this letter too. Soon after this all three members were awarded a Nobel Prize in physics in 1958. But what role played the letter idem I don't know.

5. There was firm opinion concerned Landau that he criticised severely people. Often he did not want to offend the author of criticised work with his harsh remarks. There is a characteristic history which I did not see but I heard about it very much and in details. Dau harshly blew up the work belonged to one considerable professor. He was very offended, but when Dau learned out about this he was very astonished: I have not called him "idiot", I just said that his work is idiotic. So in order to understand Landau's character it is important to distinguish his behaviour from the fact of the matter. I even was amazing with the form of his behaviour, what objective character he revealed when he didn't judge in a temper. It is known that Landau had his "merits scale" in the field of physics. It was logarithmic scale*. From physics of our century only Einstein had class 0,5. Bohr, Dirac, Heisenberg and some others belonged to the 1st class, and Landau possessed the 2nd class (here there are split, but I didn't hear his self-appraisal higher than 2.Earlier he placed himself to the class 2,5). So, One physicist, belonged to the 1st class, came out with a brilliant suggestion in 20th years but afterwards he didn't become famous and even produced the Landau's irritation and not only his with his further activity. But personality and expectancies were not taken under consideration, only achievements were appreciated. I don't know if this example and the scale will produce an impression on the readers, but I think Landau had high objective character when composed the scale. There is another example, though it will not persuade everybody. Heisenberg was given the 1st class, justly, of course—there are few people who did so much and in different physics fields. But he had no special liking in physical surroundings. Not only his political opinions were the reason for that but also his character and behaviour. I did

* Obviously common logarithms were used, for example, for class 2 suited the achievements, which were 10 times smaller than for class 1.

not know him personally so I am not going to repeat the rumours and opinions*; it is enough that Heisenberg as a person was not overfond by Landau. But in 1947—1948 Heisenberg published his articles, devoting to the attempt of building microscopical superconductivity theory. This attempt was rather lame. Landau and I had very low estimation to this theory (and it was confirmed later). But when I began to scold Heisenberg, Landau snubbed me harshly. He said: "Heisenberg is very considerable scientist and must be judged for his better works, not for bad ones. It looked like trivial thing. But actually it was a good lesson for me that I still remember. I didn't understand here anything.

Such "lessons" and developmental work with textbooks compose a substance and norms of behaviour. And different "schools" have different norms.

L.I. Mandelschtam, I.E. Tamm and L.D. Landau were absolutely different people and formed different "schools". Scientific hardness and adherence to principles, precision, connection with the experiment, width and many other things were characteristic for Landau's school (at least during his lifetime). That was impossible for Landau to be registered to somebody's work. On the contrary, sometimes his contribution to work was considerable but he refused to sign this work as co-author. There was such a moment. I asked Landau's advice while working at active field in plasma. (News AS USSR. Ser. Phys.—1944. –V. 8.—N2. P.76). As I thought that Landau had a great role in discussion of this article, so when I finished writing this work I set his hand on this work too. But when I came to Landau with this article, he refused to be co-author. Of course, I thanked him at the end of the article. Now it does not matter why he did it whether he considered that work was not big enough or was too considerable.

I know the other case when landau refused to be co-author. Physicist Z was asking advises from him about one question in optics. I also was interested in that question before, moreover even published the article, concerning it, which Z knew. He knew it but probably didn't understand it or didn't want to understand. Landau, who understood all even without my article, (I am sure that he didn't read it, but I told him it's content), explained Z the fact of the matter. Than, Z published that article and as co-author wrote

---

* But I would like to express one thought, which one famous physicist, who worked with Heisenberg and Bohr, told me. He claimed that in fact indeterminancy principle belongs to Bohr, not Heisenberg. And Heisenberg admitted that, saying that Bohr shows it hazily and he wrote in more clearly.

Landau. And Landau refused it again. So the article was published and I consider it to be a plagiarism. But certainly it was not, because Z didn't copied it from me, he just used Landau's advises.

By the way, this article by Z is quoted 2 times often than mine. And similar situation is not a rule, but is not absolute exception. Here and there some articles are quoted as pioneer, classical and so on without any reasons for that. Just these articles got into the wide range and than their supposed role fixed during the process, better expressed as "adapted by repetition". It happens sometimes by chance, very few people look through the old works, one author refered to article he came across, and than that reference began to wander from one article to another. But sometimes the unwarranted reference appearance didn't happened by chance. May be prior author "organised" this reference by himself (may be alluded to, or asked for or even demanded) , or it is known in physical surrounding that the author is influential person, may be troublesome. So the reference is being made to avoid troubles. As a matter of fact only beginners think that they can trust different prior statements and references not checking them.

6. In scientific environment the questions of priority play very important role. I wrote an article* about it and won't repeat it again. I don't remember that some material quarrels and moreover squabbles about priority had ever happened in the department which Tamm began and I have worked since 1940. I don't remember that I.E. ever mentioned his priority, I think he considered it to be beneath his dignity. So I even don't know if such questions touched him in the deep of his soul. Landau was more sensitive about it and sometimes did not hide his discontent. I don't remember the examples, but some feeling of dissatisfaction stayed with me. Landau actually read few articles (seminar served him to get known with the literature) and his own articles he didn't wrote by himself. It explained the absence of needed references in his articles. It was reasonable explanation. I think that we must not demand from people those things that you don't do by yourself, what happened to Landau sometimes. Although it is rather vexed question. Landau's works and results were better and wider known then other author's works. And he could count for more attention.

Somehow or other but I don't remember Landau dictating people in what way he must be made reference to. I will give you one recent example. A certain physicist W demands their candidates and in general "pupils" to

---

* "Who did create the theory of relativity and how?" Real collection, p. 178.

make references to him in the way like this: "as W first presented" (reference then). I think it is just indecently. If the reference is given and especially in this way it is more than enough. Landau could say: it is seen the ears of the prior person without ceremony from the word "first". If Landau allowed himself to make remarks in the questions of priority (I mean expressing his discontent), he did it because of justice and not because of aspiration for reputation. When Landau knew the matter of subject he gave others due, particularly his co-authors. Landau and I have only one joint work, it is devoted to the superconductivity theory. (ЖЭТФ.—1950. – V.20. – p.1064). And this work where I am the author or co-author occurred to be the most famous. As the name Landau is more famous then mine or may be because of the other reasons in literature this theory is made references to not as to the Ginzburg and Landau theory (our names* are in the alphabetical order in the article), but only as to Landau—Ginzburg theory or even only Landau theory. I admit that I noticed that but never draw attention of corresponding authors to it. I think only such a behaviour can be demanded, it is not important not notice it. So, Landau appreciated our work very much and mentioned it several times and always very correctly. I never made claims on Landau. I was glad because of it and , I hope, people will believe me that it was not because of shallow vanity. I just treated Landau very good and respected him. And if he had behaved differently, it would have depreciated his image for me. It is difficult to explain, but who understands—will do it.

7. Now there are some words how Landau treated Einstein. I will begin with mentioning a misunderstanding. Landau told us several times that he was speaking one time to Einstein, as I remember in Berlin, in 1930. Landau, as he said, tried to explain Einstein quantum mechanics after the seminar, but uselessly. However, Ju. B. Rumer affirms that Landau never talked to Einstein*. How this contradiction can be understood I don't know. But it is interesting, what had happen there.

* When I began to talk about names I remember that in some sense my surname is Landau, not Ginzburg. In fact my grandfather had such a surname but when he married my grandmother he took her surname because of some material views. Besides, Landau and I are probably far relatives. At first I laughed at it and told people about this but stopped doing that when Landau said that he treated me good because we were relatives.

* In 1974 I corresponded with Rumer and he let me know the next. On December 1929 Rumer and Landau met in Berlin (P. Ehrenfest introduced them) and they were sitting together (on the attic) at the colloquium where Einstein was presented. Landau said Rumer: "I will go

Now to the point. Landau put Einstein higher than any physicist our century, and this opinion is doubtless. Landau called the theory of relativity the most beautiful theory among existing ones. I don't know if this opinion is indisputable but I share it absolutely. Landau thought, as many others, that the last 30 years of his life (since 1925, after the works, devoted to Bose—Einstein statistics) Einstein was walking the wrong way. I remember the meeting of physico-mathematical department of the AS USSR (November,19,1955), which was devoted to the 50-th anniversary of foundation the theory of relativity and the memory of Einstein, who died on April 18, 1955. Tamm made opening address, than some reports were made including mine (about experimental checking of theory of relativity), and the final report Landau was devoted to Einstein, his work and life. But except this only one remained in my memory—Landau talked about "Einstein tragedy" talking about his last period of time. The matter did not concern his private tragedy (he did not have it excepting for shallow troubles and illness). The matter concerned his scientific tragedy. What caused this? Firstly, he did not accept quantum mechanics, as it is considered, he did not understand it. Secondly, he devoted many years and efforts for creating unified field theory and without success in it.

I don't agree with such conclusions and don't consider that some "scientific tragedy" existed. The subject about unified field theory is easier. Now we know that that trend was successful. I want to make reference here to Yang article** He marked those Einstein efforts to build the unified field theory*** were not successful and "some people during some period of time considered that unification idea was an obsession that seized Einstein in his old age". Further Yang writes: "Yes, it was obsession, but obsession, which suited insight that, what must fundamental structure of theoretical physics be like. And must add that exactly this understanding suits direction of physics development nowadays." So it is difficult to doubt that Einstein persuasion in importance of unification, which he defended strongly from criticism of any kind, was penetration in to the deep of problem matter."

---

downstairs and try to persuade Einstein to give up working on unified field theory." But Landau did not start the conversation there and as Rumer thinks he could not do it later too.

** Yang. C.N. // Physics Today. –1980.—V.33, N6. –P.42 (transl.: Yang // УФН.—1980 .—V.132. –P. 169).

*** The subject of Yang's article is about the last Einstein work which was published in 1955 as a application to the 5th issue of his book "The meaning of Relativity" (Transl.: M.:Science, 1966.—V. 2. – P.849).

In short, there are no any reasons for reckoning Einstein work at unified field theory among his failures. The absence of the final result in this case is naturally and can't change this conclusion.

Talking about quantum theory, it was always known that Einstein role in its development till 1926 was very big. Now a lot of new articles appeared where it is clear that his distribution is more sizeable than it was thought before*. It is interesting to know that Bohr for a long time thought about idea of light quantum (photons) sharply negatively. So in dispute between Einstein and Bohr the last was not always right, as it is believed. Talking about quantum mechanics it is not right to consider that Einstein denied it or underestimated. The point is that Einstein thought that there is something behind it that's why it is not full. Although I hold the orthodox opinions towards quantum mechanics, I was persuaded several times that the deep understanding its foundation is not widely spread and in literature the flow of debatable articles concerning this topic is not running low. Here we faced gnoseology, and in some sense go outside physics. The widely spread opinion that everything is clear in the quantum mechanics foundations is rather justly. However it is incompetently to consider all doubts about it to be obscurantism.

So I think there are no reasons in Einstein position about quantum mechanics for seeing something tragic.

Einstein was always lone person**, he worked with a few people. At the end of his life, he really was aside of "trunk-roads" of physics developing at that time. But he stayed active in social life, he corresponded a lot. His position couldn't be called isolation but homage irked him.

8. I will dwell on my relationships with Landau, not connected with our profession. I gave the article, addition to that is this text, to read different people. They made some notes and some of them I took into account, some did not change. But now I remember only one advice—to cross out text about age difference. When I asked "Why?", there was an answer; "It is unnecessary". I did not do it, but since that time I began to think that it might be a possibility to conclude that I wanted to show my close relationships with Landau.

---

* Look Pais A.// Rev. Mod. Phys. – 1979. –V. 51. – P.861. Look also: Pais A. The Science and Life of Albert Einstein. – Oxford: Oxford. Univ. Press.

** Pais writes: "If I had had to describe Einstein I would have used one word "apartness".

He addressed as "you" almost every of his students. Although he didn't address his students and physicists gotten into his sphere of influence in Moscow as "you". I met Landau in 1939 or 1940 and 15 years we addressed each other as "sir", though we communicated very often. In 1953 my wife came back to Moscow and my "home" appeared. Landau visited us often and exactly then he proposed to address each other as "you", but I resisted, it was difficult for me to say "you" to him. But Dau all the same began to say me "you". Step by step I got used to this. Such a transition was Dau's expressing of friendly relationships. I appreciated it there and still do it now. But it doesn't mean that we were friends in that sense of this word, which means close and intimate relationships. I would consider only E.M. Lifshitz among his friends. Two times, (when Landau was ill) I saw the demonstration of very warm feelings from Lifshitz side. From Landau side I didn't see such a demonstration towards anybody. Of course, it proves nothing , it happens only in emergency circumstances, and many people don't like to express their warm feelings. But I think that Landau didn't have such feelings at all.

He treated me as physicist rather positively but sensibly. He saw my merits and demerits. It was more naturally that I did not embarrass to ask him about insufficiently considered things, exposing my demerits. While estimating physicist's "class" he treated differently scientific achievements. For example, as it was noticed above, Landau did not put highly Vavilov-Cherenkov discovery. I like this effect very much. Here I appreciate my one work (published in1940) where the quantum theory of Vavilov—Cherenkov was given and it was shown, that condition of radiation appears from the lows of saving energy and impulse for the radiant particle and "photons in medium" (with energy $\hbar\omega$ and impulse $\hbar\omega n/c$ where ω--frequency and n—refractive exponent of medium). Landau considered that may be because appropriate quantum corrections are small and that's enough to use classical theory, that my mentioned work is not important. By the way, exactly due to this work Landau first heard my name and identified me in some way. That times Tamm's and Landau's departments made meetings in one or another institute. And I remember Tamm was telling Landau about my work in small room in Miusy, and Landau reacted on it rather cool.

What class could give me Landau on his logarithmic scale? I never asked him about it, I thought it would be tactlessly or may be I was afraid of getting grade outside the scale. By the way, as I remember as years passed Landau rarely made classification.

9. Except Landau I write about myself also and sometimes it occurred that Ginzburg is right and Landau and others are wrong. Unkindly reader can make conclusions uncomplimentary for me.

But I don't count on such readers, they always find something for to criticism. What about benevolent reader, if he has experience, he will understand in right way. But for inexperience youth I must explain something.

In science as in art and literature there is no dividing and it must not exist into prominent people and faceless plebs. There is the whole spectrum of achievements, abilities, standards of knowledge and so on. Outstanding physicist gets the title for his best results on his appropriate level that is unachievable for standing lower. But of course he can make poor and mistaken works. Usually outstanding physicist is rarely mistaken and often is right than physicists that possess lower classes. The existence of lower physicists is inevitable and even necessary. If they are called physicists they must have their opinion and can win the dispute with people standing higher. In short, that fact that sometimes I won the dispute with Landau this doesn't belittle his merits but shows that I am physicist and not the representative of other profession. It is so clear that it was unnecessary to explain this.

The answer for the question: "So why you gave examples where the author was right and why he is still here and has not disappeared yet*? Here, firstly I agree that it would be better if the author disappeared but secondly it is difficult to do in reminiscences. If the author writes not from words belonged to others but from his own recollections mostly he remember the episodes and facts when he talked to that person he writes about or took part in described events. I have bad memory or selecting one. I remember very well my mistakes, achievements, sometimes different unnecessary facts and names, but I don't remember verses.

So this article describes dissimilar, unequal and subjective picture. If to be afraid of critics, then you must to cut off the sizeable part of written. But I prefer the benevolent readers choose interesting parts and neglect parts, which don't have any value. Here it is important that different people find different moments to be uninteresting or curious, important or not. So it is

---

* Here there is a private matter—using of personal pronounces (I, me and so on). In scientific literature we usually don't use them (in English literature it is not so). I got into a habit not to use them and can use them in scientific article. Sometimes using different "we" and "us" in journalistic genre or popular scientific literature can be funny. All these pronounces irritate me also but I just don't know how to get rid of them.

not clear who I must be oriented at. So the most right way for author it is not to be oriented at the reader and go his own way. One example fixed this my opinion. One of my articles I end up with the "colourful", beautiful phrase. I admit that I am inclined to such a style and don't like when the manuscript comes abruptly to the end. Two well-known physicists looked through this article. One of them said that the last phrase must be eliminated, the other said, that that phrase is the best in the whole article. So who I must follow? Probably in such questions you must hear yourself first of all.

10. Now (the time this article is being written) almost 19 years passed since we have no possibility to discuss physical questions with Landau. My generation became the older generation and we left behind our teacher saying about our age. Now a lot of people work in physics, people who don't know Landau privately. But I still remember Dau and sense his absence as a big and urgent loss, if it is possible to say so. It can not be explained only by the friendly feelings towards him and his tragic end. The most important factor is that here I hold natural human feelings, who likes his profession, which plays a great role in his life. And he can not stay indifferent to the absence of person who was the bright star on the physical sky, who lived on the Olympus.

Notes to this issue.

The most complete information can be found in the collection "Recollections of L.D. Landau. (M.: Science, 1988; English issue: Landau: The physicist and the man, Recollections of L.D. Landau.—Oxford: Pergamon press, 1989) After that materials concerning Landau's imprisonment and the following shadowing appeared. Appropriate references and argument can be found in my article, published in magazine "Nature", 1993, N2, P.92. (sizeable and more precise information can be found in the article by Heifec, appeared in the "Okna" supplement to the daily Israel paper "Vesti" from December, 31, 1992).

Physics Nobel Prize Winner - 1962

Professor I. Waller

Presentation at 1962 Nobel Prize Ceremony

The winner of this year's Nobel Prize in Physics, Professor Lev Davidovic Landau at Moscow University, was born in Baku, 1908. His mathematical talents appeared at a very early age and at the age of 14 he began his studies at the University of Leningrad. After finishing them he spent one and a half years abroad, in particular with the well-known atomic physicist Niels Bohr in Copenhagen. He made a strong impression during this time thanks to his brilliant intellect and great outspokenness.

In 1930 Landau published a quantum theoretical investigation concerning the behaviour of free electrons in a magnetic field which immediately gave him international fame. This work turned out to be essential for the understanding of the properties of metals. Starting from new fruitful ideas Landau found after his return home, often in collaboration with his pupils, important results concerning the structure of magnetic substances and supraconductors and advanced fundamental theories for phase transformations and thermodynamical fluctuations.

Landau's ability to see the core of a problem and his unique physical intuition appear clearly in his investigations on liquid helium which he started after having been attached in 1937 to the Institute for Physical Problems in Moscow. The head of this institute was the famous physicist Kapitsa who then performed interesting experiments on liquid helium. The natural helium gas had earlier been liquefied by cooling to about four degrees above the absolute zero of temperature and subsequent research had shown that this fluid when further cooled about two degrees was transformed to a new state which has quite strange properties. According to a term introduced by Kapitsa it is superfluid which means that it can easily flow through very fine capillaries and slits which almost completely prevent the flow of all other liquids.

The originality in Landau's attack on the problem of explaining these phenomena was that he considered the quantized states of motion of the whole liquid instead of the states of the single atoms as other scientists had done earlier. Landau started by considering the state of the fluid at the absolute zero temperature which is its ground state. He described the excited states of the liquid by the motion of certain fictive particles called quasiparticles. Landau combined experimental results with his calculations and deduced in this way the mechanical properties of the quasi-particles. These results, from which the properties of the fluid could be calculated, were later directly confirmed by investigations on the scattering of neutrons in liquid helium. Such experiments were first performed at Atomic Energy Ltd. in Stockholm in 1957. Landau further found that there exists in liquid helium besides ordinary sound waves also waves of a "second sound". He inspired thereby a Russian scientist to confirm this phenomenon experimentally.

Natural helium consists of an isotope of atomic weight 4 apart from about one millionth of another isotope of atomic weight 3. The lighter isotope has been studied in the liquid state since about 1950. This kind of liquid helium has properties which are quite different from those of the heavier isotope because the helium nuclei of atomic weights 3 and 4 are essentially different. A satisfactory theory for the lighter helium liquid was first given by Landau in 1956 - 1958 and has many formal similarities with his above-mentioned theory for the heavier isotope. The new theory is valid only at very low temperatures, less than one tenth of a degree above absolute zero. This is, however, the most interesting temperature range. Due to the difficulty of making measurements at these low temperatures the theory was not experimentally tested until very recently. These tests have been the more favourable for the theory the more the measuring technique has been refined. Landau has also predicted a new kind of wave propagation for this liquid and has called it zero sound. He has thereby stimulated experimental scientists to great efforts aiming to detect zero sound.

The importance of Landau's investigations are apparent when one considers that an important goal of physics research is to explain the properties of liquids as completely as it has been possible to explain the properties of crystals and of rarefied gases. In their efforts to attain this goal the scientists have in general met with insurmountable difficulties. An essential exception is Landau's theories of liquid helium which therefore are an achievement of great and profound importance.

Besides his investigations on condensed matter, i.e. matter in the solid and liquid state for which he is now awarded the Nobel Prize, Landau has also made contributions of the utmost importance to other parts of physics, in particular to the theories of quantized fields and of elementary particles. He has by his original ideas and masterly investigations exercised far-reaching influence on the evolution of the atomic science of our time.

Professor Landau has unfortunately not yet fully recovered from the severe accident which he sustained at the beginning of this year. He is therefore not here to receive his Nobel Prize which is instead handed to him today by the Ambassador of Sweden in Moscow. On behalf of the Swedish Academy of Sciences I wish to express the hope that Professor Landau will soon completely recover.

PART V

THE RELEVANCE OF MATERIALS SCIENCE

The Relevance of Materials Science

Stanley L. Jaki

P.O. Box 167. Princeton N.J. 08542-01607, USA

Almost twenty five years ago, in 1977, there appeared *Principles and Applications of Ferroelectrics and Related Materials*, the first truly major monograph on the subject. The authors, M. E. Lines and A. M. Glass, both from Bell Telephone laboratories, spoke of the 1960s and 1970s as "the coming of age of ferroelectricity."[i]

The blossoming of interest in ferroelectric phenomena brought together the First International Meeting of Ferroelectricity in 1966 and prompted several regional meetings as well, such as the Japanese-Soviet Symposia on Ferroelectricity of which the first was held in Novosibirsk in 1976 and the second in Kyoto in 1980. The program, "First Principle Calculations for Ferroelectrics" had its fifth Williamsburg Workshop in 1998. The continued growth of interest in the subject is well attested by the fact that this International Meeting is already the Tenth of such Meetings.

A historian of physics will find ample material in a contrast registered by Lines and Glass. It is the difference between the theoretical understanding of the phenomenon, which, as they put it, was fairly "monotonic," and its practical applications which, to quote them again, witnessed "many ups and downs."[ii] In other words, occasional disappointments did not dampen hopes about some significant results to come. This bespoke the conviction that ferroelectric materials may play a considerable role in enhancing the broader relevance of materials science.

Of course, very little of that role was in view when, exactly eighty years ago, J. Valasek discovered that crystals of Rochelle salt exhibit hysteresis, analogous to the one characteristic of ferromagnetic materials. The phenomenon, which is due to the compound character of that salt, was first noted by Pierre de la Seignette, in 1672. As a field of study ferroelectricity is hardly more than a hundred years old, if its beginning is assigned to the observation made by F. Pockels in 1894 that Rochelle salt has an unusually high piezoelectric constant. It took another half a century before Hans Mueller of the Massachusetts Institute of Technology gave, in 1940, a combined treatment of the dielectric, piezoelectric, and elastic behavior of Rochelle salt.[iii]

Then the number of materials that exhibited ferroelectricity rapidly began to grow. Before 1950 only five crystals were known to have ferroelectric properties. In 1950 alone this number doubled. Just before 1955 the number stood at twenty. Within five more years another seventy more were added to the list and in still five more years the number increased by fifty or so. By the mid 1970s some five hundred ferroelectric materials had been known, mostly oxides. Today the number stands around a thousand, forming eight large classes. This indicates that the phenomenon is not at all a rarity in nature. Not only inorganic crystals display ferroelectric properties, but also some organic polymers and liquid crystals. Far from being a rarity in nature, the phenomenon has been recognized as a particular case of structural phase transition. The latter received a general treatment in Effective Field Approach to Phase Transitions and Some Applications to Ferroelectrics, a monograph by Prof. Julio Gonzalo, President of this Meeting.[iv]

The practical relevance of the phenomenon is amply proved through its application in optical switches, light modulators, frequency converters, and light valve arrays, to mention a few major fields of application. Not to be forgotten is the application of ferroelectric crystals in sonar devices. One more example should be recalled in some detail, because it may prove most relevant for those many millions who use computers. Ferroelectric materials are able to switch the direction of their polarization between two stable polarized states. This provides the basis for NVFRAMs, that is, non-volatile ferroelectric random access memories. Built into the hardware of a computer, thin films of NVFRAMs make possible the instantaneous storage of whatever entered on the operating system. In case of an electric blackout, only that part of information is lost that preceded by only a nanosecond the event, usually lightning, that caused the blackout.[v]

Since ferroelectricity had many workers worldwide by the late 1970s, one could expect that more than three pages would be devoted to it in Rolf E. Hummel's widely used textbook, Understanding Materials Science.[vi] But here too as in many other cases, a proportion is a relative matter. For all its vastness in its study and applications ferroelectricity may not appear vast when taken in the whole context of materials science. This science is a witness, if such a witness is needed, to the immeasurable potentialities deposited in matter. The breathtaking rate of progress in materials science today prompts the view that the sky is the only limit. Quite a difference from the expectations of Henry Ellsworth, commissioner of patents for the US government. In the report he gave to the Congress in 1843 he spoke of the "advancement of the arts from year to year" as taxing our credulity, and saw the coming of the times "when human improvement must end."[vii] On the theoretical side Lord Kelvin declared half a century later that only two small clouds darkened the clear sky of physics. One cloud was the specific heat of gases and solids as a function of their temperature, the other the null result of the Michelson-Morley experiment. Both clouds turned out to be huge storms for physics. One of them, the one relating to the specific heat of gases and solids, has completely transformed the study of matter and revealed the relevance of materials to a degree unsuspected previously.

This came about rather slowly at first. It is well to recall that only eighty years ago leading physicists thought that with protons, electrons, and neutrons the basic constituents of matter had all been known. This was already a departure from the thinking and hope of W. Crookes who in his Presidential address given in 1886 to the Chemical Section of the British Association insisted that all elements were the multiples of some primary element. The position of Crookes strongly contrasted with that of Mendeleev who in his Faraday Lecture of 1889 claimed that all the elements in his periodic table were so many individual entities, irreducible to one another. Then in 1897 J. J. Thompson discovered the electron. Twenty years later, when Rutherford discovered the neutron, physicists hoped that all matter was the composite of only three fundamental particles, the electron, the proton and the neutron. Then in the early thirties came the positron, the neutrino, and the meson. Within another decade the Pandora box of fundamental particles sprang wide open. In still another decade later Oppenheimer said that the main thing we know of fundamental particles is that none of them is fundamental. The annual issues of Particle Properties Data Booklet contain an ever larger number of entries, counted in the hundreds, of elementary particles. The standard model of fundamental particles, with its assortment of gluons, quarks (all with different flavors, colors, charms, tops, and bottoms) should appear a very complex and therefore hardly fundamental proposition indeed.

As the twentieth century turned into the twenty-first it became fashionable to take this somewhat arbitrary grouping of years into units of hundreds or centuries for an indication that with the opening of a new century many new things are bound to come. It has now become a fashion to claim that fundamental particle theories have to be wholly revamped, just because about a hundred years have gone by since the discoveries of J. J. Thompson, Rutherford, and Bohr. This may or may not be so. What is certain is that matter is enormously complex because it is very peculiar. Hence the enormous potentialities of materials science and also its great restrictions. As to the potentialities, it is not a dream, according to some, to think of flexible glass and ceramics. Materials with fantastic properties can be made and have been made to order. Strange plastic materials, undreamed of only fifty years ago, dominate everyday life. Economy is literally driven by materials science. This has always been so but especially in the last fifty years,

Prof. Hummel sounds nowhere more convincing than in introducing in his book chapter 10, "The Age of Electronic Materials." After recalling that it has become a fashion to speak of our age as the space age or the atomic age, Hummel points out that our daily lives have not been much influenced either by space probes or by nuclear reactors. Quite different has been the impact of electrical and electronic devices. Think of telephones—Hummel writes—of radio, television, refrigerators, computers, and CD players. Strangely, in 1997, he did not include mobile phones. They may be followed, for better or for worse, by hand-held TV sets. But Hummel was right in generalizing: "Life without electronics would be nearly unthinkable in many parts of the world."[viii] Therefore Prof. Hummel suggests that our age should be called the electronic age. But since he had already called earlier chapters in his book after the material that had the largest impact on past life of mankind, he looks for a corresponding material for our age, and he calls it "electronic materials age."

This, I am afraid, is a bit evasive. Prof. Hummel should have gone to the heart of the matter, and referred simply to materials that possess free electrons. Whenever a material is largely composed of atoms whose nuclei hold tight the electrons around them, the study of such a material becomes a branch of physics, called mechanics. But if materials that are made of atoms in which one or two electrons are not tightly bound, their study forms a branch of electromagnetism. Of such materials none is so abundant as iron, "ferrum" in Latin. It put the stamp on many other things, including the name ferroelectricity, which for years had been known as Seignette-electricity or Rochelle-electricity and is still called such in many French and German research papers.

It has often been observed that there is no iron or "ferrum" in materials that exhibit ferro-electricity. The name ferro-electricity is therefore a misnomer, but perhaps a most happy or telling misnomer. It is certainly a pointer to the unusually high abundance of the element ferrum or iron on the Earth. A full eighty percent of the Earth's core is made of iron and nickel. Only the planet Mercury has a proportionally larger iron content than the earth with respect to its total mass. Iron is largely missing in the Jovian planets. And iron is scarce in our Moon. This is one of the reasons why the Moon's origin has to be attributed to a collision between the quasi-solid Earth and a huge planetesimal, about the size of Mars. Owing to that collision a part of the Earth's crust, mostly of granite, was detached and eventually formed the Moon, which is very poor in iron. The Moon's magnetic dipole moment is about a thousandth of that of the Earth. Moreover, the impact seems to have largely increased the iron content of the Earth, because the core of that planetesimal may have lodged in it.

At any rate, owing to some accident, so to speak, the Earth is an abode most propitious for the study of electromagnetism, which is also an indispensable tool to investigate ferroelectric materials, although they have no iron. Here it would be tempting to wander off into some philosophical considerations about the fact that physically speaking our Earth is a very strange abode. Even taken by itself it is a peculiar body among the planets. When taken in connection with the Moon, the Earth is a very unique body. The Moon largely determines the evolution of life on the Earth, and even the rise of science there. Ptolemaic astronomy, which is the beginning or at least the forerunner of exact science, rests on the determination of the relative and absolute sizes and distances of the Earth, the Moon, and the Sun. This would have been impossible without lunar eclipses. This in turn presupposes the actual Earth-Moon system, which is certainly very unique in comparison with any moon around any other planet in the solar system.

To think that there are replicas of the Earth-Moon system elsewhere around other stars in our galaxy is surely to stretch the powers of imagination as long as imagination starts from facts and not from fantasies or wishful thinking. Our peculiar terrestrial conditions enable us to do science which is always doing some sort of cosmology. "All science is cosmology," so stated Popper decades ago,[ix] a view which by then had been a very old truth. Doing science proceeds from making some assumption about the cosmos or universe. This is surely something philosophical. It illustrates the truth of the saying that the only way to avoid philosophy is to say nothing.

But there is a reverse side of this coin. About all laws of science, about all reliable conclusions of science, the scientist assumes that they are valid throughout the universe. This is the ultimate relevance of materials science, which is science in its most tangible form. Mathematicians, it was said, can say whatever they want, a physicist must be partially sane, that is, must stand on the ground with at least one foot. The truth of theoretical constructions in physics depends on their relevance to matter, that is, to observational inferences made from theory. Further it happened time and again in the history of science that only the utilization or invention of a new material provided the tool for deciding between competing theories. Such was the case during the middle of the eighteenth century, when John Dollond utilized flint glass in making dispersion-free compound telescope objectives. This showed that both Newton's and Euler's theories of diffraction were faulty.

At other times the availability of a material surely misled physicists in their theorizing. The shoemaker's wax was such a material. An iron nail put on a piece of that wax very slowly sinks through it. Lord Kelvin took this phenomenon to argue that two very different properties (elasticity and rigidity) could coexist in one and the same matter. Thus, according to him the ether could have a very slow resistivity and also a very high degree of rigidity. The latter was needed to explain the high speed of light through the ether. The other was needed because the Earth and other huge bodies rushed through the ether with no apparent resistance.

The pitchblende was not of much use prior to Roentgen's finding that a photographic plate reproduced the image of a key because a speck of uranium lay nearby. Uranium quickly became a much sought-after element and so did pitchblende, a chief source of radium and other radioactive materials. By then silver nitrate had made photography possible, and also the making of emulsion plates for the investigation of cosmic rays. Compared with these materials, almost ridiculous may be a reference to the material called beer, which, however, prompted Glaser to think about the bubble chamber, which proved to be a chief tool in fundamental particle research and earned him the Nobel Prize.

Such and similar details may be brushed aside by the working physicist, whether in ferroelectricity or elsewhere. As one who shifted from physics to the history and philosophy of physics, I have great sympathy with physicists who would do as Gallio, the prefect in Corinth, did two thousand years ago. He could not have cared less about a particular dispute brought before him, a dispute described in the Acts of the Apostles. I even have some sympathy for the physicist who would follow one of the "proverbs of hell" as phrased by William Blake: "Drive your cart and your plow over the bones of the dead."[x] But the physicist would do well to consider the observation of F. Dyson, a noted theoretical physicist, that the history of a field should be studied if one wants to understand what is essential in it.[xi] As a historian of science I know of not a single scientist who would have tumbled on his discovery by poring over old books in order to know exactly what had happened in the past.

Still the dead remain with us. Even the greatest among scientists sees farther, because he stands on the shoulder of giants. If a giant like Newton found this by then old simile appropriate to his own case, it may not be far fetched to refer to it in a general way. The dead remain indeed with us. Without them we would not be here. The present always begins in the past and the future is always a child of the present. Pockels, Valasek, Hans Mueller were not giants in the sense in which Newton was. But they opened a new field, blazed a new path in an important branch of materials science. Science has to be about materials or it is not science but mere speculation.

Contrary to Gallio's hope, the dispute brought before him turned into a major factor in history. Past disputes in the history of science turned into vast issues, such as the wave versus particle nature of light, or whether there is medium that carries physical effects to a distance. The study of the history of such questions, which is the history of science, shows more and more a relevance of its own. It may indeed take over the role which the study of Greek and Roman classics had in general human education, half a century ago.[xii] Vacuums demand to be filled not only in physics but also in culture.

At any rate, modern economy, so a guest editorial warned in the Wall Street Journal almost exactly two years ago,[xiii] is technology based. Its progress depends on the introduction of ever new technological tools—gadgets, if you wish—and these all too often demand new forms of materials, forms provided by materials science, or the exploitation in some new form of characteristics that had been well known for some time, but not yet applied. For the past ten or so years, the urge to improve the speed of computers inspired much of that work. Only a week ago came the announcement that IBM had just succeeded in replacing silicon with carbon nanotubes for future manufacturing of chips. A carbon nanotube is a single molecule, ten times the width of an atom, and is about 500 times narrower than its silicon counterpart. Obviously, it is expected to work at much greater speed and with much less heat.

Alan Greenspan, the chairman of Federal Reserve, was right in noting the economic significance of the fast data retrieval made possible by the computer, and especially its laptop form introduced exactly twenty years ago on August 12, 1981. The computer enormously facilitated the quick tracking of inventory and thereby reduced the amount of items that had to be kept in stock for effective marketing. The nature of doing business has radically changed, with much of the deadwood of storage removed from it. What Mr. Greenspan did not note was that those computers presupposed materials science at every turn in the form of semiconductors discovered in 1948 by Bardeen, Shockley and Brattain in the Bell Physical Laboratories. Before long electronic tubes began to disappear. The discovery laid the ground for the greatest economic boom known in history, the 1990s. The boom peaked because most people who needed a PC had bought one. Today such a large number of people own laptop computers as would have been inconceivable even in 1990, when PCs were ten years old. A blue ribbon committee at that time estimated that there would be no significant increase in demand for electricity for the next ten years. But from 1990 till 2000, in a single decade, the demand grew by 10 percent. Ten percent is a relative quantity which may not be much. But as the present energy crisis of America shows, the same quantity is huge when taken in absolute terms.

This is perhaps symbolic of the fundamental role which electricity plays in everyday life and also in physics. A hundred years ago Lord Kelvin rectified his view that all was mechanical. He abandoned his vortex theory of the atom, but could suggest as replacement only the general declaration that all was electromagnetic. Lord Kelvin was too old by the time Thomson discovered the electron. Lord Kelvin did not live to see Rutherford's discovery that atoms were composed of a small nucleus with electrons orbiting around it. It took another two years before Bohr came up with his model of the hydrogen atom, in which everything was truly electromagnetic. Almost a hundred years later the magnetic spin of electrons is being tapped for further progress in computer technology as witnessed by the first "spintronics" conference held in Washington this past August.

There was a time when it was thought possible to have magnetism without electricity. Now we know that electrons generate magnetism by their rotation. A colossal form of this is the huge electric current produced by the rotating iron-nickel core of the Earth, a current responsible for the Earth's magnetic field. Permanent magnets could have originated either by the hammering of some iron oxide compounds into tools, or by lightning. But these compounds had to have electrons in their constituent atoms in order to have their own magnetic domains realigned with the Earth's magnetic field. Geological forces could help produce the displacement of electrons from their symmetrical position in the Rochelle salt and other ferroelectric compounds. As it turns out more and more, all is electrical. A proof of this is the field of ferroelectricity, with its powerful relevance for materials science.

REFERENCES

i M. E. Lines and A. M. Glass, *Principles and Applications of Ferroelectrics and Related Materials* (Oxford: Clarendon Press, 1977), p. 6.
ii Ibid., pp. 6-7.
iii Mueller's first major communication on the subject, "Properties of Rochelle Salt," appeared in 1935 in the January 15th issue of *Physical Review*, 47 (1935), pp. 175-91. Three others followed five years later under the same title in the same Journal, 57 (1940), pp. 829-41, 58 (1940), pp. 565-75 and 805-811.
iv In the series, World Scientific Lecture Notes in Physics. Volume 35. Singapore, 1998.
v See O. Auciello *et al*, "The Physics of Ferroelectric Memories," *Physics Today*, July 1998, pp. 22-27.
vi New York: Springer Verlag, 1998, pp. 213-14.
vii E. Jeffery, "Nothing Left to Invent," *Journal of the Patent Office* Society, June 1940, pp. 479-81.
viii Hummel, *Understanding Materials Science*, p. 169.
ix See K. Popper, *Conjectures and Refutations* (New York: Harper and Row, 1968), p. 136.
x "The Marriage of Heaven and Hell," see *Blake. Complete Writings*, with variant readings, ed. by G. Keynes (Oxford University Press, 1966), p. 150.
xi See Dyson's review of volume II of E. T. Whittaker's *A History of Theories of Ether and Electricity*, in *Scientific American*, March 1954, pp. 92-95.
xii H. Butterfield, "The History of Science and the Study of History," *Harvard Literary Bulleting* 13 (1959), p. 331.
xiii G. Melloan, "Yes, America has a 'New Economy': Technology," *Wall Street Journal*, September 21, 1999, p. A27.

The Nobel Prize in Solid State Physics. Laureates

The Nobel Foundation

2001. Eric A. Cornell, Wolfgang Ketterle, Carl E. Wieman "for the achievement of Bose-Einstein condensation in dilute gases of alkali atoms, and for early fundamental studies of the properties of the condensates"

Presented by: Professor Sune Svanberg*

Three quarters of a century ago, Bose and Einstein made theoretical predictions that this year's Nobel Laureates in Physics succeeded in implementing 70 years later in the form of an exciting new state of matter. The so-called Bose-Einstein condensate they achieved has the same relationship to a piece of ordinary matter as a laser beam has to the light from a bulb.

The world consists of interacting particles and waves. In our ordinary perception of reality, we are fairly confident that we can distinguish easily between waves and particles, but in fact they can sometimes switch identities. Light waves may be regarded as a stream of massless particles, also called photons, and particles of matter sometimes have a wavelike nature. Generally speaking, the corresponding wavelength is extremely

* Member of The Royal Swedish Academy of Sciences and of The Nobel Committee for Physics.

short, but for atoms that move slowly, it may be observable. Einstein predicted that if a gas is cooled to very low temperatures, all the atoms should gather in the lowest energy state. Matter waves of the individual atoms then merge into a single wave; indeed, they can be said to "sing in unison." Thousands of atoms behave like one big superatom. This is Bose-Einstein condensation. However, the phenomenon can only occur with a special type of particles called bosons. Electrons are an example of another type of particles. They are not at all "sociable"; they absolutely do not wish to be in the same state and behave in the same way, but instead they arrange themselves in the increasingly complex electron shells of atoms, building up the periodic system of the elements.

To achieve Bose-Einstein condensation, a gas must be cooled to less than one millionth of a degree above absolute zero. The 1997 Nobel Laureates had developed effective laser-based methods for cooling and trapping atomic gases, but the path to Bose-Einstein condensation remained extremely difficult. Carl Weiman pointed toward a successful method: laser cooling of alkali atoms in a so-called magneto-optical atom trap and then continued reduction of speed through evaporative cooling, which is a way of systematically getting rid of the fastest atoms. After extensive efforts by the research team in Colorado, Eric Cornell solved the last remaining problem that prevented condensation. The successful experiments using rubidium atoms were reported in 1995.

Wolfgang Ketterle worked independently of Cornell and Wieman, and 4 months after them he reported large condensates of laser-cooled sodium atoms. Ketterle was able to demonstrate that the condensate actually behaved as a single coherent wave. He did an experiment similar to the one where two stones are thrown simultaneously at a calm water surface and their wave patterns roll into each other, strengthening and weakening each other in a systematic way. This is in greater contrast to what happens when uncoordinated matter, for example two fistfuls of sand, are thrown on the water surface. Ketterle was also able to extract a beam of coherent matter from the condensate, thus achieving the first atom laser. An ordinary laser yields coherent radiation, an atom laser a stream of coherent matter.

When a gas consisting of uncoordinated atoms turns into a Bose-Einstein condensate, it is like when the various instruments of an orchestra with their different tones and timbres, after warming up individually, all join in the same tone.

After the pioneering experiments of this year's Laureates, condensates have now been achieved by more than twenty additional research teams. Many fascinating applications are conceivable. Precision measurements utilizing slow atoms may offer big surprises: perhaps what we today call natural constants are completely constant. The new control over matter that Bose-Einstein condensation provides may have far-reaching practical applications, for example in lithography and nanotechnology.

Professor Cornell, Professor Ketterle, Professor Wieman. Your groundbreaking work on Bose-Einstein condensation has opened up a very fruitful area of research and potential applications. On behalf of the Royal Swedish Academy of Sciences, I wish to congratulate you on your great accomplishments. I now ask you to step forward to receive your Nobel Prizes from the hands of His Majesty the King.

2000. Zhores I. Alferov, Herbert Kroemer, "for developing semiconductor heterostructures used in high-speed- and opto-electronics" and Jack S. Kilby "for his part in the invention of the integrated circuit"

Presented by: Professor Tord Claeson*

Information technology (IT) influences our lives at many levels. We use it to collect, process, communicate and present information. IT controls high-tech processes as well as medical diagnostic instruments and everyday home appliances. Computers are linked in a global network and only a few years from now, they will number one billion. The performance of microelectronic circuits seems to be increasing one hundred-fold every ten years - at unchanged prices. IT is viewed as a prime mover in the economic upswing our society has experienced over the past decade.

This year's Nobel Prize in Physics rewards contributions to the early developments of microelectronics and photonics, focusing on the integrated circuit, or "chip," as well as semiconductor heterostructures for lasers and high-speed transistors.

The transistor was invented around Christmas 1947 and the discoverers of the transistor effect were awarded the 1956 Nobel Prize in Physics. Ten years after that discovery, transistors had replaced vacuum tubes on a large

* Member of The Royal Swedish Academy of Sciences and Chairman of The Nobel Committee for Physics.

scale. Beaches were being flooded by pop music, and one of the inventors is said to have exclaimed: "If only I had never invented that transistor."

Discrete transistors were soldered on circuit boards together with other components. But the emerging computers required ten thousands of transistors on the same board, a time-consuming and error-prone task.

As a newly hired engineer, Jack Kilby did not get his two weeks of vacation in the summer of 1958. Instead he had the privilege of thinking undisturbed on working time. He designed a circuit out of components made from a single semiconductor material that had been processed in different ways. This had already been suggested, but it was in conflict with the prevailing industrial practice of producing parts in the cheapest available material. On September 12, he was able to demonstrate that an integrated circuit worked - the birth date of the integrated circuit is one of the most important birth dates in the history of technology. Since then, things have moved fast. Chips being made today contain nearly a billion bits of memories or logic gates in processors - the brains of computers.

What was needed was not only more, smaller and cheaper transistors, but also faster ones. Early transistors were relatively slow. Semiconductor heterojunctions were proposed as a way of increasing amplification and achieving higher frequencies and power. Such a heterostructure consists of two semiconductors whose atomic structures fit one another well, but which have different electronic properties. A carefully worked out proposal was published in 1957 by Herbert Kroemer. Today, high-speed transistors are found in mobile (cellular) phones and in their base stations, in satellite dishes and links. There they are part of devices that amplify weak signals from outer space or from a faraway mobile telephone without drowning in the noise of the receiver itself.

Semiconductor heterostructures have been at least equally important to the development of photonics - lasers, light emitting diodes, modulators and solar panels, to mention a few examples. The semiconductor laser is based upon the recombination of electrons and holes, emitting particles of light, photons. If the density of these photons becomes sufficiently high, they may begin to move in rhythm with each other and form a phase-coherent state, that is, laser light. The first semiconductor lasers had low efficiency and could only shine in short pulses.

Herbert Kroemer and Zhores Alferov suggested in 1963 that the concentration of electrons, holes and photons would become much higher if they were confined to a thin semiconductor layer between two others - a double heterojunction. Despite a lack of the most advanced equipment, Alferov and his co-workers in Leningrad (now St. Petersburg) managed to produce a laser that effectively operated continuously and that did not require troublesome cooling. This was in May 1970, a few weeks earlier than their American competitors.

Lasers and light emitting diodes (LEDs) have been further developed in many stages. Without the heterostructure laser, today we would not have had optical broadband links, CD players, laser printers, bar code readers, laser pointers and numerous scientific instruments. LEDs are used in displays of all kind, including traffic signals. Perhaps they will entirely replace light bulbs. In recent years, it has been possible to make LEDs and lasers that cover the full visible wavelength range, including blue light.

I have emphasized the technical consequences of these discoveries, since these are easier to explain than the spectacular scientific breakthroughs that they have also led to. Challenging problems and matching resources have led to large-scale basic research. The advanced materials and tools of microelectronics are being used for studies in nanoscience and of quantum effects. Scientific experiments and computations are, of course, highly computerized.

Semiconductor heterostructures can be regarded as laboratories of two-dimensional electron gases. The 1985 and 1998 Nobel Prizes in physics for quantum Hall effects were based on such confined geometries. They can be reduced further to form one-dimensional quantum channels and zero-dimensional quantum dots for future studies.

Drs. Alferov, Kilby and Kroemer, I have briefly described some consequences of your discoveries and inventions. Few have had such a beneficial impact on mankind as yours. I also predict that there will be continued development, as we may be only halfway through the information technology revolution. New effects may appear as a result of basic research. When, what and where we cannot say, but we can be sure they will come.

On behalf of the Royal Swedish Academy of Sciences, I would like to convey the warmest congratulations to you and ask you to step forward to receive the Nobel Prize from the hands of His Majesty the King.

1999. Gerardus 't Hooft, Martinus J.G. Veltman "for elucidating the quantum structure of electroweak interactions in physics."

1998. Robert B. Laughlin, Horst L. Störmer, Daniel C. Tsui "for their discovery of a new form of quantum fluid with fractionally charged excitations."

1997. Steven Chu, Claude Cohen-Tannoudji, William D. Phillips "for development of methods to cool and trap atoms with laser light."

1996. David M. Lee, Douglas D. Osheroff, Robert C. Richardson "for their discovery of superfluidity in helium-3."

1995. Martin L. Perl "for the discovery of the tau lepton."
Frederick Reines "for the detection of the neutrino."

1994. Bertram N. Brockhouse "for the development of neutron spectroscopy" and **Clifford G. Shull** "for the development of the neutron diffraction technique"

Presented by: Professor Carl Nordling*

This year's Nobel Prize in Physics has been awarded to Bertram Brockhouse and Clifford Shull for their pioneering contributions to the development of neutron scattering techniques for the study of liquid and solid matter. In simple terms, one could say that Shull answered the question of where atoms "are," while Brockhouse answered the question of what they "do."

At the end of World War II, research conditions underwent a radical change, especially in the United States. For some years, every single neutron emitted by a radioactive source, produced in an accelerator or released in a nuclear reactor had been employed for one single purpose: to produce the first atomic bomb. Suddenly a major new resource was being placed in the service of peaceful research. Neutrons could henceforth perform other tasks

* Member of The Royal Swedish Academy of Sciences.

besides splitting atomic nuclei. Words like TOP SECRET were no longer automatically stamped across the cover of each research report.

Researchers had been familiar with neutrons as building blocks in the atomic nucleus for more than a decade. They had also done some thinking and conducted experiments concerning the properties of neutrons as free particles outside the nucleus. For example, they knew that neutrons possessed a dual nature that was characteristic of their tiny world: the ability to behave both as particles and as waves. In their latter guise, neutrons had been reflected against the atomic planes of a crystal in the same way X-rays previously had.

This provided a hint that some day, neutrons might become a tool for studying the microstructure of matter at the atomic level. The door was already ajar, but had not yet been opened wide.

Brockhouse and Shull followed their own individual strategies, both with the aim of gaining new knowledge about liquid and solid materials, otherwise called "condensed matter."

Shull took advantage of the fact that the wavelength of neutrons from a reactor may be roughly equal to the distance between the atoms in a solid body or a liquid. When the neutrons bounce against atomic nuclei, they do not lose energy, but their scattering is concentrated in directions that are determined by the structure in which the atoms are arranged. Shull revealed that neutrons could answer questions that the X-ray diffraction method had failed to answer, such as where the atoms of the light element hydrogen are located in an ice crystal.

Another breakthrough concerned magnetic structures. Neutrons themselves are small magnets and can interact very efficiently with the atoms in a magnetic material. Shull demonstrated how neutrons can reveal the magnetic properties of metals and alloys. The X-ray method had been powerless to accomplish this task as well.

While Shull was studying elastic neutron scattering, that is, scattering that occurs without energy changes, Brockhouse was concentrating on inelastic scattering. In the latter, neutrons lose part of their energy to the material or pick up energy from it.

Brockhouse designed ingenious instruments with which he managed to record the energy spectrum of the scattered neutrons. This enabled him to gather new information about such phenomena as atomic vibrations in crystals, diffusion movements in liquids and fluctuations in magnetic material. As a consequence, the study of these types of phenomena underwent a renaissance.

Over the years since Brockhouse and Shull made the contributions for which they are now being awarded the Nobel Prize, their methods have found widespread applications. Thousands of researchers are using neutron scattering to study the structure and dynamics of the new ceramic superconductors, molecule movements on surfaces for catalytic exhaust emission control, the interaction between proteins and the genetic material of viruses, the connection between the structure and elastic properties of polymers, the rapidly fading memory of the atomic structure of a metallic melt and much more. The pioneers of this broad field of research are the recipients of this year's Nobel Prize in Physics.

Professor Brockhouse, Professor Shull,

You have been awarded the 1994 Nobel Prize in Physics for your pioneering contributions to the development of neutron scattering in condensed matter research. It is my privilege to convey to you the heartiest congratulations of the Royal Swedish Academy of Sciences, and I now ask you to receive the Prize from the hands of His Majesty the King.

1993. Russell A. Hulse, Joseph H. Taylor Jr. "for the discovery of a new type of pulsar, a discovery that has opened up new possibilities for the study of gravitation"

1992. Georges Charpak "for his invention and development of particle detectors, in particular the multiwire proportional chamber"

1991. Pierre-Gilles de Gennes "for discovering that methods developed for studying order phenomena in simple systems can be generalized to more complex forms of matter, in particular to liquid crystals and polymers"

Presented by: Professor Ingvar Lindgren*

This year's Nobel Prize in Physics has been awarded to Pierre-Gilles de Gennes, College de France, Paris, for his investigations of *liquid crystals*

* Member of The Royal Swedish Academy of Sciences and Chairman of The Nobel Committee for Physics.

and polymers. De Gennes has shown that mathematical models, developed for studying simpler systems, are applicable also to such complicated systems. De Gennes has discovered relations between different, seemingly quite unrelated, fields of physics - connections which nobody has seen before.

Liquid crystals and polymers can be regarded as intermediate states between *order and disorder*. A simple crystal, such as ordinary salt, is an example of almost perfect order - its atoms or ions are located in exact positions relative to each other. An ordinary liquid is an example of the opposite, complete disorder, its atoms or ions seem to move in completely irregular fashion. These examples represent two extremes of the concept order-disorder. In nature, there are more subtle forms of order and a liquid crystal is an example of that. It can be well ordered in one dimension but completely disordered in another. De Gennes has generalized the description of order for media of this type and been able to see analogies with, e.g. magnetic and superconducting materials.

The discovery of the remarkable substances we now call liquid crystals was made by the Austrian botanist *Friedrich Reinitzer* slightly more than a hundred years ago. In studying plants, he found that a substance related to the cholesterol had two distinct melting points. At the lower temperature, the substance became liquid but opaque and at the higher temperature completely transparent. Earlier, similar properties had been found in stearin. The German physicist *Otto Lehmann* found that the material was completely uniform between these temperatures with properties characteristic of a liquid as well as a crystal. Therefore, he named it "liquid crystal".

All of us have seen liquid crystals in the display of digital watches and pocket calculators. Most likely, we shall shortly see them also on the screen of our TV sets. Applications of this kind depend upon the unique optical properties of the liquid crystals and the fact that these can easily be changed, e.g. by an electric field.

It has been known for a long time that liquid crystals scatter light in an exceptional way, but all early explanations of this phenomenon failed. De Gennes found the explanation in the special way the molecules of a liquid crystal are ordered. One of the phases of a liquid crystal, called *nematic*, can be compared with a ferromagnet, where the atoms, which are themselves tiny magnets, are ordered so that they point in essentially the same direction - with slight variations. These variations follow a strict mathematical rule, which near the so-called *critical temperature*, where the magnet ceases to be magnetic, attains a very special form. In the liquid crystal the molecules are ordered in a similar way *at every temperature*, which explains its remarkable optical properties.

Another large field, where de Gennes has been very active, is that of *polymer physics*. A polymer consists of a large number of molecular fragments, monomers, which are linked together to form long chains or other structures. These molecules can be formed in a countless number of ways, giving the polymer materials a great variety of chemical and physical properties. We are quite familiar with some of the applications, which range from plastic bags to parts of automobiles and aircraft.

Also in these materials, de Gennes has found analogies with critical phenomena appearing in magnetic and superconducting materials. For instance, the size of the polymer in a solution increases by a certain power of the number of monomers, which is mathematically analogous to the behavior near a critical temperature of a magnet. This had led to the formulation of scaling laws, from which simple relations between different properties of polymers can be deduced. In this way, predictions can be made about unknown properties - predictions which later in many cases have been confirmed by experiments.

Major progress in science is often made by transfering knowledge from one discipline to another. Only few people have sufficiently deep insight and sufficient overview to carry out this process. De Gennes is definitely one of them.

Professor de Gennes,

You have been awarded the 1991 Nobel Prize in Physics for your outstanding contributions to the understanding of liquid crystals and polymers. It is my privilege to convey to you the heartiest congratulations of the Royal Swedish Academy of Sciences, and I now ask you to receive the Prize from the hands of His Majesty the King.

1990. Jerome I. Friedman, Henry W. Kendall, Richard E. Taylor "for their pioneering investigations concerning deep inelastic scattering of electrons on protons and bound neutrons, which have been of essential importance for the development of the quark model in particle physics"

1989. Norman F. Ramsey, "for the invention of the separated oscillatory fields method and its use in the hydrogen maser and other atomic clocks", and **Hans G. Dehmelt, Wolfgang Paul** "for the development of the ion trap technique."

1988. Leon M. Lederman, Melvin Schwartz, Jack Steinberger "for the neutrino beam method and the demonstration of the doublet structure of the leptons through the discovery of the muon neutrino"

1987. J. Georg Bednorz, K. Alexander Müller "for their important breakthrough in the discovery of superconductivity in ceramic materials"

Presented by: Professor Gösta Ekspong*

The Nobel Prize for Physics has been awarded to Dr. Georg Bednorz and Professor Dr. Alex Müller by the Royal Swedish Academy of Sciences "for their important breakthrough in the discovery of superconductivity in ceramic materials". This discovery is quite recent - less than two years old - but it has already stimulated research and development throughout the world to an unprecedented extent. The discovery made by this year's laureates concerns the transport of electricity without any resistance whatsoever and also the expulsion of magnetic flux from superconductors.

Common experience tells us that bodies in motion meet resistance in the form of friction. Sometimes this is useful, occasionally unwanted. One could save energy, that is to say fuel, by switching off the engine of a car when it had attained the desired speed, were it not for the breaking effect of friction. An electric current amounts to a traffic of a large number of electrons in a conductor. The electrons are compelled to elbow and jostle among the atoms which usually do not make room without resistance. As a consequence some

---

* Member of The Royal Swedish Academy of Sciences.

energy is converted into heat. Sometimes the heat is desirable as in a hot plate or a toaster, occasionally it is undesirable as when electric power is produced and distributed and when it is used in electromagnets, in computers and in many other devices.

The Dutch scientist Heike Kamerlingh-Onnes was awarded the Nobel Prize for Physics in 1913. Two years earlier he had discovered a new remarkable phenomenon, namely that the electric resistance of solid mercury could completely disappear. Superconductivity, as the phenomenon is called, has been shown to occur in some other metals and alloys.

Why hasn't such an energy saving property already been extensively applied? The answer is, that this phenomenon appears only at very low temperatures; in the case of mercury at -269 degrees Celsius, which means 4 degrees above the absolute zero. Superconductivity at somewhat higher temperatures has been found in certain alloys. However, in the 1970's progress seemed to halt at about 23 degrees above the absolute zero. It is not possible to reach this kind of temperatures without effort and expense. The dream of achieving the transport of electricity without energy losses has been realized only in special cases.

Another remarkable phenomenon appears when a material during cooling crosses the temperature boundary for superconductivity. The field of a nearby magnet is expelled from the superconductor with such force that the magnet can become levitated and remain floating in the air. However, the dream of frictionless trains based on levitated magnets has not been realisable on a large scale because of the difficulties with the necessarily low temperatures.

Dr. Bednorz and Professor Müller started some years ago a search for superconductivity in materials other than the usual alloys. Their new approach met with success early last year, when they found a sudden drop towards zero resistance in a ceramic material consisting of lanthanum-barium-copper oxide. Sensationally, the boundary temperature was 50 % higher than ever before, as measured from absolute zero. The expulsion of magnetic flux, which is a sure mark of superconductivity, was shown to occur in a following publication.

When other experts had overcome their scientifically trained sceptiscism and had carried out their own control experiments, a large number of scientists decided to enter the new line of research. New ceramic materials were synthesized with superconductivity at temperatures such that the cooling suddenly became a simple operation. New results from all over the world flooded the international scientific journals, which found difficulties in coping with the situation. Research councils, industries and politicians are busily considering means to best promote the not so easy development work in order to benefit from the promising possibilities now in sight.

Scientists strive to describe in detail how the absence of resistance to the traffic of electrons is possible and to find the traffic rules, i. e. the laws of nature, which apply. The trio of John Bardeen, Leon Cooper and Robert Schrieffer found the solution 30 years ago in the case of the older types of superconductors and were awarded the Nobel Prize for Physics in 1972. Superconductivity in the new materials has reopened and revitalized the scientific debate in this field.

Herr Dr Bednorz und Herr Professor Müller:

In Ihren bahnbrechenden Arbeiten haben Sic einen neuen, sehr erfolgreichen Weg fir die Erforschung und die Entwicklung der Supraleitung angegeben. Sehr viele Wissenschaftler hohen Ranges sind zurzeit auf dem Gebiet tätig, das Sie eröffnet haben.

Mir ist die Aufgabe zugefallen, Ihnen die herzlichsten Glückwünsche der Küniglich Schwedischen Akademie der Wissenschaften zu übermitteln. Darf ich Sie nun bitten vorzutreten um Ihren Preis aus der Hand Seiner Majestät des Königs entgegenzunehmen.

1986. Ernst Ruska,"for his fundamental work in electron optics, and for the design of the first electron microscope" and **Gerd Binnig, Heinrich Rohrer** "for their design of the scanning tunneling microscope"

Presented by: Professor Sven Johansson*

The problem of the basic structure of matter has long interested man but it was not until the time of the Greek philosophers that the problem took on a scientific character. These ideas reached their culmination in Democritos' theory which postulated that atoms were the building blocks of matter. All this was, however, mere speculation, and it was first the early science and technology of Western Europe which made it possible to tackle the problem experimentally.

The first major breakthrough came with the invention of the microscope. The significance of the microscope in the fields of, for example, biology and

* Member of The Royal Swedish Academy of Sciences.

medicine is well known, but it did not provide a means of studying the basic nature of matter. The reason is that there is a limit to the amount of detail one can see in a microscope. This is connected with the wave nature of light. In the same way as ocean waves are not affected, to any great degree, by small objects, but only by larger ones, for example a breakwater, light will not produce a picture of an object that is too small. The limit is set by the wavelength of light which is about 0.0005 mm. We know that an atom is 1000 times smaller. It is clear, therefore, that something radically new was needed in order to be able to see an atom.

This new development was the electron microscope. The electron microscope is based on the principle that a short coil of a suitable construction, carrying an electric current, can deflect electrons in the same way that a lens deflects light. A coil can therefore give an enlarged image of an object that is irradiated with electrons. The image can be registered on a fluorescent screen or a photographic film. In the same way that lenses can be combined to form a microscope, it was found that an electron microscope could be constructed of coils. As the electrons used in an electron microscope have a much shorter wavelength than light, it is thus possible to reach down to much finer details. Several scientists, among them Hans Busch, Max Knoll, and Bodo von Borries, contributed to the development of the instrument, but Ernst Ruska deserves to be placed foremost. He built in 1933 the first electron microscope with a performance significantly better than that of an ordinary light microscope. Developments since then have led to better and better instruments. The importance, in many areas of research, of the invention of the electron microscope should, by now, be well known.

The microscope can be regarded as an extension of the human eye. But sight is not the only sense we use to orientate us in our surroundings, another is feeling. With modern technology it is possible to construct equipment that is based on the principle of feeling, using, for example, a sort of mechanical linger. The "finger" may be a very fine needle which is moved across the surface of the structure to be investigated. By registering the needle's movements in the vertical direction as it traverses the surface, a sort of topographical map is obtained, which, in principle, is equivalent to the image obtained in a electron microscope. It is clear that this is a rather coarse method of microscopical investigation and no one had expected any revolutionary developments in this field. However, two basic improvements led to a breakthrough. The most important of these was that a method for keeping the tip of the needle at a very small and exact constant distance from the surface was developed, thus eliminating the mechanical contact between the needle and the surface, which was a limiting factor. This was achieved using the so-called tunnelling effect. This involves applying a potential between the needle tip and the surface so that an electric current flows between the needle and the surface without actually touching them, provided that the tip of the needle and the surface are close enough together. The magnitude of the current is strongly dependent on the distance, and can therefore be used to keep the needle a certain distance above the surface with the aid of a servo mechanism, typically 2-3 atomic diameters. It was also decisive that it turned out to be possible to produce extremely fine needles so that the tip consists of only a few atoms. It is clear that if such a fine tip is moved across a surface at a height of a few atomic diameters the finest atomic details in the surface structure can be registered. It is as if one were feeling the surface with an infinitely fine finger. A crystal surface which appears completely flat in a microscope is seen with this instrument to be a plain on which atoms rise like hills in a regular pattern.

Attempts by Russell Young and co-workers to realize these ideas revealed enormous experimental difficulties. The scientists who finally mastered these difficulties were Gerd Binnig and Heinrich Rohrer. Here it was a question of moving the needle over the surface of the sample and registering its vertical position, with great precision and without disturbing vibrations. The data obtained arc then printed out, in the form of a topographic map of the surface, by a computer. The investigation may be concerned with a crystal surface, whose structure is of interest in microelectronic applications. Another example is the investigation of the adsorption of atoms on a surface. It has also been found to be possible to study organic structures, for example, DNA molecules and viruses. This is just the beginning of an extremely promising and fascinating development. The old dream from antiquity of a visible image of the atomic structure of matter is beginning to look like a realistic possibility, thanks to progress in modern microscopy.

Professor Ruska, Dr Binnig, Dr Rohrer!

In Ihrer bahnbrechenden Arbeit haben Sie den Grund fir die entscheidenden Entwicklungen moderner Mikroskopie gelegt. Es ist jetzt möglich, die kleinsten Einzelheiten der Struktur von Materie zu erkennen. Dies ist von grösster Bedeutung - nicht nur in der Physik, sondern such in vielen anderen Bereichen der Wissenschaft.

Es gereicht mir zur Ehre und Freude, Ihnen die herzlichsten Glöckwünsche der Königlich Schwedischen Akademie der Wissenschaften zu iibermitteln. Darf ich Sic nun bitten vorzutreten, urn Ihren Preis aus der Hand Seiner Majestät des Königs entgegenzunehmen.

1985. Klaus von Klitzing "for the discovery of the quantized Hall effect"

Presented by: Professor Stig Lundqvist*

This year's Nobel Prize for Physics has been awarded to Professor Klaus von Klitzing for the discovery of the quantized Hall effect.

This discovery is an example of these unexpected and surprising discoveries that now and then take place and which make research in the sciences so exciting. The Nobel Prize is sometimes an award given to large projects, where one has shown great leadership and where one with ingenuity combined with large facilities and material resources has experimentally verified the correctness of theoretical models and their predictions. Or, one has succeeded through creation of new theoretical concepts and methods to develop theories for fundamental problems in physics that resisted all theoretical attempts over a long period of time. However, now and then things happen in physics that no one can anticipate. Someone discovers a new phenomenon or a new fundamental relation in areas of physics where no one expects anything exciting to happen.

This was exactly what happened when Klaus von Klitzing in February 1980 was working on the Hall effect at the Hochfelt-Magnet-Labor in Grenoble. He discovered from his experimental data that a relation which had been assumed to hold only approximately seemed to hold with an exceptionally high accuracy and in this way the discovery of the quantized Hall effect was made.

The discovery by von Klitzing has to do with the relation between electric and magnetic forces in nature and has a long history. Let us go back to 1820, when the Danish physicist H.C. Ørsted found that an electric current in a wire influenced a compass needle and made it change its direction. He discovered this phenomenon in a class with his students. No one had seen a relation between electric and magnetic forces before. More than 50 years later a young American physicist, E.H. Hall, speculated that the magnetic

* Member of The Royal Swedish Academy of Sciences.

force might influence the charge carriers in a metallic wire placed in a magnetic field and give rise to an electric voltage across the wire. He was able to show that when sending an electric current through a strip of gold there was a small voltage across the wire in a direction perpendicular both to the current and the magnetic field. That was the discovery of the Hall effect.

The Hall effect is now a standard method frequently used to study semiconductor materials of technical importance, and the effect is described in all textbooks in solid state physics. The experiment is in principle very simple and requires only a magnetic field plus instruments to measure current and voltage. If one varies the magnetic field, the current and voltage will change in a completely regular way and no surprising effects are expected to happen.

von Klitzing studied the Hall effect under quite extreme conditions. He used an extremely high magnetic field and cooled his samples to just a couple of degrees above the absolute zero point of temperature. Instead of the regular change one would expect, he found some very characteristic steps with plateaus in the conductivity. The values at these plateaus can with extremely high accuracy be expressed as an integer times a simple expression that just depends of two fundamental constants: the electric elementary charge and Planck's constant which appear everywhere in quantum physics.

The result represents a quantization of the Hall effect - a completely unexpected effect. The accuracy in his results was about one part in ten million, which would correspond to measuring the distance between Stockholm and von Klitzing's home station Stuttgart with an accuracy of a few centimeters. The discovery of the quantized Hall effect is a beautiful example of the close interrelation between the highly advanced technology in the semiconductor industry and fundamental research in physics. The samples used by von Klitzing were relined versions of a kind of transistor we have in our radios. His samples, however, had to satisfy extremely high standards of perfection and could only be made by using a highly advanced technique and refined technology.

The quantized Hall effect can only be observed in a two-dimensional electron system. Two-dimensional electron systems do not occur in nature. However, the development in semiconductor technology has made possible the realization of a two-dimensional electron system. In the kind of transistor that von Klitzing used, some of the electrons are bound to the interface between two parts of the transistor. At sufficiently low temperature the electrons can move only along the interface and one has effectively a two-dimensional electron system.

von Klitzing's discovery of the quantized Hall effect attracted immediately an enormous interest. Because of the extremely high accuracy the effect can be used to define an international standard for electric resistance. The metrological possibilities are of great importance and have been subject to detailed studies at many laboratories all over the world.

The quantized Hall effect is one of the few examples, where quantum effects can be studied in ordinary macroscopic measurements. The underlying detailed physical mechanisms are not yet fully understood. Later experiments have revealed completely new and unexpected properties and the study of two-dimensional systems is now one of the most challenging areas of research in physics.

Professor von Klitzing,

On behalf of the Royal Swedish Academy of Sciences I wish to convey our warmest congratulations and ask you to receive your prize from the hands of His Majesty the King.

1984. Carlo Rubbia, Simon van der Meer "for their decisive contributions to the large project, which led to the discovery of the field particles W and Z, communicators of weak interaction"

1983. Subramanyan Chandrasekhar "for his theoretical studies of the physical processes of importance to the structure and evolution of the stars" and **William Alfred Fowler** *"for his theoretical and experimental studies of the nuclear reactions of importance in the formation of the chemical elements in the universe"*

1982. Kenneth G. Wilson "for his theory for critical phenomena in connection with phase transitions"

Presented by: Professor Stig Lundqvist*

---

* Member of The Royal Swedish Academy of Sciences.

The development in physics is on the whole characterized by a close interaction between experiment and theory. New experimental discoveries lead often rapidly to the development of theoretical ideas and methods that predict new phenomena and thereby stimulate further important experimental progress. This close interaction between theory and experiment keeps the frontiers of physics moving forward very rapidly.

However, there have been a few important exceptions, where the experimental facts have been well known for a long time but where the fundamental theoretical understanding has been lacking and where the early theoretical models have been incomplete or even seriously in error. I mention here three classical examples from the physics of the twentieth century, namely superconductivity, critical phenomena and turbulence. Superconductivity was discovered in the beginning of this century, but in spite of great theoretical efforts by many famous physicists, it took about fifty years until a satisfactory theory was developed. The theory of superconductivity was awarded the Nobel Prize in physics exactly ten years ago. The critical phenomena occur at phase transitions, for example between liquid and gas. These phenomena were known even before the turn of the century, and some simple but incomplete theoretical models were developed at an early stage. In spite of considerable theoretical efforts over many decades, one had to wait until the early seventies for the solution. The problem was solved in an elegant and profound way by Kenneth Wilson, who developed the theory which has been awarded this year's Nobel Prize in physics. The third classical problem I mentioned, namely turbulence, has not yet been solved, and remains a challenge for the theoretical physicists.

From daily life we know that matter can exist in different phases and that transitions from one phase to another may occur if we change, for example, the temperature. A liquid goes over into gas phase when sufficiently heated, a metal melts at a certain temperature, a permanent magnet loses its magnetization above a certain critical temperature, just to give a few examples. Let us consider the transition between liquid and gas. When we come close to the critical point, there will appear fluctuations in the density of the liquid at all possible scales. These fluctuations take the forms of drops of liquids mixed with bubbles of gas. There will be drops and bubbles of all sizes from the size of a single molecule to the volume of the system. Exactly at the critical point the scale of the largest fluctuations becomes infinite, but the role of the smaller fluctuations can by no means be ignored. A proper theory for the critical phenomena must take into account the entire spectrum of length scales. In most problems in physics one has to deal with only one length scale. This problem required the development of a new type of theory capable of describing phenomena at all possible length scales, for example, from the order of a centimeter down to less than one millionth of a centimeter.

Wilson succeeded in an ingenious way to develop a method to solve the problem, published in two papers from 1971. A frontal attack on this problem is impossible, but he found a method to divide the problem into a sequence of simpler problems, in which each part can be solved. Wilson built his theory on an essential modification of a method in theoretical physics called renormalization group theory.

Wilson's theory gave a complete theoretical description of the behaviour close to the critical point and gave also methods to calculate numerically the crucial quantities. During the decade since he published his first papers we have seen a complete breakthrough of his ideas and methods. The Wilson theory is now also successfully applied to a variety of problems in other areas of physics.

Professor Wilson,

You are the first theoretical physicist to develop a general and tractable method, where widely different scales of length appear simultaneously. Your theory has given a complete solution to the classical problem of critical phenomena at phase transitions. Your new ideas and methods seem also to have a great potential to attack other important and up to now unsolved problems in physics.

I am very happy to have the privilege of expressing the warmest congratulations of the Royal Swedish Academy of Sciences. I now ask you to receive your Nobel Prize from the hands of His Majesty the King.

1981. Nicolaas Bloembergen, Arthur Leonard Schawlow"for their contribution to the development of laser spectroscopy" and **Kai M. Siegbahn** "for his contribution to the development of high-resolution electron spectroscopy"

Presented by: Professor Ingvar Lindgren*

* Member of The Royal Swedish Academy of Sciences.

This year's Nobel prize in physics is shared between three scientists - Nicolaas Bloembergen and Arthur Schawlow, both from the United States, and Kai Siegbahn from Sweden - for their contributions to the development of two important spectroscopic methods - laser spectroscopy and electron spectroscopy.

Both of these methods are based upon early discoveries of Albert Einstein. One of the major problems for physicists of the last century was to explain, with the "classical" concepts, the so-called photoelectric effect, i.e. the emission of electrons from a metal surface irradiated with light of short wavelength. In 1905 Einstein explained this phenomenon in a simple and elegant way, using the quantum hypothesis introduced by Max Planck five years earlier. According to this model, light is a wave motion, but it is quantized, i.e. it is emitted in small pieces - light quanta or photons - which in some respects behave like particles. This discovery was the first foundation stone to be laid in the building of the "new" physics - the quantum physics - which was to develop rapidly during the first decades of this century.

The photoelectric effect is the basis of the spectroscopy which Kai Siegbahn has developed together with his collaborators in Uppsala. When a photon of high energy - e.g. from an X-ray tube - hits an atom, it can penetrate deeply into the atom and expel an electron. By analyzing the electrons expelled in this way, it is possible to extract valuable information about the interior of the atom. Early experiments of this kind where performed in the second decade of this century, but the method was not sufficiently developed to probe the atomic structure until the 1950's. At that time Kai Siegbahn had for a number of years developed more and more sophisticated instruments for analyzing electrons emitted at the decay of certain radioactive nuclei - so called beta decays. When he and his collaborators applied this technique to analyze the electrons emitted in the photoelectric process, the new era of electron spectroscopy was born. With this spectroscopy it became possible to determine the binding energy of atomic electrons with higher accuracy than was previously possible. This was of great importance for testing new atomic models and computation schemes, which were being developed at the same time, partly due to the simultaneous rapid development of computers. It was furthermore found that the electronic binding energy was to some extent dependent upon the chemical environment of the atom, and this led to a new method for chemical analysis - ESCA - which stands for "Electron Spectroscopy for

Chemical Analysis". Nowadays, this method is being applied at hundreds of laboratories all over the world, particularly in the investigation of surface reactions, such as corrosion and catalytic reactions, i.e. reactions where a substance may initiate or stimulate a chemical reaction, seemingly without taking part in it. Such reactions are of vital importance for the process industry, and the spectroscopy developed by Siegbahn and his collaborators can be of great help in our efforts to understand processes of this kind.

The second form of spectroscopy which is awarded with this year's Nobel prize - the laser spectroscopy - is based upon another early discovery of Einstein. It had been known for a long time that atoms and molecules could absorb light as well as spontaneously emit light of certain wave lengths. In 1917 Einstein found that light could also *stimulate* atoms or molecules to emit light of the same kind. This is the basic process in the laser. Photons emitted in such a stimulated process have not only the same wave length but they also oscillate in phase with each other. We call light of this kind *coherent*.

Coherent light could be compared with a marching military troop - where the soldiers correspond to the photons - while non-coherent light in this model could be compared with people on a busy shopping street on a Saturday morning. Those who have done their military service know that marching soldiers should keep in step. However, there is one occasion, namely when the troop crosses a small bridge, when it is necessary to break step, otherwise the strong, coherent vibrations of the troop could break the bridge. The situation is similar for the coherent light. Due to the fact that the photons oscillate in phase, such light will have a much stronger effect than incoherent light on the irradiated material, and this gives laser light its very special character.

Coherent radiation was first produced in the microwave region, using an instrument we call maser (MASER=Microwave Amplification by means of Stimulated Emission of Radiation). The idea of the maser was conceived in the middle 1950's by the American Charles Townes and by Basov and Prokhorov from the Soviet Union, who shared the Nobel prize in Physics in 1964. Townes and Arthur Schawlow extended the idea of the maser to the optical region - i.e. for visible light - and this led to the construction of the laser two years later (LASER= Light Amplification by means of Stimulated Emission of Radiation). At Stanford University Schawlow has led a research group, which has developed a number of advanced methods, where the laser is used to study the properties of atoms and molecules with extreme accuracy. This has stimulated the development of new theoretical models and improved appreciably our knowledge of these building blocks of matter.

Nicolaas Bloembergen has contributed to the development of laser spectroscopy in a different way. Laser light is sometimes so intense that, when it is shone on to matter, the response of the system could not be described by existing theories. Bloembergen and his collaborators have formulated a more general theory to describe these effects and founded a new field of science we now call *non-linear optics*. Several laser spectroscopy methods are based upon this phenomenon, particularly such methods where two or more beams of laser light are mixed in order to produce laser light of a different wave length. Such methods can be applied in many fields, for instance, for studying combustion processes. Furthermore, it has been possible in this way to generate laser light of shorter as well as longer wave lengths, which has extended the field of application for laser spectroscopy quite appreciably.

Professor Bloembergen, Professor Schawlow, Professor Siegbahn: you have all contributed significantly to the development of two spectroscopic methods, namely the laser spectroscopy and the electron spectroscopy. These methods have made it possible to investigate the interior of atoms, molecules and solids in greater detail than was previously possible. Therefore, your work has had a profound effect on our present knowledge of the constitution of matter.

On behalf of the Royal Swedish Academy of Sciences I wish to extend to you the heartiest congratulations and I now invite you to receive this year's Nobel prize in Physics from the hands of His Majesty the King!

1980. James Watson Cronin, Val Logsdon Fitch "for the discovery of violations of fundamental symmetry principles in the decay of neutral K-mesons"

1979. Sheldon Lee Glashow, Abdus Salam, Steven Weinberg "for their contributions to the theory of the unified weak and electromagnetic interaction between elementary particles, including, inter alia, the prediction of the weak neutral current"

1978. Pyotr Leonidovich Kapitsa "for his basic inventions and discoveries in the area of low-temperature physics". **Arno Allan Penzias, Robert Woodrow Wilson** "for their discovery of cosmic microwave background radiation"

Presented by: Professor Lamek Hulthén*

This year's prize is shared between Peter Leonidovitj Kapitza, Moscow, "for his basic inventions and discoveries in the area of low-temperature physics" and Arno A. Penzias and Robert W. Wilson, Holmdel, New Jersey, USA, "for their discovery of cosmic microwave background radiation".

By low temperatures we mean temperatures just above the absolute zero, -273°C, where all heat motion ceases and no gases can exist. It is handy to count degrees from this zero point: "degrees Kelvin" (after the British physicist Lord Kelvin) E.g. 3 K (K = Kelvin) means the same as -270°C.

Seventy years ago the Dutch physicist Kamerlingh-Onnes succeeded in liquefying helium, starting a development that revealed many new and unexpected phenomena. In 1911 he discovered *superconductivity* in mercury: the electric resistance disappeared completely at about 4 K. 1913 Kamerlingh-Onnes received the Nobel prize in physics for his discoveries, and his laboratory in Leiden ranked for many years as the Mekka of low temperature physics, to which also many Swedish scholars went on pilgrimage.

In the late twenties the Leiden workers got a worthy competitor in the young Russian Kapitza, then working with Rutherford in Cambridge, England. His achievements made such an impression that a special institute was created for him: the Royal Society Mond Laboratory (named after the donor Mond), where he stayed until 1934. Foremost among his works from this period stands an ingenious device for liquefying helium in large quantities - a pre-requisite for the great progress made in low temperature physics during the last quarter-century.

Back in his native country Kapitza had to build up a new institute from scratch. Nevertheless, in 1938 he surprised the physics community by the discovery of the *superfluidity* of helium, implying that the internal friction (viscosity) of the fluid disappears below 2.2 K (the so-called lambda-point of helium). The same discovery was made independently by Allen and Misener at the Mond Laboratory. Later Kapitza has pursued these investigations in a

* Member of The Royal Swedish Academy of Sciences.

brilliant way, at the same time guiding and inspiring younger collaborators, among whom we remember the late Lev Landau, recipient of the physics prize 1962 "for his pioneering theories for condensed matter, especially liquid helium". Among Kapitza's accomplishments we should also mention the method he developed for producing very strong magnetic fields.

Kapitza stands out as one of the greatest experimenters of our time, in his domain the uncontested pioneer, leader and master.

We now move from the Institute of Physical Problems, Moscow, to Bell Telephone Laboratories, Holmdel, New Jersey, USA. Here Karl Jansky, in the beginning of the thirties, built a large movable aerial to investigate sources of radio noise and discovered that some of the noise was due to radio waves coming from the Milky Way. This was the beginning of radio astronomy that has taken such an astounding development after the second World War - as an illustration let me recall the discovery of the pulsars, honoured with the physics prize 1974.

In the early 1960ies a station was set up in Holmdel to communicate with the satellites Echo and Telstar. The equipment, including a steerable horn antenna, made it a very sensitive receiver for microwaves, i.e. radio waves of a few cm wavelength. Later radio astronomers Arno Penzias and Robert Wilson got the chance to adapt the instrument for observing radio noise e.g. from the Milky Way. They chose a wave length c. 7 cm where the cosmic contribution was supposed to be insignificant. The task of eliminating various sources of errors and noise turned out to be very difficult and time-consuming, but by and by it became clear that they had found a background radiation, equally strong in all directions, independent of time of the day and the year, so it could not come from the sun or our Galaxy. The strength of the radiation corresponded to what technicians call an antenna temperature of 3 K.

Continued investigations have confirmed that this background radiation varies with wave length in the way prescribed by well-known laws for a space, kept at the temperature 3 K. Our Italian colleagues call it "la luce fredda" - the cold light.

But where does the cold light come from? A possible explanation was given by Princeton physicists Dicke, Peebles, Roll and Wilkinson and published together with the report of Penzias and Wilson. It leans on a cosmological theory, developed about 30 years ago by the Russian born physicist George Gamow and his collaborators Alpher and Herman. Starting from the fact that the universe is now expanding uniformly, they concluded that it must have been very compact about 15 billion years ago and ventured to assume that the universe was born in a huge explosion the "Big Bang". The temperature must then have been fabulous: 10 billion degrees, perhaps more. At such temperatures lighter chemical elements can be formed from existing elementary particles, and a tremendous amount of radiation of all wave lengths is released. In the ensuing expansion of the universe, the temperature of the radiation rapidly goes down. Alpher and Herman estimated that this radiation would still be left with a temperature around 5 K. At that time, however, it was considered out of the question, that such a radiation would ever be possible to observe. For this and other reasons the predictions were forgotten.

Have Penzias and Wilson discovered "the cold light from the birth of the universe"? It is possible - this much is certain that their exceptional perseverance and skill in the experiments led them to a discovery, after which cosmology is a science, open to verification by experiment and observation.

Piotr Kapitsa, Arno Penzias, Robert Wilson,

In accordance with our tradition I have given a brief account in Swedish of the achievements, for which you share this year's Nobel prize in Physics. It is my privilege and pleasure to congratulate you on behalf of the Royal Swedish Academy of Sciences and ask you to receive your prizes from the hands of His Majesty the King!

1977. Philip Warren Anderson, Sir Nevill Francis Mott, John Hasbrouck van Vleck "for their fundamental theoretical investigations of the electronic structure of magnetic and disordered systems"

1976. Burton Richter, Samuel Chao Chung Ting "for their pioneering work in the discovery of a heavy elementary particle of a new kind"

1975. Aage Niels Bohr, Ben Roy Mottelson, Leo James Rainwater "for the discovery of the connection between collective motion and particle motion in atomic nuclei and the development of the theory of the structure of the atomic nucleus based on this connection"

1974. Sir Martin Ryle, Antony Hewish "for their pioneering research in radio astrophysics: Ryle for his observations and inventions, in particular of the aperture synthesis technique, and Hewish for his decisive role in the discovery of pulsars"

1973. Leo Esaki, Ivar Giaever"for their experimental discoveries regarding tunneling phenomena in semiconductors and superconductors, respectively" **and Brian David Josephson** "for his theoretical predictions of the properties of a supercurrent through a tunnel barrier, in particular those phenomena which are generally known as the Josephson effects"

Presented by: Professor Stig Lundqvist*

The 1973 Nobel Prize for physics has been awarded to Drs. Leo Esaki, Ivar Giaever and Brian Josephson for their discoveries of tunnelling phenomena in solids.

The tunnelling phenomena belong to the most direct consequences of the laws of modern physics and have no analogy in classical mechanics. Elementary particles such as electrons cannot be treated as classical particles but show both wave and particle properties. Electrons are described mathematically by the solutions of a wave equation, the Schrödinger equation. An electron and its motion can be described by a superposition of

* Member of The Royal Swedish Academy of Sciences.

simple waves, which forms a wave packet with a finite extension in space. The waves can penetrate a thin barrier, which would be a forbidden region if we treat the electron as a classical particle. The term tunnelling refers to this wave-like property - the particle "tunnels" through the forbidden region. In order to get a notion of this kind of phenomenon let us assume that you are throwing balls against a wall. In general the ball bounces back but occasionally the ball disappears straight through the wall. In principle this could happen, but the probability for such an event is negligibly small.

On the atomic level, on the other hand, tunnelling is a rather common phenomenon. Let us instead of balls consider electrons in a metal moving with high velocities towards a forbidden region, for example a thin insulating barrier. In this case we cannot neglect the probability of tunneling. A certain fraction of the electrons will penetrate the barrier by tunnelling and we may obtain a weak tunnel current through the barrier.

The interest for tunnelling phenomena goes back to the early years of quantum mechanics, i.e. the late twenties. The best known early application of the ideas came in the model of alpha-decay of heavy atomic nuclei. Some phenomena in solids were explained by tunnelling in the early years. However, theory and experiments often gave conflicting results, no further progress was made and physicists lost interest in solid state tunnelling in the early thirties.

With the discovery of the transistor effect in 1947 came a renewed interest in the tunnelling process. Many attempts were made to observe tunnelling in semiconductors, but the results were controversial and inconclusive.

It was the young Japanese physicist Leo Esaki, who made the initial pioneering discovery that opened the field of tunnelling phenomena for research. He was at the time with the Sony Corporation, where he performed some deceptively simple experiments, which gave convincing experimental evidence for tunnelling of electrons in solids, a phenomenon which had been clouded by questions for decades. Not only was the existence of tunnelling in semiconductors established, but he also showed and explained an unforeseen aspect of tunnelling in semiconductor junctions. This new aspect led to the development of an important device, called the tunnel diode or the Esaki diode.

Esaki's discovery, published in 1958, opened a new field of research based on tunnelling in semiconductors. The method soon became of great importance in solid state physics because of its simplicity in principle and the high sensitivity of tunnelling to many finer details.

The next major advance in the field of tunnelling came in the field of superconductivity through the work of Ivar Giaever in 1960. In 1957, Bardeen, Cooper and Schrieffer had published their theory of superconductivity, which was awarded the 1972 Nobel Prize in physics. A crucial part of their theory is that an energy gap appears in the electron spectrum when a metal becomes superconducting. Giaever speculated that the energy gap should be reflected in the current-voltage relation in a tunnelling experiment. He studied tunnelling of electrons through a thin sandwich of evaporated metal films insulated by the natural oxide of the film first evaporated. The experiments showed that his conjecture was correct and his tunnelling method soon became the dominating method to study the energy gap in superconductors. Giaever also observed a characteristic fine structure in the tunnel current, which depends on the coupling of the electrons to the vibrations of the lattice. Through later work by Giaever and others the tunnelling method has developed into a new spectroscopy of high accuracy to study in detail the properties of superconductors, and the experiments have in a striking way confirmed the validity of the theory of superconductivity.

Giaver's experiments left certain theoretical questions open and this inspired the young Brian Josephson to make a penetrating theoretical analysis of tunnelling between two superconductors. In addition to the Giaever current he found a weak current due to tunelling of coupled electron pairs, called Coopers pairs. This implies that we get a supercurrent through the barrier. He predicted two remarkable effects. The first effect is that a supercurrent may flow even if no voltage is applied. The second effect is that a high frequency alternating current will pass through the barrier if a constant voltage is applied.

Josephson's theoretical discoveries showed how one can influence supercurrents by applying electric and magnetic fields and thereby control, study and exploit quantum phenomena on a macroscopic scale. His discoveries have led to the development of an entirely new method called quantum interferometry. This method has led to the development of a rich variety of instruments of extraordinary sensitivity and precision with application in wide areas of science and technology.

Esaki, Giaever and Josephson have through their discoveries opened up new fields of research in physics. They are closely related because the pioneering work by Esaki provided the foundation and direct impetus for Giaever's discovery and Giaever's work in turn provided the stimulus which led to Jo- sephson's theoretical predictions. The close relation between the abstract concepts and sophisticated tools of modern physics and the practical applications to science and technology is strongly emphasized in these discoveries. The applications of solid state tunnelling already cover a wide range. Many devices based on tunneling are now used in electronics. The new quantum interferometry has already been used in such different applications as measurements of temperatures near the absolute zero, to detect gravitational waves, for ore prospecting, for communication through water and through mountains, to study the electromagnetic field around the heart or brain, to mention a few examples.

Drs. Esaki, Giaever and Josephson,

In a series of brilliant experiments and calculations you have explored different aspects of tunelling phenomena in solids. Your discoveries have opened up new fields of research and have given new fundamental insight about electrons in semiconductors and superconductors and about macroscopic quantum phenomena in superconductors.

On behalf of the Royal Academy of Sciences I wish to express our admiration and convey to you our warmest congratulations. I now ask you to proceed to receive your prizes from the hands of his Majesty the King.

1972. John Bardeen, Leon Neil Cooper, John Robert Schrieffer "for their jointly developed theory of superconductivity, usually called the BCS-theory"

Presented by: Professor Stig Lundqvist*

The 1972 Nobel Prize for physics has been awarded to Drs John Bardeen, Leon N. Cooper and J. Robert Schrieffer for their theory of superconductivity, usually referred to as the BCS-theory.

Superconductivity is a peculiar phenomenon occurring in many metallic materials. Metals in their normal state have a certain electrical resistance, the magnitude of which varies with temperature. When a metal is cooled its resistance is reduced. In many metallic materials it happens that the electrical resistance not only decreases but also suddenly disappears when a

* Member of The Royal Swedish Academy of Sciences.

certain critical temperature is passed which is a characteristic property of the material.

This phenomenon was discovered as early as 1911 by the Dutch physicist Kamerlingh Onnes, who was awarded the Nobel Prize for physics in 1913 for his discoveries.

The term superconductivity refers to the complete disappearance of the electrical resistance, which was later verified with an enormous accuracy. A lead ring carrying a current of several hundred ampères was kept cooled for a period of 2 1/2 years with no measurable change in the current.

An important discovery was made in the thirties, when it was shown that an external magnetic field cannot penetrate a superconductor. If you place a permanent magnet in a bowl of superconducting material, the magnet will hover in the air above the bowl, literally floating on a cushion of its own magnetic field lines. This effect may be used as an example for the construction of friction-free bearings.

Many of the properties of a metal change when it becomes superconducting and new effects appear which have no equivalent in the former's normal state. Numerous experiments have clearly shown that a fundamentally new state of the metal is involved.

The transition to the superconductive state occurs at extremely low temperatures, characteristically only a few degrees above absolute zero. For this reason practical applications of the phenomenon have been rare in the past and superconductivity has been widely considered as a scientifically interesting but exclusive curiosity confined to the low temperature physics laboratories. This state of affairs is rapidly changing and the use of superconducting devices is rapidly increasing. Superconducting magnets are often used for example in particle accelerators. Superconductivity research has in recent years resulted in substantial advances in measuring techniques and an extensive used in the computer field is also highly probable. Advanced plans for the use of superconductivity in heavy engineering are also in existence. By way of an example, it may be mentioned that the transport of electric energy to the major cities of the world with the use of superconductive lines is being planned. Looking further ahead one can see, for example, the possibility of building ultrarapid trains that run on superconducting tracks.

Superconductivity has been studied experimentally for more than sixty years. However, the central problem, the question of the physical mechanism responsible for the phenomenon remained a mystery until the late fifties.

Many famous physicists tackled the problem with little success. The difficulties were related to the very special nature of the mechanism sought. In a normal metal the electrons more around individually at random, somewhat similar to the atoms in a gas, and the theory is, in principle, fairly simple. In superconductive metals the experiments suggested the existence of a collective state of the conduction electrons-a state in which the electrons are strongly coupled and their motion correlated so that there is a gigantic coherent state of macroscopic dimension containing an enormous number of electrons. The physical mechanism responsible for such a coupling remained unknown for a long time. An important step towards the solution was taken in 1950 when it was discovered simultaneously on theoretical and experimental grounds that superconductivity must be connected with the coupling of the electrons to the vibrations of the atoms in the crystal lattice. The conduction electrons are coupled to each other via these vibrations. Starting from this fundamental coupling of the electrons Bardeen, Cooper and Schrieffer developed their theory of superconductivity, published in 1957, which gave a complete theoretical explanation of the phenomenon of superconductivity.

According to their theory the coupling of the electrons to the lattice oscillations leads to the formation of bound pairs of electrons. These pairs play a fundamental role in the theory. The complete picture of the mechanism of superconductivity appeared when Bardeen, Cooper and Schrieffer showed that the motion of the different pairs is very strongly correlated and that this leads to the formation of a gigantic coherent state in which a large number of electrons participate. It is this ordered motion of the electrons in the superconductive state in contrast to the random individual motion in a normal crystal that gives superconductivity its special properties.

The theory developed by Bardeen, Cooper and Schrieffer together with extensions and refinements of the theory, which followed in the years after 1957, succeeded in explaining in considerable detail the properties of superconductors. The theory also predicted new effects and it stimulated intense activity in theoretical and experimental research which opened up new areas. These latter developments have led to new important discoveries which are being used in a number of interesting ways especially in the sphere of measuring techniques.

Developments in the field of superconductivity during the last fifteen years have been greatly inspired by the fundamental theory of superconductivity and have strikingly verified the validity and great range of the concepts and ideas developed by Bardeens, Cooper and Schrieffer.

Drs. Bardeen, Cooper and Schrieffer,

You have in your fundamental work given a complete theoretical explanation of the phenomenon of superconductivity. Your theory has also predicted new effects and stimulated an intensive activity in theoretical and experimental research. The further developments in the field of superconductivity have in a striking way confirmed the great range and validity of the concepts and ideas in your fundamental paper from 1957.

On behalf of the Royal Academy of Sciences I wish to convey to you the warmest congratulations and I now ask you to receive your prizes from the Hands of His Royal Highness the Crown Prince.

1971. Dennis Gabor "for his invention and development of the holographic method"

1970. Hannes Olof Gösta Alfvén "for fundamental work and discoveries in magneto-hydrodynamics with fruitful applications in different parts of plasma physics" and Louis Eugène Félix Néel "for fundamental work and discoveries concerning antiferromagnetism and ferrimagnetism which have led to important applications in solid state physics"

Presented by: Professor Torsten Gustafsson*

About two thousand years ago, the first magnetic compass was made in China by stroking a piece of iron with a lump of magnetite. Such a compass always arouses much surprise, from the child who asks about the invisible force which aligns it along the north-south axis, to the scientist, who here confronts one of the very difficult problems of physics.

Three states of magnetism have long been recognised, die-, para- and ferromagnetism. In the two former, the elementary magnets of the atoms behave independently of one another when subjected to a magnetic field. However, in ferromagnetism, which is many times stronger, they are aligned collectively, which makes the understanding of the physics much more difficult.

---

* Member of The Royal Swedish Academy of Sciences.

The first scientist who tried to explain magnetism was Ampere with his hypothesis about elementary currents. In 1907, Pierre Weiss found that there must be a special kind of force which aligned the elementary magnets, although he could not identify it. In his doctor's thesis in 1911. Niels Bohr showed that magnetism could not be caused by currents originating from the classical motion of electrical charges, but that something completely new was needed. Using the new ideas of atomic physics, Heisenberg in 1928 was able to give a qualitative explanation of the aligning force occurring in ferromagnetics. To these three types of magnetism, Néel in 1932 added a fourth, anti-ferromagnetism. He found that for certain crystals adjacent elementary magnets could align themselves anti-parallel and not parallel as in ferro-magnetic materials. He deduced the existence of anew variant of the force postulated by Weiss and presented a model for crystals which are built up from two interlaced lattices with equally strong magnetic fields acting in opposite directions. Anti-ferromagnetism is an ordered state with important properties. Thus, Néel showed that the magnetic state should disappear at a temperature now known as the Néel point, in analogy with the Curie point. Similarly, other remarkable observations in the physics of the solid state were explained.

In 1948, Néel made another fundamental discovery with his explanation of the strong magnetism found in the ferrite materials, of which magnetite is one. He generalized his earlier assumption by assuming that the lattices could be of different strengths and could produce external fields. In magnetite, with three atoms of iron and four of oxygen, the effects of two of the iron atoms cancel out while the third gives rise to the magnetic field. It is remarkable that magnetite which in the hands of the Chinese was used to produce the first compass, is in fact not ferromagnetic, but, in Néel's terminology, ferrimagnetic.

Néel could present an accurate description of the behaviour of the new synthetic magnetic materials and so explain hitherto puzzling experimental observations. These developments have been of considerable technical importance, *e.g.* in computer memories and in high-frequency techniques.

Néel has made many other contributions, such as investigations in the theory of magnetic domains and the discovery of the effect found in small particles, called super-paramagnetism.

Professor Néel. I have attempted to describe your major discoveries which follow in the great French tradition of studies of magnetic phenomena.

I have particularly emphasised your discoveries of anti-ferro- and ferrimagnetism which have been of such importance in the shaping of modern theories of magnetism.

I have the pleasure and the honour to convey to you the most sincere congratulations of the Royal Academy of Science.

Professor Alfvén, Professor Néel. I invite you to receive the Nobel Prize in Physics from the hands of His Majesty the King.

1969. Murray Gell-Mann "for his contributions and discoveries concerning the classification of elementary particles and their interactions"

1968. Luis Walter Alvarez "for his decisive contributions to elementary particle physics, in particular the discovery of a large number of resonance states, made possible through his development of the technique of using hydrogen bubble chamber and data analysis"

1967. Hans Albrecht Bethe "for his contributions to the theory of nuclear reactions, especially his discoveries concerning the energy production in stars"

1966. Alfred Kastler "for the discovery and development of optical methods for studying Hertzian resonances in atoms"

1965. Sin-Itiro Tomonaga, Julian Schwinger, Richard P. Feynman "for their fundamental work in quantum electrodynamics, with deep-ploughing consequences for the physics of elementary particles"

1964. Charles Hard Townes, Nicolay Gennadiyevich Basov, Aleksandr Mikhailovich Prokhorov "for fundamental work in the field of quantum electronics, which has led to the construction of oscillators and amplifiers based on the maser-laser principle"

Presented by: Professor B. Edlén*

The Nobel Prize for physics is in this year given for the invention of the maser and the laser. "Maser" stands for "microwave amplification by stimulated emission of radiation", and the word "laser" is obtained by replacing "microwave" by "light".

The key to the invention is the concept of stimulated emission which was introduced by Einstein already in 1917. By a theoretical analysis of the Planck radiation formula he found that the well-known process of absorption must be accompanied by a complementary process implying that received radiation can stimulate the atoms to emit the same kind of radiation. In this process lies a potential means for amplification. However, the stimulated emission was long regarded as a purely theoretical concept which never could be put to work or even be observed, because the absorption would be the completely dominating process under all normal conditions. An amplification can occur only if the stimulated emission is larger than the absorption, and this in turn requires that there should be more atoms in a high energy state than in a lower one. Such an unstable energy condition in matter is called an inverted population. An essential moment in the invention of the maser and the laser was, therefore, to create an inverted population under such circumstances that the stimulated emission could be used for amplification.

The first papers about the maser were published 10 years ago as a result of investigations carried out simultaneously and independently by Townes and co-workers at Columbia University in New York and by Basov and Prochorov at the Lebedev Institute in Moscow. In the following years there were designed a number of masers of widely different types, and many people made important contributions to this development. In the type that is now being mostly used the maser effect is obtained by means of the ions of certain metals imbedded in a suitable crystal. These masers work as extremely sensitive receivers for short radiowaves. They are of great

---

* Member of The Nobel Committee for Physics.

importance in radio astronomy and are being used in space research for recording the radio signals from satellites.

The optical maser, that is, the laser, dates from 1958, when the possibilities of applying the maser principle in the optical region were analysed by Schawlow and Townes as well as in the Lebedev Institute. Two years later the first laser was operating.

The step from the microwaves to visible light means a 100000-fold increase in frequency and causes such changes in the operation conditions that the laser may be regarded as an essentially new invention. In order to achieve the high radiation density required for the stimulated emission to become dominating, the radiating matter is enclosed between two mirrors that force the light to traverse the matter many times. During this process the stimulated radiation grows like an avalanche until all the atoms have given up their energy to the radiation. The fact that the stimulated and stimulating radiation have exactly the same phase and frequency is essential for the result of the process. By virtue of resonance all parts of the active medium combine their forces to give one strong wave. The laser emits what is called coherent light, and this is the decisive difference between the laser and an ordinary light source where the atoms radiate quite independent of each other.

Lasers have now been made in many different shapes. The first, and still most frequently used, type consists of a ruby rod, a few inches long, with the polished and silvered end faces serving as mirrors. The radiation leaves eventually the crystal through one of the end faces which is made slightly transparent. The ruby consists of aluminium oxide with a small admixture of chromium. The chromium ions give to the ruby its red colour, and they are also responsible for the laser effect. The inverted population is produced by the light from a xenon flash lamp. This is absorbed by the ions, putting them in such a condition that they can be stimulated to emit a red light with a welldefined wavelength.

Normally, a large number of successive pulses of laser light is emitted during the time of one flash from the lamp, but by retarding the release until the stored energy has reached a maximum all the energy can be put into one big pulse. The power of the emitted light can then reach more than a hundred million watts. Since, moreover, the emerging ray bundle is strictly parallel, the whole energy can be concentrated by means of a lens on a very small area, producing an enormous power per unit area. From a scientific point of view it is especially interesting that the electrical field strength produced in the light wave may amount to some hundred million volts/cm and thus surpass the forces that keep the electron shells of the atoms together. The high photon density opens up quite new possibilities for studying the interaction of radiation and matter.

Another type of laser, in which the light is emitted from a gas excited by an electric discharge, produces continuously a radiation with a very sharply defined wavelength. This radiation can be used for measurements of lengths and velocities with a previously unattainable precision.

The invention of the laser has provided us with a powerful new tool for research in many fields, the exploitation of which has only just started. Its potential technical applications have been much publicised and are therefore well known. Regarding, especially, the extreme power concentration obtainable with a laser, it should be noted that this effect is limited to short time intervals and very small volumes and therefore attains its main importance for micro-scale operations. It should be emphasized, finally, that the use of a laser beam for destructive purposes over large distances is wholly unrealistic. The "death ray" is and remains a myth.

Dr. Townes, Dr. Basov and Dr. Prochorov. By your ingenious studies of fundamental aspects of the interaction between matter and radiation you have made the atoms work for us in a new and most remarkable way. These magic devices called maser and laser have opened up vast new fields for research and applications which are being exploited with increasing intensity in many laboratories all over the world. On behalf of the Royal Swedish Academy of Sciences I extend to you our warm congratulations and now ask you to receive the Nobel prize from the hands of His Majesty the King.

1963. Eugene Paul Wigner"for his contributions to the theory of the atomic nucleus and the elementary particles, particularly through the discovery and application of fundamental symmetry principles" and **Maria Goeppert-Mayer, J. Hans D. Jensen** "for their discoveries concerning nuclear shell structure"

1962. Lev Davidovich Landau "for his pioneering theories for condensed matter, especially liquid helium"

1961. Robert Hofstadter "for his pioneering studies of electron scattering in atomic nuclei and for his thereby achieved discoveries concerning the stucture of the nucleons" and **Rudolf Ludwig Mössbauer** "for his researches concerning the resonance absorption of gamma radiation and his discovery in this connection of the effect which bears his name"

1960. Donald Arthur Glaser "for the invention of the bubble chamber"

1959. Emilio Gino Segrè, Owen Chamberlain "for their discovery of the antiproton"

1958. Pavel Alekseyevich Cherenkov, Il´ja Mikhailovich Frank, Igor Yevgenyevich Tamm "for the discovery and the interpretation of the Cherenkov effect"

1957. Chen Ning Yang, Tsung-Dao Lee "for their penetrating investigation of the so-called parity laws which has led to important discoveries regarding the elementary particles"

1956. William Bradford Shockley, John Bardeen, Walter Houser Brattain "for their researches on semiconductors and their discovery of the transistor effect"

Presented by: Professor E.G. Rudberg*

In these days 250 years have elapsed since Benjamin Franklin was born: the printer and educator, the statesman, the pioneer in the field of electricity. It was Franklin who strung a high-tension line from a thundercloud to a green pasture in idyllically rural Philadelphia. He showed that the cloud held electric energy. A kite drew energy out of the cloud. The kite string was drenched by the rain and therefore conducted the charge down to a key, which gave off sparks when approached too closely. Franklin had tied one end of a silk ribbon to the key; he clutched the other end of the ribbon as he stooped under a cowshed to keep his silk insulator dry.

---

* Member of The Nobel Committee for Physics.

A conductor and an insulator was what Franklin needed for his power line. Electrical engineering would have been unthinkable today if Nature had not presented the material in these two extreme classes, metallic conductors and insulators. Mobile carriers of charge are almost entirely lacking in an insulator, but a good conductor has plenty of them, about one for each atom. As early as 100 years ago they carried the current in the first Atlantic cable from the Old World to the New - in a fraction of a second. A group of charged carriers enters at the European end, and immediately afterwards carriers emerge from the American end - but not the same group. Over the entire length of the cable, carriers are standing tightly packed. The emigrants must push to make room for themselves at the very entrance. This push darts as a shock with the speed of light down the long line of carriers finally ejecting those that are standing next to the exit in America. Charge is therefore transported with lightning speed, although each carrier only moves a short distance. In the old days, carriers were thought to be of two kinds, positive and negative, moving in opposite directions. Franklin held that only one kind was needed. Franklin's contention was supported by the great discoveries around the year 1900. The carriers in metallic conductors are electrons, and they all carry the same negative charge.

If the Easter pilgrims in Piazza San Pietro were to represent the carriers in a metal, then an insulator would resemble the Antarctic with one solitary traveller. In the abundance of carriers there is an enormous gap between conductors and insulators. In this gap it is now possible to place the semiconductors, with carriers about as numerous as the longshoremen in a harbour when a loaded freighter has just arrived. The semiconductors now in use are artificial products made from elements such as germanium or silicon. The pure element has very few carriers. Through small additions of certain contaminants, however, it is possible to alter the supply of carriers. Every atom of phosphorus, forced as a lodger on silicon, donates one carrier to the house, a negative electron. A few parts in 100,000 make a good semiconductor. Still more remarkable is that a guest atom of boron provides a carrier of the opposite kind - positive. This the guest manages to accomplish by stealing an electron which his host, silicon, had kept locked up. Where that electron was, a hole is now left. This hole can migrate in the semiconductor, and it then acts as a carrier of positive charge.

It is possible to have both holes and electrons as carriers in a semiconductor at the same time. Donors and thieves are lodged in such proportions that one kind of carrier, or the opposite kind, will prevail. Much of the technical importance of semiconductors stems from the interplay of holes and electrons. The idea of two kinds of carriers is contrary to Franklin's views. This idea was put forward in the 1930's, at a time when rectifiers based on semiconductors began to find important uses. Attempts were made to control these rectifiers by means of an extra electrode, just as a radio valve is controlled by the grid - without success. Finally, in 1948, the discovery of transistor action gave Shockley, Bardeen, and Brattain the key to the control mechanism and, in addition, a new weapon for tackling the semiconductor problems.

The description must now borrow a picture from the classical books of adventure. To place a negative electrode against a semiconductor with negative carriers - this is like bringing a ship up to a quay in the Orient, with the yellow flag of the plague hoisted. The place becomes deserted by its carriers. Unloading - current - is blocked. But exchange that negative flag of pestilence for a positive sign and the carriers will return, the contact becoming conducting. Electrically this is called rectification. In those seafaring tales it was perhaps possible to induce the carriers to return, without striking the flag, merely by throwing some gold coins on the quay, thus positively destroying the insulation. It is possible to destroy the blockade in the semiconductor in a similar fashion by throwing in some positive holes around which the negative carriers will gather. This is transistor action. It is a fine thing that the carriers' strike can be broken up by rather few holes, which do not cost much energy. Thus the current in the rectifier is controlled through the injection of holes. A transistor functions much like a radio valve. But it is smaller, and it does not require current to heat a filament. Hearing aids, computing machines, telephone stations and many others are in need of just such a device.

The physicists at Murray Hill decided to map out that region, poor in carriers, near a negative electrode, using a movable probe at the surface of the semiconductor. This is done in the same fashion as electric prospecting for ore, but the scale is a different one. Bardeen and Brattain moved their tiny probe under the microscope, using a micrometer screw. When the probe was made positive quite close to the electrode they found that the blockade was lifted. The probe acted as an injector of holes. Shockley and his collaborators hastened to utilize this injector in a series of ingeniously conceived experiments, which then disclosed many properties of holes: how fast they travel, how long they live and other characteristics. With new tools such as these, semiconductor physics is today a seething field of research.

From Philadelphia's old pasture to today's Murray Hill is not many miles - but 200 years. Evidently there is more than the geographical proximity that connects Franklin's work with the discoveries of his latter-day country men.

Doctor Shockley, Doctor Bardeen, Doctor Brattain. The summit of Everest was reached by a small party of ardent climbers. Working from an advance base, they succeeded. More than a generation of mountaineers had toiled to establish that base. Your assault on the semiconductor problem was likewise launched from a high-altitude camp, contributed by many scientists. Yours, too, was a supreme effort - of foresight, ingenuity and perseverence, excercised individually and as a team. Surely, supreme joy befalls the man to whom those breathtaking vistas from the summit unfold. You must have felt it, overwhelmingly. This joy is now shared by those who laboured at the base. Shared, too, is the challenge of untrodden territory, now seen for the first time, calling for a new scientific attack.

Thus salutes you, Nobel Laureates, the Royal Academy of Sciences.

And now, my solemn duty, nay, my treasured privilege: to invite you to receive your award from the hands of His Majesty the King.

1955. Willis Eugene Lamb "for his discoveries concerning the fine structure of the hydrogen spectrum" and **Polykarp Kusch** "for his precision determination of the magnetic moment of the electron"

1954. Max Born "for his fundamental research in quantum mechanics, especially for his statistical interpretation of the wavefunction" and **Walther Bothe** "for the coincidence method and his discoveries made therewith"

1953. Frits (Frederik) Zernike "for his demonstration of the phase contrast method, especially for his invention of the phase contrast microscope"

1952. Felix Bloch, Edward Mills Purcell "for their development of new methods for nuclear magnetic precision measurements and discoveries in connection therewith"

Presented by: Professor E. Hulthén*

For the man in the street I suppose the compass needle is the most familiar magnetic instrument. But when and where the compass was first used is a much-debated question, where we grope between Chinese records from the year 2,600 B.C. and ship's logs made by the Norsemen in their Icelandic voyages in the 12th and 13th centuries. It is typical of all such records, whether of gunpowder or the compass, that they refer to inventions that had long been in use. The very idea of invention, of having been the first, had doubtless not the same significance formerly that it has today. As a matter of fact, the scientific study of magnetism in the sense in which we understand it, was begun only with the publication in London of Gilbert's work De Magnete in the year 1,600 A.D. The subsequent investigation and classification of magnetic substances led to their division into three categories: the ferromagnetics or strong magnetics such as iron, cobalt and nickel; the paramagnetic or weak magnetics, including chiefly crystals and fluids; and finally the diamagnetics, with their magnetic repulsion, a property intrinsic in all substances. A compass needle made of a diamagnetic substance turns at right angles to the magnetic lines of force, and thus comes to point in an east-westerly direction. Fortunately, diamagnetism is too weak to cause shipwreck in this way. This wealth of magnetic is today joined by a fourth category, the nuclear magnetism deriving from the atomic nucleus.

The magnetic field radiating from the infinitesimally tiny atomic nucleus is so feeble that its existence was still scarcely more than divined only fifteen or twenty years ago. Thus when Bloch and Purcell, this year's Nobel Prize winners in Physics, are able to register nuclear magnetism with a precision exceeding almost all other measurements in physics, one supposes that this must be thanks to the use of special methods and accessories. But what interest or useful purpose may conceivably be served by such subtleties?

If we consider the methods that have been employed, we soon recognize the idea that runs through all more advanced measurements of a body's magnetic moments. Thus the celebrated German mathematician and

* Member of The Nobel Committee for Physics.

physicist, Karl Friedrich Gauss, determined in 1836 the magnetic moment of thc compass needle in relation to its moment of inertia, simply by observing the oscillations of the needle in a magnetic field of known strength.

Now the electrons or the atomic nucleus do not, it is true, behave in quite the same way as the compass needle in the magnetic field, but rather in the manner of the top, the gyroscope, which spins and precesses about the perpendicular. But the electronic and nuclear spins are just as characteristic for these particles as are their electric charge and mass (the atomic weights), so that the deep import of a determination of their gyromagnetic indices becomes immediately obvious.

Now what possibilities exist for the observation and measurement of the frequencies of the electrons and the atomic nucleus in the magnetic field? This is where the new phase in the development comes in. In this connection I need only remind you of the resonance between our radio apparatuses and radio waves. The comparison is actually quite justified, as the electronic and atomic nuclear frequencies in the magnetic field fall precisely within the region for the short-wave radio with wavelengths varying between some tens of meters and the centimeterwaves employed in radar technique.

These atomic frequencies in the magnetic field are so characteristic for each element and its isotopes that they are more undisturbed and regular than the balance-wheel, pendulum and vibrating quartz-crystal in our modern chronometers.

The method for the determination of the nuclear magnetic moment through resonance with radio waves has long been well-known, and was rewarded by the Academy of Sciences with the Nobel Prize for the year 1944 to Rabi. It was with similar methods that the paramagnetism of crystals deriving from the electronic spin was investigated by Gorter in Leiden.

Rabi carried out his investigations on nuclear magnetic moments according to the molecular-ray method, an artificial method which has, certainly, the inestimable advantage that the investigated substance is in a state of very high rarefaction, though at the same time this limits its application. The methods of Purcell and Bloch imply a great simplification and generalization in this respect, which enables their application to solid, liquid and gaseous substances. This brings us to the useful purposes which may be served. Since each kind of atom and its isotopes have a sharply defined and characteristic nuclear frequency, we can in any object placed between the poles of an electromagnet seek out and examine with radio waves all the various kinds of atom and isotopes present in the object in question, and, *which is the essential point*, this without in any perceptible way affecting the same, its form, crystalline structure, etc. This form of analysis *in situ* is therefore probably not parallelled in any other known methods of analysis. Its extraordinary sensitiveness also makes it particularly well-adapted as a micro-method in many scientific and technical fields.

Professor Purcell. As far I have been able to follow your activities since you stopped working at the great Radiation Laboratory at M.I.T. at the end of the War, and up to your development of the excellent method of *nuclear resonance absorption* for which you have been awarded your Nobel Prize, you have happily realized man's old dream of beating the sword into a ploughshare. Your wide experience in electronics and the deep interest you early showed in paramagnetic phenomena may thus conceivably have contributed to the invention of your method, which through its extraordinary sensitiveness gives us a deep insight into the constitution of crystals and fluids, and the interactions, so-called relaxations, between the tiniest particles of matter.

In part with this method, and in part without it, you and your collaborators have made a number of important discoveries, among which I would like particulary to stress the three following:

Your method for studying nuclear magnetic resonance in weak magnetic field produced according to the solenoid method, which is of great value for the absolute determination of nuclear magnetic moments.

In the very interesting experiment which you performed together with Dr. Pound, you have produced with paramagnetic resonance the rather unique situation in which the state of the atomic nucleus corresponds to negative temperatures in the absolute-temperature scale.

Finally, as a quite spectacular discovery I may mention your observation with Dr. Ewen in 1951 of a line in the galactic radiospectrum caused by atomic hydrogen, an important contribution to radioastronomy.

Please accept our congratulations, and receive your Nobel Prize from the hands of His Majesty.

Professor Bloch. It would be difficult in the few minutes at my disposal to try to give the main features of the nuclear induction method for which you have been awarded your Nobel Prize. It would be still more difficult for me to give an exhaustive account of the ways that led you to this invention.

You began your career as a theoretical physicist, well-known for your fundamental contributions to the theory of metals.

When, quite unexpectedly, you went over to experimental research, this must have been, I feel, with deliberation and assurance. For you had in your kitbag a tool of extraordinary value, the method for the magnetic polarization of a beam of neutrons. The inestimable value of possessing a good idea, of indefatigably testing and perfecting it, is best illustrated by your precision-measurements of the magnetic moment of the neutron, one of the most difficult and at the same time most important tasks in nuclear physics.

But ideas give birth to new ideas, and it was, as I understand, in this way that you hit upon the excellent notion of eliminating the difficult absolute determination of the magnetic field by a direct measurement of the neutron moment in units of the proton cycle (the nuclear magneton). According to your own account it was this solution which finally led you to the nuclear induction method.

In congratulating you I now beg you to receive your Nobel Prize from the hands of His Majesty.

1951. Sir John Douglas Cockcroft, Ernest Thomas Sinton Walton "for their pioneer work on the transmutation of atomic nuclei by artificially accelerated atomic particles"

1950. Cecil Frank Powell "for his development of the photographic method of studying nuclear processes and his discoveries regarding mesons made with this method"

1949. Hideki Yukawa "for his prediction of the existence of mesons on the basis of theoretical work on nuclear forces"

1948. Patrick Maynard Stuart Blackett "for his development of the Wilson cloud chamber method, and his discoveries therewith in the fields of nuclear physics and cosmic radiation"

1947. Sir Edward Victor Appleton "for his investigations of the physics of the upper atmosphere especially for the discovery of the so-called Appleton layer"

1946. Percy Williams Bridgman "for the invention of an apparatus to produce extremely high pressures, and for the discoveries he made therewith in the field of high pressure physics"

1945. Wolfgang Pauli "for the discovery of the Exclusion Principle, also called the Pauli Principle"

1944. Isidor Isaac Rabi "for his resonance method for recording the magnetic properties of atomic nuclei"

1943. Otto Stern "for his contribution to the development of the molecular ray method and his discovery of the magnetic moment of the proton"

1942. The prize money was with 1/3 allocated to the Main Fund and with 2/3 to the Special Fund of this prize section.

1941. The prize money was with 1/3 allocated to the Main Fund and with 2/3 to the Special Fund of this prize section.

1940. The prize money was with 1/3 allocated to the Main Fund and with 2/3 to the Special Fund of this prize section.

1939. Ernest Orlando Lawrence "for the invention and development of the cyclotron and for results obtained with it, especially with regard to artificial radioactive elements"

1938. **Enrico Fermi** "for his demonstrations of the existence of new radioactive elements produced by neutron irradiation, and for his related discovery of nuclear reactions brought about by slow neutrons"

1937. Clinton Joseph Davisson, George Paget Thomson "for their experimental discovery of the diffraction of electrons by crystals"

Presented by: Professor H. Pleijel*

The Nobel Prize for Physics for the year 1937 will today be delivered to Dr. C.J. Davisson and Professor G.P. Thomson for their discovery of the interference phenomena arising when crystals are exposed to electronic beams.

The study of the dispersion and diffraction phenomena produced by beams of electrons impinging on crystal surfaces was begun already in 1922 by Davisson and his collaborator Kunsman. These investigations soon obtained special actuality in connection with the theory of mechanical waves pronounced in 1923 by the Nobel Prize winner Prince de Broglie. According to this theory material particles are always linked with a system of travelling

* Chairman of the Nobel Committee for Physics of The Royal Swedish Academy of Sciences.

waves, a «wave-packet», forming the constituent parts of matter and determining its movements. We might get a popular picture of the relation between a material particle and the associated mechanical waves, if we assume space filled with wave systems travelling with somewhat different velocities. In general these waves neutralize one another, but at certain points it happens that a great number of waves are in such a position as to reinforce one another and form a marked wave crest. This wave crest then corresponds to a material particle. Since, however, the waves travel with different velocity they will part from one another, and the wave crest disappears to be found again at a nearby point. The material particle has moved. The wave crest will thus travel, but the velocity with which this is done is quite different from the one with which the underlying wave systems move. The material particle in general moves at right angles to the surfaces of the mechanical waves, just as a ray of light is, as a rule, directed at right angles to the surface planes of the light waves.

The theory of de Broglie derived from analogies between the laws ruling the movement of a material particle and those applying in the case of the passage of a ray of light.

A great number of phenomena observed in optics can neither be explained nor described by the aid of rays of light, and this holds true especially of the diffraction and dispersion phenomena produced when light passes through a narrow slit or by a sharp edge. To explain those phenomena it is necessary to have recourse to the hypothesis of the propagation of light by means of waves.

In recent times, the existence of diffraction and interference phenomena has settled a dispute regarding the nature of a certain radiation. This time the X-rays were concerned. The question was whether these rays consist of particles ejected with great velocity or of electromagnetic waves.

The mechanical grids utilized for studying interference phenomena in optics let through the X-rays without diffraction. This might be due to the wavelength of these rays being so short that the grids became too wide. The Nobel Prize winner von Laue then got the ingenious idea to use as grids, crystals, the regularly arranged atoms of which could serve as diffraction centres. It was also stated that the X-rays in those grids gave rise to diffraction and interference phenomena; the X-rays consequently consisted of waves.

The mechanical waves of de Broglie now correspond to the waves of light and the path of the material particle to the passage of the ray of light.

In his theory de Broglie found a simple relation between the velocity of the material particle and the wavelength of the «wave-packet» associated with this particle. The greater the velocity of the particle the shorter is the wavelength. If the velocity of the particle is known, it is then possible to calculate, by means of the formula indicated by de Broglie, the wavelength and *vice versa*.

The theory of de Broglie of mechanical waves and the development of wave mechanics have been of radical importance to modern atom theory.

It is therefore quite natural that this revolutionary theory should become the object of assiduous research as to its consequences and of efforts to prove experimentally the existence of mechanical waves.

As has already been mentioned, Davisson had, together with his collaborator Kunsman, in the year before the theory of de Broglie was presented, started a series of experiments on the diffraction phenomena produced when a beam of electrons impinges with a certain velocity on the surface of a crystal. These experiments which were continued during the following years, gave, however, at the beginning results rather strange and hard to explain, probably due to the great experimental difficulties connected with the apparatus arrangement. In 1928, however, the investigations met with such a success that Davisson and his collaborator Germer were able to present the incontestable evidence, reached by experiments, of the existence of mechanical waves and of the correctness of the theory of de Broglie. Four months later Professor Thomson, who had been studying the same problem independently of Davisson and by the aid of a different apparatus equipment for his experiments, also confirmed de Broglie's theory.

For their experiments Davisson and Germer availed themselves of a cubic nickel crystal. Here the atoms are symmetrically arranged in planes parallel to the end surfaces of the crystal, the atoms forming a quadratic network in the planes. However, as radiation surface was not used the end plane of the cube but the triangular plane obtained, if an angle of the cube is symmetrically cut off. The atoms in this plane form a triangular network.

A minute bundle of electrons of determined velocity were emitted perpendicularly upon this plane. If we assume the incoming electrons replaced by mechanical waves, the planes of which are thus parallel to the surface of the crystal, these mechanical waves will strike the atoms lying in the surface simultaneously, and these atoms as centres will, in their turn, emit new mechanical waves in all directions. The waves going out in a certain direction can be studied and measured by the aid of a so-called

Faraday chamber placed in this direction. In this chamber the mechanical waves cause the same effect as the corresponding electrons. In order to describe better how the outgoing radiation arises, let us suppose the receiving device placed so as to capture the waves going out parallel to the crystal plane and at right angles to one of the sides of the triangle. Parallel to this side the atoms lie in parallel rows with a certain distance between the rows, this distance having been determined beforehand by the aid of X-ray investigations. Every row now emits its wave. But the waves from the inner rows arrive later, due to the longer way they have to pass to reach the edge of the triangle. As a rule an irregular system of waves is thus obtained in which the waves neutralize each other, and consequently no outgoing wave is produced. If on the other hand the mechanical waves should be of such a wavelength that the distance between the rows of atoms becomes equal to the wavelength or to a multiple thereof, all the outgoing waves will be in phase and reinforce one another. In this case a wave system going out in the direction indicated is obtained or, if preferable, a bundle of outgoing electronic beams.

The experiments now showed at what velocities of the incoming electrons outgoing beams are produced, and these have, according to what has been stated above, a wavelength equal to the distance between the rows of atoms. Since thus the wavelength of the mechanical waves had been found and since the velocity of the corresponding electron was known, it was possible to check the formula of de Broglie. Davisson found that the theory agreed with the experiments except for 1 to 2%. Davisson and Germer examined the reflection of the electronic beams in various directions and obtained results which agreed with the wave theory.

During his experiments Davisson used electron beams with rather a low velocity corresponding to the one obtained when an electron is made to pass a voltage between 50 and 600 volts.

Thomson, on the other hand, for his experiments availed himself of swift electrons with a velocity corresponding to voltages between 10,000 and 80,000 volts. These swift electrons have afterwards proved to be of great use in connection with studies on the structure of matter.

For his experiments Thomson made use of exceedingly thin films of celluloid, gold, platinum, or aluminium. He made the electron beam fall perpendicularly upon the film and examined the diffraction figures produced on a fluorescent screen placed behind the film, or else had them reproduced on a photographic plate. The thickness of the films used for the experiments amounted to between 1/10,000 and 1/100,000 of a millimetre. Such a film now consists of innumerable small crystals of various directions. In accordance with what the theory indicates, there is generally obtained on the screen a series of concentric rings corresponding to the various directions of the planes in a crystal where a regularly arranged network of atoms can be found. From the diametre of a ring, the wavelength of the mechanical wave can be determined, and to make possible the production of a ring this wavelength must be in accordance with the spacing of the planes in the system of planes to which the ring corresponds. A similar method has been applied previously by Debye-Scherrer for X-rays analysis of the structure of crystals. Thomson found very good agreement with the theory of de Broglie. He further found that a magnetic field influencing the beams having passed the film produced a lateral movement of the image on the screen, which shows that these beams consist of bundles of electrons.

For the above-mentioned experiments electrons have been employed as matter; later investigations have confirmed the correctness of de Broglie's theory also for such cases where beams of molecules, atoms, and atom nuclei have been used.

The purpose of the said experiments was to verify the theory of de Broglie, and to this end was utilized the knowledge of the arrangement of the atoms in a crystal, this knowledge having been previously acquired as a result of investigations by means of X-rays. Now that the law of de Broglie has become known and acknowledged, the opposite way has been taken. From the law of de Broglie we know the wavelength of the mechanical waves accompanying an electronic beam with a certain velocity of the electrons. By changing this velocity we can then obtain electronic waves with known wavelengths. By application of one or the other of the investigation methods mentioned above we can find the distances between the various atom planes within the crystal and thus also the structure of the crystal. The procedure is here the same as the one previously applied to determine the structure of crystals by means of X-rays. We have thus obtained a new method for such investigation, but the two methods have found very different fields of application due to the different nature of the beams employed. The X-rays are pure electromagnetic rays like the rays of light, and they therefore influence but slightly the atoms of the crystal, and owing to this circumstance easily traverse the crystal structure. From the same reason the diffracted rays are comparatively feeble, and many hours' exposure is therefore required to record X-ray diagrams. The mechanical waves, on the other hand, are associated with electrical charges which are very strongly influenced by the charges of the crystal atoms. The mechanical waves will therefore be rapidly absorbed in the crystal, and the interference figures obtained only come from an exceedingly thin surface layer. In return the intensity of the diffracted or reflected bundles of electrons becomes very great, and the time of exposure required is consequently extremely short, in many cases only a fraction of a second. These properties of the electronic beams make them an exceedingly important complement to the X-rays as far as researches on the structure of matter are concerned. At the important investigations of the structure of surfaces good results can be attained only by the new method, since the images of the X-rays are influenced by the matter lying behind the surface layer. By the aid of electronic beams it has thus been possible to explain how the structure of the surfaces of metals is changed by various mechanical, thermal, or chemical treatment. It has also been possible to ascertain the properties of thin layers of gases and powder. On account of the rapid exposure which the electronic beams permit, we can follow the course of the changes occurring in connection with the oxidization of metals and also observe the corrosion phenomenon in iron and steel for various thermal treatment as well as the chemical process ensuing when metals are attacked by corrosive substances. The intensity of radiation is so great that one can easily carry out investigations of the structure of crystals with a mass of less than a millionth of a gram. This has made it possible to discover in certain substances exceedingly minute crystalline structures, which it would not have been possible to find by means of X-ray investigations.

It would bring us too far here to enter upon the multitude of experimental results furnished by the method with electronic beams, especially as new fields of application of the electron beam are incessantly being opened up within the spheres of physical and chemical research.

Dr. Davisson. When you found that electron beams touching crystals give rise to phenomena of diffraction and interference, this signified in itself a discovery that widened essentially our knowledge of the nature of electrons. But this discovery has proved to be of still greater importance. Your researches concerning these phenomena resulted in your presenting the first positive, experimental evidence of the wave nature of matter. The investigation methods that you and Professor Thomson have elaborated and the further research work carried out by both of you have provided science with a new, exceedingly important instrument for examining the structure of matter, an instrument constituting a very valuable complement to the earlier method which makes use of the X-ray radiation. The new investigations have already furnished manifold new, significant results within the fields of physics and chemistry and of the practical application of these sciences.

On behalf of the Royal Swedish Academy of Sciences I congratulate you on your important discoveries, and I now ask you to receive your Nobel Prize from the hands of His Majesty.

The Royal Swedish Academy of Sciences much regrets that Professor Thomson has not had the opportunity of being present on this occasion to receive in person his Nobel Prize. The prize will now instead be delivered to His Excellency the Minister of Great Britain.

Your Excellency. Permit me to request you to receive on behalf of Professor Thomson the Nobel Prize for Physics from the hands of His Majesty.

1936. Victor Franz Hess "for his discovery of cosmic radiation" and **Carl David Anderson** "for his discovery of the positron"

1935. James Chadwick "for the discovery of the neutron"

1934. The prize money was with 1/3 allocated to the Main Fund and with 2/3 to the Special Fund of this prize section

1933. Erwin Schrödinger, Paul Adrien Maurice Dirac "for the discovery of new productive forms of atomic theory"

1932. Werner Karl Heisenberg "for the creation of quantum mechanics, the application of which has, inter alia, led to the discovery of the allotropic forms of hydrogen"

1931. The prize money was allocated to the Special Fund of this prize section

1930. Sir Chandrasekhara Venkata Raman "for his work on the scattering of light and for the discovery of the effect named after him"

Presented by: Professor H. Pleijel*

The Academy of Sciences, has resolved to award the Nobel Prize in Physics for 1930 to Sir Venkata Raman for his work on the scattering of light and for the discovery of the effect named after him.

The diffusion of light is an optical phenomenon, which has been known for a long time. A ray of light is not perceptible unless it strikes the eye directly. If, however, a bundle of rays of light traverses a medium in which extremely fine dust is present, the ray of light will scatter to the sides and the path of the ray through the medium will be discernible from the side. We can represent the course of events in this way; the small particles of dust begin to oscillate owing to electric influence from the ray of light, and they form centres from which light is disseminated in all directions. The wavelength, or the number of oscillations per second, in the light thus diffused is here the same as in the original ray of light. But this effect has different degrees of strength for light with different wavelengths. It is stronger for the short wavelengths than for the long ones, and consequently it is stronger for the blue part of the spectrum than for the red part. Hence if a ray of light containing all the colours of the spectrum passes through a medium, the yellow and the red rays will pass through the medium without appreciable scattering, whereas the blue rays will be scattered to the sides. This effect has received the name of the "Tyndall effect".

Lord Rayleigh, who has made a study of this effect, has put forward the hypothesis that the blue colours of the sky and the reddish colouring that is observed at sunrise and sunset is caused by the diffusion of light owing to the fine dust or the particles of water in the atmosphere. The blue light from the sky would thus be light-scattered to the sides, while the reddish light would be light that passes through the lower layers of the atmosphere and which has become impoverished in blue rays owing to scattering. Later, in 1899, Rayleigh threw out the suggestion that the phenomenon in question might be due to the fact that the molecules of air themselves exercised a scattering effect on the rays of light.

In 1914 Cabannes succeeded in showing experimentally that pure and dustless gases also have the capacity of scattering rays of light.

---

* Chairman of the Nobel Committee for Physics of The Royal Swedish Academy of Sciences.

But a closer examination of scattering in different substances in solid, liquid, or gaseous form showed that the scattered light did not in certain respects exactly follow the laws which, according to calculation, should hold good for the Tyndall effect. The hypothesis which formed the basis of this effect would seem to involve, amongst other things, that the rays scattered to the sides were polarized. This, however, did not prove to be exactly the case.

This divergence from what was to be expected was made the starting point of a searching study of the nature of scattered light, in which study Raman was one of those who took an active part. Raman sought to find the explanation of the anomalies in asymmetry observed in the molecules. During these studies of his in the phenomenon of scattering, Raman made, in 1928, the unexpected and highly surprising discovery that the scattered light showed not only the radiation that derived from the primary light but also a radiation that contained other wavelengths, which were foreign to the primary light.

In order to study more closely the properties of the new rays, the primary light that was emitted from a powerful mercury lamp was filtered in such a way as to yield a primary light of one single wavelength. The light scattered from that ray in a medium was watched in a spectrograph, in which every wavelength or frequency produces a line. Here he found that, in addition to the mercury line chosen, there was obtained a spectrum of new sharp lines, which appeared in the spectrograph on either side of the original line. When another mercury line was employed, the same extra spectrum showed itself round it. Thus, when the primary light was moved, the new spectrum followed, in such a way that the frequency distance between the primary line and the new lines always remained the same.

Raman investigated the universal character of the phenomenon by using a large number of substances as a scattering medium, and everywhere found the same effect.

The explanation of this phenomenon, which has received the name of the "Raman effect" after its discoverer, has been found by Raman himself, with the help of the modern conception of the nature of light. According to that conception, light cannot be emitted from or absorbed by material otherwise than in the form of definite amounts of energy or what are known as "light quanta". Thus the energy of light would possess a kind of atomic character. A quantum of light is proportionate to the frequency of rays of light, so that in the case of a frequency twice as great, the quanta of the rays of light will also be twice as great.

In order to illustrate the conditions when an atom emits or absorbs light energy, we can, according to Bohr, picture to ourselves the atom as consisting of a nucleus, charged with positive electricity round which negative electrons rotate in circular paths at various distances from the centre. The path of every such electron possesses a certain energy, which is different for different distances from the central body.

Only certain paths are stable. When the electron moves in such a path, no energy is emitted. When, on the other hand, an electron falls from a path with higher energy to one with lower energy - that is to say, from an outer path to an inner path - light is emitted with a frequency that is characteristic of these two paths, and the energy of radiation consists of a quantum of light. Thus the atom can give rise to as many frequencies as the number of different transitions between the stable paths. There is a line in the spectrum corresponding to each frequency.

An incoming radiation cannot be absorbed by the atom unless its light quantum is identical with one of the light quanta that the atom can emit.

Now the Raman effect seems to conflict with this law. The positions of the Raman-lines in the spectrum do not correspond, in point of fact, with the frequencies of the atom itself, and they move with the activating ray. Raman has explained this apparent contradiction and the coming into existence of the lines by the effect of combination between the quantum of light coming from without and the quanta of light that are released or bound in the atom. If the atom, at the same time as it receives from without a quantum of light, emits a quantum of light of a different magnitude, and if the difference between these two quanta is identical with the quantum of light which is bound or released when an electron passes from one path to another, the quantum of light coming from without is absorbed. In that case the atom will emit an extra frequency, which either will be the sum of or the difference between the activating ray and a frequency in the atom itself. In this case these new lines group themselves round the incoming primary frequency on either side of it, and the distance between the activating frequency and the nearest Raman-lines will be identical with the lowest oscillation frequencies of the atom or with its ultrared spectrum.What has been said as to the atom and its oscillations also holds good of the molecule.

In this way we get the ultrared spectrum moved up to the spectral line of the activating light. The discovery of the Raman-line has proved to be of extraordinarily great importance for our knowledge of the structure of molecules.

So far, indeed, there have been all but insuperable difficulties in the way of studying these ultrared oscillations, because that part of the spectrum lies so far away from the region where the photographic plate is sensitive. Raman's discovery has now overcome these difficulties, and the way has been opened for the investigation of the oscillations of the nucleus of the molecules. We choose the primary ray within that range of frequency where the photographic plate is sensitive. The ultrared spectrum, in the form of the Raman-lines, is moved up to that region and, in consequence of that, exact measurements of its lines can be effected.

In the same way the ultraviolet spectrum can be investigated with the help of the Raman effect. Thus we have obtained a simple and exact method for the investigation of the entire sphere of oscillation of the molecules.

Raman himself and his fellow-workers have, during the years that have elapsed since the discovery was made, investigated the frequencies in a large number of substances in a solid, liquid, and gaseous state. Investigations have been made as to whether different conditions of aggregation affect atoms and molecules, and the molecular conditions in electrolytic dissociation and the ultrared absorption spectrum of crystals have been studied.

Thus the Raman effect has already yielded important results concerning the chemical constitution of substances; and it is to foresee that the extremely valuable tool that the Raman effect has placed in our hands will in the immediate future bring with it a deepening of our knowledge of the structure of matter.

Sir Venkata Raman. The Royal Academy of Sciences has awarded you the Nobel Prize in Physics for your eminent researches on the diffusion of gases and for your discovery of the effect that bears your name. The Raman effect has opened new routes to our knowledge of the structure of matter and has already given most important results.

I now ask you to receive the prize from the hands of His Majesty.

1929. Prince Louis-Victor Pierre Raymond de Broglie "for his discovery of the wave nature of electrons"

1928. Owen Willans Richardson "for his work on the thermionic phenomenon and especially for the discovery of the law named after him"

Presented by: Professor C.W. Oseen*

Among the great problems that scientists conducting research in electrotechnique are today trying to solve, is that of enabling two men to converse in whatever part of the world each may be. In 1928 things had reached the stage when we could begin to establish telephonic communication between Sweden and North America. On that occasion there was a telephone line of more than 22,000 kilometres in length between Stockholm and New York. From Stockholm, speech was transmitted via Berlin to England by means of a cable and overhead lines; from England by means of wireless to New York; then, via a cable and lines by land, over to Los Angeles and back to New York, and from there by means of a new line to Chicago, returning finally to New York. In spite of the great distance, the words could be heard distinctly and this is explained by the fact that there were no fewer than 166 amplifiers alone the line. The principle of construction of an amplifier is very simple. A glowing filament sends out a stream of electrons. When the speech waves reach the amplifier, they oscillate in tune with the sound waves but are weakened. The speech waves are now made to put the stream of electrons in the same state of oscillation as they have themselves. So exactly does the stream of electrons adapt itself to the speech waves that the amplification could be repeated 166 times without the distinctness of speech being lost.

I should like to give another example of what has recently been attained in that department. On the 16th of February 1928, there was a conference between the American Institute of Electrical Engineers in New York and the Institution of Electrical Engineers in London. The various speeches could be heard in both places by means of loud-speakers.

Most people here present will certainly be able to call to mind those anxious days, when news of the missing Nobile expedition was awaited all over the world. Everyone will no doubt remember that the first word of the

---

* Chairman of the Nobel Committee for Physics of The Royal Swedish Academy of Sciences.

lost expedition was picked up by a wireless amateur. I think that on this occasion it was clear to many people that wireless is not only a means of diversion and as such, one of the more prominent - but also one of the most valuable expedients in the struggle against that sort of Nature which is still unconquered.

Every owner of a valve receiving-set knows the importance of the valve in the apparatus - the valve, the essential part of which is the glowing filament.

At the Jubilee, held in the twenty-fifth year of the reign of King Oscar II, our medical men were enabled to take up the struggle against the tuberculosis, thanks to the Jubilee Fund. At the Jubilee held on Your Majesty's 70th birthday, the fight against cancer was taken up in the same manner. We all know that Röntgen rays are one of the keenest weapons employed in this struggle. But we know, too, that this weapon is double-edged. The rays cannot only do good but also do harm. All depends on the accurate regulation of their strength and intensity. Quite recently, a change has taken place in this department. Röntgen rays are obtained when rapidly moving electrons collide with a solid body. By using a glowing filament in order to produce the electron stream, the means of regulating accurately the strength and intensity of Röntgen rays has been obtained.

Behind the progress which has here been briefly pointed out, lies the work of many men. But we have seen that they all have one thing in common. A "red thread" connects them - the glowing filament.

As early as 1737, a French scientist, Du Fay by name, found out that air in proximity to a glowing body is a conductor of electricity. Valuable researches concerning the character of this conductivity was made by Elster and Geitel, two German scientists. Their investigations were continued by Mr. J.J. Thomson, the Grand Old Man of English Physics of today. By these researches they have found it probable that the conductivity of air in proximity to a glowing metal depends on electrons in the air, which have been made free in some way or another. So far had the researches advanced when Mr. O. W. Richardson appeared and devoted himself to it. He began by laying down a theory for the phenomenon. According to this theory the phenomenon is bound up with the electrical conductivity of metals. The latter depends on the fact that there are free electrons in a metal. At higher temperatures these cannot, according to Mr. Richardson, be retained by the body but they are emitted according to a fixed law. But a theory alone does not give any knowledge of reality. That can be obtained only by means of experimental research. So Mr. Richardson proceeded to do this. The point was to find out if the theory was really right. The strenuous work of twelve years was necessary to settle this question. So hard was the struggle that even so late as in the twelfth year, there was a time when it was uncertain whether Mr. Richardson's theory was not completely wrong, and if the origin of the phenomenon was not quite different, being, for instance, chemical reactions between the metal and impurities in it. But in the end, Mr. Richardson's theory proved to be correct in all essential points. The most important fact was that Mr. Richardson's opinion about the thermion-phenomenon with fixed laws was totally confirmed. Through this fact a solid basis was obtained for the practical application of the phenomenon. Mr. Richardson's work has been the starting-point and the prop of the technical activity which has]ed to the progress of which I have just spoken.

Professor Richardson. You are a happy man. You possess the very thing that gives life its chief value. You can devote yourself with all your strength to the activity that you love. We constantly see the results of this activity come to light. Besides this, you are fortunate enough to see the harvest ripen to the benefit of mankind in the fields you tilled in your youth. For one who is so rich it is but a little thing to receive the greatest prize which the Royal Academy of Sciences has at its disposal as a reward for a scientific discovery. I ask you, however, to receive from our King's hand the Nobel Prize for Physics for the year 1928.

1927. Arthur Holly Compton "for his discovery of the effect named after him" and Charles Thomson Rees Wilson "for his method of making the paths of electrically charged particles visible by condensation of vapour"

1926. Jean Baptiste Perrin "for his work on the discontinuous structure of matter, and especially for his discovery of sedimentation equilibrium"

1925. James Franck, Gustav Ludwig Hertz "for their discovery of the laws governing the impact of an electron upon an atom"

1924. Karl Manne Georg Siegbahn *"for his discoveries and research in the field of X-ray spectroscopy"*

1923. Robert Andrews Millikan "for his work on the elementary charge of electricity and on the photoelectric effect"

1922. Niels Henrik David Bohr "for his services in the investigation of the structure of atoms and of the radiation emanating from them"

1921. Albert Einstein "for his services to Theoretical Physics, and especially for his discovery of the law of the photoelectric effect"

Presented by: Professor S. Arrhenius*

There is probably no physicist living today whose name has become so widely known as that of Albert Einstein. Most discussion centres on his theory of relativity. This pertains essentially to epistemology and has therefore been the subject of lively debate in philosophical circles. It will be no secret that the famous philosopher Bergson in Paris has challenged this theory, while other philosophers have acclaimed it wholeheartedly. The theory in question also has astrophysical implications which are being rigorously examined at the present time.

---

* Chairman of the Nobel Committee for Physics of The Royal Swedish Academy of Sciences.

Throughout the first decade of this century the so-called Brownian movement stimulated the keenest interest. In 1905 Einstein founded a kinetic theory to account for this movement by means of which he derived the chief properties of suspensions, i.e. liquids with solid particles suspended in them. This theory, based on classical mechanics, helps to explain the behaviour of what are known as colloidal solutions, a behaviour which has been studied by Svedberg, Perrin, Zsigmondy and countless other scientists within the context of what has grown into a large branch of science, colloid chemistry.

A third group of studies, for which in particular Einstein has received the Nobel Prize, falls within the domain of the quantum theory founded by Planck in 1900. This theory asserts that radiant energy consists of individual particles, termed "quanta", approximately in the same way as matter is made up of particles, i.e. atoms. This remarkable theory, for which Planck received the Nobel Prize for Physics in 1918, suffered from a variety of drawbacks and about the middle of the first decade of this century it reached a kind of impasse. Then Einstein came forward with his work on specific heat and the photoelectric effect. This latter had been discovered by the famous physicist Hertz in 1887. He found that an electrical spark passing between two spheres does so more readily if its path is illuminated with the light from another electrical discharge. A more exhaustive study of this interesting phenomenon was carried out by Hallwachs who showed that under certain conditions a negatively charged body, e.g. a metal plate, illuminated with light of a particular colour - ultraviolet has the strongest effect - loses its negative charge and ultimately assumes a positive charge. In 1899 Lenard demonstrated the cause to be the emission of electrons at a certain velocity from the negatively charged body. The most extraordinary aspect of this effect was that the electron emission velocity is independent of the intensity of the illuminating light, which is proportional only to the number of electrons, whereas the velocity increases with the frequency of the light. Lenard stressed that this phenomenon was not in good agreement with the then prevailing concepts.

An associated phenomenon is photo-luminescence, i.e.phosphorescence and fluorescence. When light impinges on a substance the latter will occasionally become luminous as a result of phosphorescence or fluorescence. Since the energy of the light quantum increases with the frequency, it will be obvious that a light quantum with a certain frequency can only give rise to the formation of a light quantum of lower or, at most, equal frequency. Otherwise energy would be created. The phosphorescent or fluorescent light hence has a lower frequency than the light inducing the photo-luminescence. This is Stokes' rule which was explained in this way by Einstein by means of the quantum theory.

Similarly, when a quantum of light falls on a metal plate it can at most yield the whole of its energy to an electron there. A part of this energy is consumed in carrying the electron out into the air, the remainder stays with the electron as kinetic energy. This applies to an electron in the surface layer of the metal. From this can be calculated the positive potential to which the metal can be charged by irradiation. Only if the quantum contains sufficient energy for the electron to perform the work of detaching itself from the metal does the electron move out into the air. Consequently, only light having a frequency greater than a certain limit is capable of inducing a photo-electric effect, however high the intensity of the irradiating light. If this limit is exceeded the effect is proportional to the light intensity at constant frequency. Similar behaviour occurs in the ionisation of gas molecules and the so-called ionisation potential may be calculated, provided that the frequency of the light capable of ionising the gas is known.

Einstein's law of the photo-electrical effect has been extremely rigorously tested by the American Millikan and his pupils and passed the test brilliantly. Owing to these studies by Einstein the quantum theory has been perfected to a high degree and an extensive literature grew up in this field whereby the extraordinary value of this theory was proved. Einstein's law has become the basis of quantitative photo-chemistry in the same way as Faraday's law is the basis of electro-chemistry.

1920. Charles-Edouard Guillaume "in recognition of the service he has rendered to precision measurements in Physics by his discovery of anomalies in nickel steel alloys"

1919. Johannes Stark "for his discovery of the Doppler effect in canal rays and the splitting of spectral lines in electric fields"

1918. Max Karl Ernst Ludwig Planck "in recognition of the services he rendered to the advancement of Physics by his discovery of energy quanta"

1917. Charles Glover Barkla *"for his discovery of the characteristic Röntgen radiation of the elements"*

1916. The prize money was allocated to the Special Fund of this prize section

1915. Sir William Henry Bragg, William Lawrence Bragg "for their services in the analysis of crystal structure by means of X-rays"

Presented by: Professor G. Granqvist*

Von Laue's epoch-making discovery of the diffraction of the X-rays in crystals, on the one hand established wave motion as the essential quality of those rays and, on the other, afforded the experimental proof of the existence of molecular gratings in the crystals. The problem, however, of calculating the crystal structures from von Laue's formulae was an exceedingly complicated one, in as much as not only the space lattices, but also the wavelengths and the intensity-distribution over the various wavelengths in the spectra of the X-rays, were unknown quantities. It was consequently a discovery of epoch-making significance when W.L. Bragg found out that the phenomenon could be treated mathematically as a reflection by the successive parallel planes that may be placed so as to pass through the lattice points, and that in this way the ratio between the wavelengths and the

---

* Chairman of the Nobel Committee for Physics of The Royal Swedish Academy of Sciences.

distances of the said planes from each other can be calculated by a simple formula from the angle of reflection.

It was only by means of that simplification of the mathematical method that it became possible to attack the problem of the crystal structures, but to attain the end in view it was further necessary that the photographic method employed by von Laue should be replaced by an experimental one, based on the reflection principle, which admitted of a definite, even though at first unknown, wavelength being made use of. The instrument requisite for the said purpose, the so-called X-ray spectrometer, was constructed by Professor W.H. Bragg, W.L. Bragg's father, and it has been with the aid of that instrument that father and son have carried out, in part conjointly, in part each on his own account, a series of extremely important investigations respecting the structure of crystals.

If a number of cubes are laid on and beside each other in such a way that one cube face coincides in every case with the face of an adjoining cube, whereby consequently eight vertices always meet in one point, those angular points give a visual picture of the lattice points in the so-called simple cubic lattice. If again a lattice point is placed so as to coincide with the central point of each cube face, the so-called face-centred cubic lattice is obtained, whereas the centred cubic lattice has one lattice point in every cube-centre. With the exception of these three cases there is no cubic lattice that fulfils the condition that parallel planes placed in any direction whatever so as to pass through all the lattice points, shall also be at a constant distance from each other. The space lattice in the regular or cubic system must therefore coincide with one of those three, or constitute combinations of them. In such lattice combinations, on the other hand, in which the condition just mentioned is not fulfilled, where consequently parallel planes placed to pass through all the lattice points in certain directions are not equidistant, that circumstance is revealed by an abnormal intensity distribution among spectra of different orders, when the reflection takes place by those planes.

From crystallographical data it is always known how the face of a cube is situated in any given regular crystal, and there is consequently no difficulty in fixing the crystal on the spectrometer table in such a way that the reflection shall take place by planes with any prescribed orientation.

The rays falling on the crystal were produced by X-ray tubes, platinum being at first used for the anticathode. The characteristic X-radiation of the metals consists, as is well known, of a few strong lines or narrow bands, and the very first experiments with the spectrometer revealed the X-radiation that is characteristic of platinum. However, in the research undertaken to find out the nature of complicated space lattices, in which an abnormal intensity distribution among spectra of varying orders constitutes one of the most important of the results observed, it soon proved desirable to have available an X-radiation of approximately half the wavelength of the strongest platinum-line. From theoretical considerations W.H. Bragg regarded it as probable that a metal whose atomic weight was somewhere near the figure 100, would give a characteristic radiation of the desired wavelength. Accordingly anticathodes of palladium and rhodium were produced, which fully answered the purpose in view, so that spectra ev en of the fifth order could be obtained and measured. In order to take practical advantage, however, of those results, it was essential to have a method for calculating the intensity in the case of a complicated space lattice, that would prove simpler than the one given by von Laue's theory, and W.L. Bragg developed one.

The above is a brief sketch of the methods discovered by the two Braggs for investigating crystal structures. The results of their investigations embrace a large number of crystals belonging to various systems and can only be cursorily summarized in this place.

To begin with, the two investigators applied themselves to the simplest types of the regular system, represented by the alkaline haloid salts. It then proved that potassium bromide and potassium iodide showed the spectra that are characteristic of a face-centred cubic lattice, while the spectra of potassium chloride represented a simple cubic lattice, sodium chloride occupying an intermediate position. As it must be assumed, on the strength of the analogy of these salts, both in a chemical and a crystallographical sense, that they are possessed of a corresponding space lattice, which could also be corroborated in another way, it was proved by those researchers that the lattice of the crystals in question consists of two face-centred cubic lattices corresponding to the two atoms, which interpenetrate in such a way that they together constitute one single cubic lattice.

From these investigations it follows that a metal atom in the crystals of the alkaloid salts is situated at one and the same distance from the six haloid atoms nearest to it, and vice versa - a relationship that was found to prevail, mutatis mutandis, in all the crystals examined. That means the exceedingly important discovery, both for molecular physics and chemistry, that the crystals consist of atomic lattices and not, as has been always imagined, of molecular ones.

Two face-centred cubic lattices can also interpenetrate in such a way that every point belonging to the one lattice is at the centre of gravity of a tetrahedron whose vertices are points belonging to the other lattice. That structure was found by the two Braggs in the diamond, and afforded an experimental support for the tetrahedral arrangement that chemists postulate for the four-coordinate carbon. On the other hand, the explanation became evident of why crystallographers have not been able to agree regarding the class in the regular system to which the diamond should be referred.

It would carry us too far and be quite too complicated a proceeding to give an account here of the further investigations into the space lattices of the crystals. It will suffice to add that, in the course of their investigations, the two Braggs have also discovered important relations between the amplitude and the phase difference of the diffracted rays on the one hand and the atomic weights on the other, and have also shown experimentally the influence of heat on the space lattice.

Finally it may be mentioned that the two investigators have also determined the wavelengths of the X-rays and the distances between the successive planes placed to pass through the lattice points with such exactitude, that the error, if any, is probably a t most some few units per cent and is more due to the general physical constant entering into the calculations than to the measurements themselves.

Thanks to the methods that the Braggs, father and son, have devised for investigating crystal structures, an entirely new world has been opened and has already in part been explored with marvellous exactitude. The significance of these methods, and of the results attained by their means, cannot as yet be gauged in its entirety, however imposing its dimensions already appear to be. In consideration of the great importance that these methods possess for research in the realm of physics, the Swedish Royal Academy of Sciences decided that the 1915 Nobel Prize in Physics should be divided between Professor W.H. Bragg and his son W.L. Bragg, in recognition of their services in promoting the investigation of crystal structures by means of X-rays.

1914. Max von Laue "for his discovery of the diffraction of X-rays by crystals"

Presented by: Professor G. Granqvist*

Seldom indeed can a discovery in the field of physics have given rise to such intensive research work as did that of Röntgen in 1896, when he proved the existence of a new form of rays which had hitherto been unknown and which, owing to their remarkable characteristics, have since achieved a position of the greatest importance, not only in the field of pure physics but also in connection with research work throughout the other sciences.

Notwithstanding the considerable number of tests which have been carried out since their discovery and directed toward investigation of the true nature of X-rays, it was not until over a decade had passed that their true nature had finally been elucidated .

Already during the first tests it was established that not even the strongest magnetic fields were able to alter the direction of the rays. It was equally impossible to prove the existence of a refraction on transfer of the rays from one medium to another. If the X-rays were of a corpuscular nature they could not, therefore, be carriers of an electrical charge, as is the case with other known rays of corpuscular nature. If, therefore, we wish to disregard matter which has no electrical charge, it is necessary to assume that the particles, whose motion is characteristic for the X-rays, bear two charges of opposite sign, one of which neutralizes the other. On the other hand, from the fact that there was no evidence of refraction of the X-rays, it was possible to assume that, should they consist of a transverse wave motion - as is the case with light waves - the relevant wavelength would have to be very small, as for very small wavelengths, according to the theory of dispersion of light, the refractive index would approach unity.

After hurriedly discarding an hypothesis which had been expounded initially, according to which X-rays were believed to consist of longitudinal wave motions in ether, opinions as to their actual nature were divided according to the above two alternatives. Nevertheless an objective presentation could only describe them as a type of impulse of an unknown nature.

On the basis of an hypothesis expounded as early as 1896 by Stokes and Wiechert this impulse was believed to consist of a disturbance which occurs

---

* Chairman of the Nobel Committee for Physics of The Royal Swedish Academy of Sciences.

in the ether when the cathode-ray particle, i.e. a forward-rushing electron, is impeded on colliding with molecules of matter. This disturbance or impulse was believed to propagate in all directions at the speed of light from the ether surrounding the electron. In each part of the space this disturbance was maintained for a period of identical duration to that in which the electron was impeded. This period of time, multiplied by the speed of light, was described as the impulse width, a quantity which, if the nature of the X-rays were the same as that of the light rays, would coincide with the wavelength.

According to that theory the X-ray impulse, which originates perpendicular to the cathode-ray bundle by which it is excited, is alleged to be completely polarized. The evidence of this type of polarization was first produced by Barkla in 1905, but, contrary to the theory, the polarization was not complete but only partial. While it was possible to explain the causative factors of this aberration the characteristics of the polarization were not adequate to prove the existence of a transverse undulation.

Once Dorn had succeeded, in 1897, in determining the fraction of the energy of the impeded electrons which is converted to X-rays, W. Wien was able to calculate the impulse width which, according to his figures, amounted to approximately 10^{-10} cm, or only one hundred-thousandth of the shortest known wavelengths of light. The short impulse width thus determined could explain the lack of success with previous diffraction tests which had been carried out on slits with X-rays, for even with the narrowest slit the diffraction phenomenon, which is produced by such small impulse widths or wavelengths, would have to lie just about at the limits of possible observation. And it may, in actual fact, only be said even of the most accurate of these tests conducted by Walter and Pohl that they render diffraction highly probable. From the research carried out by these scientists it would meanwhile seem to follow that the upper limit for the impulse width of X-rays lies at 4×10^{-9}.

This was the situation when von Laue placed a research medium of the highest import at the disposal of science by virtue of his epoch-making discovery of the interference of X-rays and, at the same time, proved that X-rays, as is the case with light rays, consist of progressive transversal waves.

Previous research had indicated, as is mentioned in the foregoing, that it was highly probable that, if X-rays are wave motions of the same type as light rays, then their wavelengths would have to be of an order of 10^{-9} cm. In order to obtain clear interference phenomena of the same type as those which are caused when light rays pass a grating it was necessary for the distance between the grating slits to be of an order of 10^{-8} cm. But this is approximately the distance between the molecules of a solid body and it was in this manner that von Laue arrived at the idea of employing, as a diffraction grating, a solid body with regularly-arranged molecules, e.g. a crystal. As early as 1850 Bravais had introduced into crystallography the assumption that the atoms composing the various crystals are arranged in regular groups, so-called three-dimensional lattices or space-lattices, whose constants could be calculated with the aid of crystallographic data.

However, the theoretical basis of a space-lattice was unknown and thus it was first necessary for von Laue to develop this theory if else the investigation were to have a value. This he did mainly according to the same approximations as those conventional to the science of optics as applied to normal one-dimensional lattices.

Von Laue left the execution of the experimental work in the hands of W. Friedrich and P. Knipping. The apparatus which they employed consisted of a lead box into which they admitted a thin bundle of X-rays which they directed so as to fall upon a precisely oriented crystal. Sensitized film was positioned both behind and at the sides of the crystal. Already the preparatory tests showed that the intensity maxima which had been anticipated by von Laue became evident in the form of blackened spots on the film positioned behind the crystal.

From the grouping shown by these intensity maxima in accordance with the requirements of the theory, as established, for such photograms of various crystals and from the degree of clarity with which they have been reproduced, it follows that they are an interference phenomenon. Absorption tests have shown that the rays which give rise to the points of interference are actually X-rays, and from this von Laue has deduced with a high degree of certainty that the X-rays which cause intensity maxima on irradiation of a crystal have the character of a wave motion. However, the same is required also for those rays employed for irradiation purposes, for, as he says, were they of a corpuscular nature, coherent oscillations could only arise from those atoms set into motion by the identical corpuscle and these atoms would have to form together one whole agglomerate whose dimensions would tee largest in the direction of radiation. However, contrary to what was indicated in the experiment, this would result in the intensity maxima consisting of irregular concentric circles.

As a result of von Laue's discovery of the diffraction of X-rays in crystals proof was thus established that these light waves are of very small wavelengths. However, this discovery also resulted in the most important discoveries in the field of crystallography. It is now possible to determine the position of atoms in crystals and much important knowledge has been gained in this connection. We can anticipate further discoveries of equal note in the future. It is thus rendered likely that experimental research into the influence of temperature upon diffraction will provide the solution to the question of a zero-point energy, or will at least be of some assistance in arriving at a solution to this problem, as the temperature factor assumes a different value according to whether a zero-point energy exists or not. However, the direct results of this discovery of diffraction are of no less importance: it is now possible to subject the X-ray spectra to direct examination, their line spectra can even be photographed, and science has thus been enriched by a method of research whose full implications can not yet be fully appreciated.

If it is permissible to evaluate a human discovery according to the fruits which it bears then there are not many discoveries ranking on a par with that made by von Laue. If one reflects further on the fact that only a few years have passed since his discovery was first published it may surely be said that, when awarding the Nobel Prize for Physics, the Royal Academy of Sciences will presumably seldom, if ever, be in a position of such close agreement with the letter of the Testament as on this occasion in deciding to award the Nobel Prize for Physics for the year 1914 to Professor Max von Laue, for his discovery of the diffraction of X-rays in crystals.

1913. Heike Kamerlingh Onnes "for his investigations on the properties of matter at low temperatures which led, inter alia, to the production of liquid helium"

Presented by: former Councillor Th. Nordstrom*

At its meeting on the 11th November the Royal Academy of Sciences decided to award the Nobel Prize for Physics for the year 1913 to Dr. Heike Kamerlingh Onnes, Professor at the University of Leyden "for his investigations on the properties of matter at low temperatures which led, inter alia, to the production of liquid helium".

As early as 100 years ago research into the behaviour of gases at various pressures and temperatures gave a great impetus to physics. Since this time the study of the connection between the pressure, the volume and the

---

* President of The Royal Swedish Academy of Sciences.

temperature of gases has played a very important part in physics, and particularly in thermodynamics - one of the most important disciplines of modern physics.

In the years 1873 and 1880 Van der Waals presented his famous laws governing gases which, owing to their great importance for thermodynamics, were rewarded by the Royal Academy of Sciences in 1910 with the Nobel Prize for Physics.

The thermodynamic laws of Van der Waals were laid down on atheoretical basis under the assumption that certain properties could be attributed to molecules and molecular forces. In the case of gases the properties of which are changed by pressure and temperature, or in one way or another do not agree with Van der Waals' hypothesis, deviations from these laws occur.

A systematic experimental study of these deviations and the changes they undergo due to temperature and the molecular structure of the gas must therefore contribute greatly to our knowledge of the properties of the molecules and of the phenomena associated with them.

It was for this research that Kamerlingh Onnes set up his famous laboratory at the beginning of the 1880's, and in it he designed and improved, with unusual success, the physical apparatus needed for his experiments.

It is impossible to report briefly here on the many important results of this work. They embrace the thermodynamic properties at low temperatures of a series of monatomic and diatomic gases and their mixtures, and have contributed to the development of modern thermodynamics and to an elucidation of those associated phenomena which are so difficult to explain. They have also made very important contributions to our knowledge of the structure of matter and of phenomena related to it.

Whilst important on its own account, this research has gained greater significance because it has led to the attainment of the lowest temperatures so far reached. These lie in the vicinity of so-called absolute zero, the lowest temperature in thermodynamics.

The attainment of low temperatures in general was not possible until we learnt to condense the so-called permanent gases, which, since Faraday's pioneer work in this field in the middle of the 1820's, has been one of the most important tasks of thermodynamics.

After Olszewski, Linde, and Hampson had prepared liquid oxygen and air in a variety of ways, and after Dewar, having overcome great experimental difficulties, had succeeded in condensing hydrogen, all temperatures down to -259°C, i.e. all temperatures down to 14° from absolute zero, could be attained.

At these low temperatures all known gases can easily be condensed, except for helium, which was discovered in the atmosphere in the year 1895.

Thus, by condensing this it would be possible to reach still lower temperatures. After both Olszewski and Dewar, Travers, and Jacquerod had tried in vain to prepare liquid helium, using a variety of met hods it was generally assumed that it was impossible.

The question was solved in 1908, however, by Kamerlingh Onnes, who then prepared liquid helium for the first time.

I should have to cover too much ground if I were to report here on the experimental equipment with which Kamerlingh Onnes was at last successful in liquefying helium, and on the enormous experimental difficulties which had to be overcome. I would only mention here that the liquefaction of helium represented a continuation of the long series of investigations into the properties of gases and liquids at low temperatures which Kamerlingh Onnes has carried out in so praiseworthy a manner. These investigations finally led to the determination of the so-called isotherms of helium and the knowledge gained here was the first step towards the liquefaction of helium. Kamerlingh Onnes has constructed cold baths with liquid helium which permit research to be done into the properties of substances at temperatures which lie between 4,3° and 1,15° from absolute zero.

The attainment of these low temperatures is of the greatest importance to physics research, for at these temperatures both the properties of the substances and also the course followed by physical phenomena, are generally quite different from those at our normal and higher temperatures, and a knowledge of these changes is of fundamental importance in answering many of the questions of modern physics.

Let me mention one of these particularly here.

Various principles borrowed from gas thermodynamics have been transferred to the so-called theory of electrons, which is the guiding principle in physics in explaining all electrical, magnetic, optical, and many heat phenomena.

The laws which have been arrived at in this way also seem to be confirmed by measurements at our normal and higher temperatures. That the situation is at very low temperatures not the same, however, has, amongst other things, been shown by Kamerlingh Onnes' experiments on resistance to electrical conduction at helium temperatures and by the determinations which Nernst and his students have carried out in relation to specific heat at liquid temperatures.

It has become more and more clear that a change in the whole theory of electrons is necessary. Theoretical work in this direction has already been begun by a number of research workers, particularly by Planck and Einstein.

In the meantime new supports had to be created for these investigations. These could only be obtained by a continued experimental study of the properties of substances at low temperatures, particularly at helium temperatures, which are the most suitable for throwing light upon phenomena in the world of electrons. Kamerlingh Onnes' merit lies in the fact that he has created these possibilities and at the same time opened up a field of the greatest consequence and significance to physical science.

Owing to the great importance which Kamerlingh Onnes' work has been seen to have for research in physics, the Royal Academy of Sciences has found ample grounds for bestowing upon him the Nobel Prize for Physics for the year 1913.

1912. Nils Gustaf Dalén "for his invention of automatic regulators for use in conjunction with gas accumulators for illuminating lighthouses and buoys"

1911. Wilhelm Wien "for his discoveries regarding the laws governing the radiation of heat"

1910. Johannes Diderik van der Waals "for his work on the equation of state for gases and liquids"

1909. Guglielmo Marconi, Carl Ferdinand Braun "in recognition of their contributions to the development of wireless telegraphy"

1908. Gabriel Lippmann "for his method of reproducing colours photographically based on the phenomenon of interference"

1907. Albert Abraham Michelson "for his optical precision instruments and the spectroscopic and metrological investigations carried out with their aid"

1906. Sir Joseph John Thomson "in recognition of the great merits of his theoretical and experimental investigations on the conduction of electricity by gases"

1905. Philipp Eduard Anton von Lenard "for his work on cathode rays"

1904. Lord (John William Strutt) Rayleigh "for his investigations of the densities of the most important gases and for his discovery of argon in connection with these studies"

1903. Antoine Henri Becquerel "in recognition of the extraordinary services he has rendered by his discovery of spontaneous radioactivity" and **Pierre Curie, Marie Curie** "in recognition of the extraordinary services they have rendered by their joint researches on the radiation phenomena discovered by Professor Henri Becquerel"

1902. Hendrik Antoon Lorentz, Pieter Zeeman "in recognition of the extraordinary service they rendered by their researches into the influence of magnetism upon radiation phenomena"

1901. Wilhelm Conrad Röntgen "in recognition of the extraordinary services he has rendered by the discovery of the remarkable rays subsequently named after him"
